High-Pressure Shock Compression of Condensed Matter

High-Pressure Shock Compression of Condensed Matter

L.L. Altgilbers, M.D.J. Brown, I. Grishnaev, B.M. Novac, I.R. Smith, I. Tkach, and Y. Tkach : Magnetocumulative Generators

T. Antoun, D.R. Curran, G.I. Kanel, S.V. Razorenov, and A.V. Utkin : Spall Fracture

J. Asay and M. Shahinpoor (Eds.) : High-Pressure Shock Compression of Solids

S.S. Batsanov : Effects of Explosion on Materials: Modification and Synthesis Under High-Pressure Shock Compression

R. Cherét : Detonation of Condensed Explosives

L. Davison, D. Grady, and M. Shahinpoor (Eds.) : High-Pressure Shock Compression of Solids II

L. Davison, Y. Horie, and T. Sekine (Eds.) : High-Pressure Shock Compression of Solids V

L. Davison, Y. Horie, and M. Shahinpoor (Eds.) : High-Pressure Shock Compression of Solids IV

L. Davison and M. Shahinpoor (Eds.) : High-Pressure Shock Compression of Solids III

A.N. Dremin : Toward Detonation Theory

Y. Horie, L. Davison, and N.N. Thadhani (Eds.) : High-Pressure Shock Compression of Solids VI

R. Graham : Solids Under High-Pressure Shock Compression

J.N. Johnson and R. Cherét (Eds.) : Classic Papers in Shock Compression Science

V.F. Nesterenko : Dynamics of Heterogeneous Materials

M. Sućeska : Test Methods of Explosives

J.A. Zukas and W.P. Walters (Eds.) : Explosive Effects and Applications

G.I. Kanel, S. V. Razorenov, and V.E. Fortov : Shock-Wave Phenomena and the Properties of Condensed Matter

V.E. Fortov, L.V. Altshuler, R.F. Trunin, and A.I. Funtikov : High-Pressure Shock Compression of Solids VII

Lalit C. Chhabildas Lee Davison
Yasuyuki Horie (Eds.)

High-Pressure Shock Compression of Solids VIII

The Science and Technology of High-Velocity Impact

With 272 Figures

Springer

Dr. Lalit C. Chhabildas
Solid Dynamics
and Energetic Materials Department
Sandia National Laboratories
Department 1647, MS1181
Albuquerque, New Mexico 87185-1181, USA
Email: lcchhab@sandia.gov

Dr. Lee Davison
39 Cañoncito Vista Road
Tijeras, NM 87059, USA
Email: leedavison@aol.com

Dr. Yasuyuki Horie
Los Alamos National Laboratory
MS F699
Los Alamos, NM 87545, USA
Email: horie@lanl.gov

Editors-in-Chief:

Lee Davison
Yasuyuki Horie

Library of Congress Control Number: 2004110191

ISBN 3-540-22866-7 Springer Berlin Heidelberg New York

Springer is a part of Springer Science+Business Media

springeronline.com

Printed in Germany

Typesetting: by the author and TechBooks using a Springer LaTeX macro package
Cover design: *design & production* GmbH, Heidelberg

Printed on acid-free paper 54/3141/jl 5 4 3 2 1 0

Preface

This is the eighth in a series of books published on *High-Pressure Shock Compression of Solids*. In this book, the emphasis is on the science and technology of high velocity impact, a subject that is relatively new. It is a direct outgrowth of World War II research and the manned spaceflight program for well-controlled impact studies to understand material response to intense dynamic loading. This book is based on plenary lectures presented at the 54th Aeroballistic Range Association (ARA) Meeting held in Santa Fe, New Mexico, October, 20–24, 2003. The many chapters were prepared by leading researchers and fellows of the Association who have made significant contributions to the field.

The ARA is an international organization of experimental and computational range facilities, operating in research or commercial mode that requires projectiles or macro-objects to be launched at high velocities. Studies include projectiles' aeroballistic characteristics or hypervelocity impact mechanics or phenomena that arise as a result of their impact on other bodies. The association holds annual meetings to discuss topics pertaining to state-of-the-art research and applied programs on ballistic studies, hypervelocity launch techniques and applications, diagnostics development, and safety issues associated with the operation of range facilities. The topics of discussion are pulished informally as ARA Proceedings that are held proprietary to its members. This philosophy is adopted to foster frank discussions, to stimulate technical advances, protect intellectual property rights, and allow open disclosures of embarrassing accident descriptions or unsuccessful trials – all of which are necessary for advanced technological development. The ARA has succeeded in promoting the safe and efficient operation of range facilities, the development of many of the test methods now routinely used, and the formation of a contributing commercial industry that supply the needed research tools.

One of the primary motivations for this book is to document the history of the development of high velocity launchers and the relevant high-speed and high-time-resolution diagnostics. The examples presented in this book are strong evidence of the significant contribution the development of smoothbore research guns has had in the field of high velocity impact. The tools have been extremely useful in obtaining well-controlled and well-characterized projectile launches, not only of flat plates for the material property studies generally referred to as shock physics, but also for arbitrarily-shaped projectiles for both aerodynamic and impact studies.

The first chapter, by Hallock F. Swift, highlights the history of two-stage light-gas gun technology development, beginning in the late 1940s and continuing to the present. He discusses the motivating factors and the program needs that led to the focused development of the two-stage gun technology. The second chapter, by Henri Bernier, describes in detail the theory and technical methods used to design and operate two-stage light-gas guns. Projectile launch velocities approaching 9 km/s represent the technological limit of conventional two-stage light-gas guns, and programs are underway to devise means for achieving higher launch velocities. One such effort is the development of a plasma drag accelerator described in Chap. 3 by Eduard Igenbergs and Martin Rott. A high velocity, plasma "drags" small diameter particles to velocities well over 10 km/s. This technology was driven in part by the need to investigate micrometeoroid impacts on spacecraft. Chapter 4, by Lalit Chhabildas and Marcus Knudson, describes the technical conditions that must be met to launch projectile plates to velocities approaching 30 km/s. Early study in the 1990s enhanced the capability of two-stage guns to launch plates at velocities as high as 16 km/s. More recently, pulsed-electrical-power techniques have produced magnetic pressures sufficient to launch plates at velocities approaching 30 km/s.

In Chap. 5, Alois Stilp discusses the development of sabots to launch arbitrarily-shaped projectiles. This technology has allowed evaluation of the lethality and vulnerability of many conventional weapon systems. The technology needs to be precise in that the sabot package must not only protect the projectile during launch but also ensure a proper and stable flight path after it exits the muzzle.

To assist with the interpretation of ballistic events, explosions, etc., imaging methods such as high-speed photography and flash x-radiography have proven to be indispensable. Chapter 6, by Vernon Draxler, highlights many of the applications and provides examples of the use of these imaging techniques. In Chap. 7, Peter Fuller traces the historical development of high-speed photography over the past 150 years. It should be emphasized that obtaining images of projectiles moving at velocities of from a few hundreds of meters per second to ten kilometers per second imposes stringent conditions balancing the accuracy requirements with the exposure time necessary to prevent blurring. This is addressed in these chapters. With the advent of lasers, the diagnostics have recently been extended to include imaging holograms of ballistics events, as discussed in Chap. 8, prepared by Gray Hough. In many respects, this is a revolutionary development since construction of three-dimensional holographic images offers the potential for making this a powerful tool to promote improved understanding of ballistic events.

Complementing the high-speed imaging techniques are nanosecond high time-resolution time-resolved test methodologies that use interferometry, piezoresistive, or piezoelectric gauges to infer the deformation features resulting from shock structures in impacted solids. William Isbell discusses this subject in Chap. 9. The final chapter by Jim Asay and Marcus Knudson highlights recent developments in which pulsed magnetic fields were used to investigate isentropic compression in materials. When combined with shock Hugoniot data and isotherms obtained using static techniques, this offers the potential for determining more comprehensive equations of state, over wider temperature, pressure and specific volume ranges, than any heretofore available.

We would like to acknowledge all the authors for their contribution to this book and in particular their personal contribution to the science and technology of high velocity impact. As indicated, this is a dynamic, diverse, mature, and challenging field ranging from material science studies to ballistic impact investigations. The development of new technology, even to this date, continues to be inspiring and educational.

Albuquerque, New Mexico — *Lalit C. Chhabildas*
Tijeras, New Mexico — *Lee Davison*
Los Alamos, New Mexico — *Yasuyuki Horie*

Contents

List of Contributors

James R. Asay
Institute for Shock Physics
Washington State University
Pullman, WA 99163
USA
jrasay@wsu.edu

Henri Bernier
66Av. François Coppée
93250 Villemomble
henri.bernier@free.fr

Lalit C. Chhabildas
Solid Dynamics and Energetic Materials Department
Sandia National Laboratories
Albuquerque, NM 87185-1181
USA
lcchhab@sandia.gov

Vernon C. Draxler
13416 Island View Drive
Elk River, MN 55330
USA

Peter W.W. Fuller
47 Maderia Avenue
Bromley, Kent BR1 4AR
United Kingdom
beegpete@btinternet.com

Gary Hough
Southwest Research Institute
San Antonio, Texas 78228
USA

Eduard Igenbergs
Techniche Universitat München
Fachgebeit Raumfahrttechnik
85748 Garching
Germany
e.igenbergs@lrt.mw.tum.de

William M. Isbell
ATA Associates
600 Via Hierba
Santa Barbara, CA 93110
USA
bvisbell@cox.net

Marcus D. Knudson
Shock and Z-Pinch Physics Department
Sandia National Laboratories
Albuquerque, NM 87185
USA
mdknuds@sandia.gov

Martin Rott
Techniche Universitat München
Fachgebeit Raumfahrttechnik
85748 Garching
Germany
m.rott@lrt.mw.tum.de

Alois J. Stilp
Bussardweg 7
D-79110 Freiburg i.Br.
Germany

Hallock Swift
Physics Applications, Inc.
7635 Wilmington Pike
Dayton OH 45458
USA
physapp@att.net

1 Light-Gas Gun Technology: A Historical Perspective

Hallock F. Swift

1.1 Introduction

Since their inception during 1948, two-stage light-gas guns have evolved into the primary laboratory devices for launching macro-size models to velocities above 2.5 km/s (8200 ft/s). Today, light-gas guns are used routinely to launch projectiles with sizes ranging from spheres and cylinders 0.1 mm in diameter and weighing a few micrograms to complex models with transverse dimensions up to 175 mm that weigh several kilograms. Some of the larger units have been launched successfully with active electromechanical systems aboard. Large and complex models can achieve launch velocities above 7.0 km/s (23,000 ft/s) without experiencing damage. Smaller and more rugged models made from a variety of plastics, metals and ceramics are launched intact to velocities exceeding 9.0 km/s (29,500 ft/s). Bore fitting plastic cylinders with limited aspect ratios have achieved velocities above 11.0 km/s (36,000 ft/s). Finally, disks of low-density titanium-alloys have been launched to velocities up to 16 km/s (52,500 ft/s) by using light-gas guns in conjunction with impact-driven velocity amplifiers mounted at their muzzles.

Two-stage light-gas guns have emerged as virtually the only devices available for launching models in support of a wide range of laboratory studies that require high-speed flights. Fortunately, such gas guns have proven to be reliable and easily used laboratory devices. Hypersonic aerodynamic phenomena associated with atmospheric reentry of space-traveling vehicles have been simulated, as have aerothermal phenomena. These phenomena have been produced in laboratory equipment where they can be studied closely and measured with high precision. Hypervelocity impacts producing local stress levels up to nearly 1.0 TPa (150 million lb/in^2) during planar impact experiments have proven invaluable for studying details of the response of materials to ultra high-pressure environments. Armor systems for protecting space vehicles from natural meteoroid impacts and from similar encounters with man-made particles have been developed almost exclusively using results from gas gun impact studies. The effectiveness of supervelocity artillery projectiles impacting modern armor systems has been studied extensively using light-gas guns. Aeroelastic phenomena that ultimately control the aiming accuracy of very high velocity slender projectiles as they fly through the atmosphere have been simulated. Many other research areas ranging from fabrication of new materials to evaluating modern methods for processing food have also benefited from modern light-gas gun technology.

The occasionally convoluted story of light-gas gun technology development over the past half century provides significant lessons about how to, and not to, exploit modern technological breakthroughs. The story of this development effort is often entertaining as well.

Organization of this chapter is necessarily complex because many of the points discussed occurred simultaneously or were superimposed upon oneanother as they evolved over an extended period of time. We start with a description of the early history of light-gas gun technology development that extends up to about 1980 when modern development began in earnest. Then the technology is treated by individual subject areas. The chapter concludes with a discussion of recent history (1980 to the present) and an abbreviated estimate of what the near future may hold for the technology.

1.2 Early History

1.2.1 The Post World War II Technology Situation

The story of two-stage light-gas gun development extends back well before their original invention in 1948. It is widely agreed that the Allies won World War II mainly through supplying their fighting troops with more and generally superior equipment than was available to troops of the Axis Powers. These materials were produced mainly in the United States. One notable exception to allied superiority in material quality over the Axis powers was gunnery extending in size from larger small arms to heavy artillery. The Germans, in particular, produced weapons with higher velocity capabilities and better workability than many of their allied counterparts. A prime example is the famous German 88-mm field gun. It was initially distributed as an antiaircraft canon but it was found to be a superb antitank gun and, finally, it served admirably as the main armament for the famous German Tiger Tank. Among the Allies, the USSR also produced several important guns that outperformed their American counterparts.

After World War II, farsighted officers in the United States Army and Navy became concerned about this perceived technology gap, especially as the USSR emerged as a potential military opponent during the early postwar years. Research and development projects were quietly initiated to lay groundwork for improving American ordnance capabilities. At the same time, technology from the German V2 rocket project during World War II was being married to nuclear weapons development to create modern ballistic missile weapons systems in both the United States and the USSR. Rocketry development also triggered engineering interest in space exploration in both camps. Development of space-traveling missiles and research vehicles required laboratory capability for launching a variety of models to velocities up to

6.0 km/s (21,500 ft/s) for ICBM technology development and to much higher velocities for satellite and deep space vehicle development.

No technology was available during early post-war years for producing the required model velocities under laboratory conditions. The nearest approach came from single stage solid propellant guns (powder guns) whose muzzle velocities were limited to approximately 2.75 km/s (9000 ft/s). Thus, the stage was set for fundamental new thinking in the area of launching macro-projectiles to high velocities.

1.2.2 Two-Stage Light-Gas Gun Invention

Professor E.J. Workman and his research group at the New Mexico Institute of Mining and Technology (NMIMT) approached the technical staff at the United States Army Ballistic Research laboratory during early 1948 with a revolutionary proposal for extending launch velocity capabilities of guns. He pointed out that launch velocities of powder guns had been maximized at levels near 2.75 km/s (9000 ft/s) by minimizing launch mass, lengthening the launch tube, evacuating the launch tube to eliminate back pressure on the projectile face, increasing the mass and volume of the propellant charge, and making propellant burn rates correspond to gun launch needs. Prof. Workman asserted that velocity limits of solid propellant (powder) guns could be traced to the nature of the propellant burn products that formed the driver gas charge for these devices. Only a finite amount of energy is released by combustion of the propellant to form the burn products that expand to accelerate the projectile. At least part of this energy must be used to accelerate the mass of the local burn product gas to the instantaneous velocity of a projectile as it is being launched along the gun tube. As projectile velocities approach 2.75 km/s, essentially all the available energy is used up accelerating the driver gas so little or none is left for further acceleration of the projectile. This factor causes a pressure gradient to develop along the gas column in the launch tube. This gradient results in the gas pressure at the base of the projectile decreasing to values well below the pressure in the propellant chamber as the projectile achieves high velocities during its travel along the launch tube.

Professor Workman proposed to alleviate this problem by replacing propellant burn product gas with a column of light-gas for projectile launches. The new gas charge would have a much lower molecular weight than propellant burn products while retaining or even exceeding energy densities of propellant burn products. In this way, similar energy densities within the driver gas could produce much higher peak flow velocities that would reduce the effects of gas pressure gradients along the gun tube and allow higher projectile velocities to be attained. Typical propellant burn products consist of mixtures of H_2O, CO_2, and N_2 with average molecular weights near 30 g/mole. Substituting helium as a driver gas lowers the gas molecular weight to 4.0 g/mole (only ~1/8 that of the gas from propellant burn products.) The

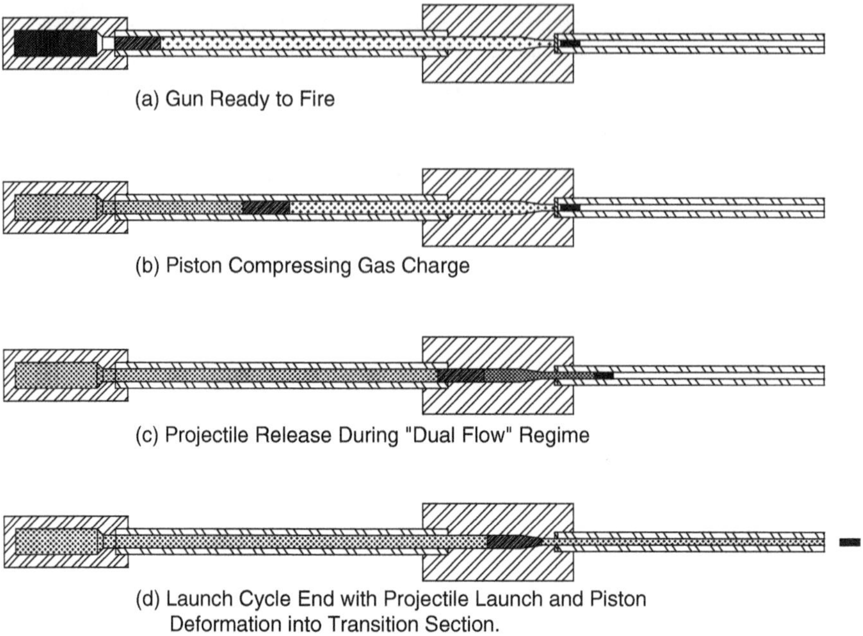

Fig. 1.1. Light-gas gun sketches showing various stages in the firing process

problem then becomes "How is energy to be added to a column of helium gas to develop energy densities comparable to (or greater than) those contained by propellant burn products [1]? The technique chosen was adiabatic compression of helium gas originally contained within the tube of a smoothbored gun byfigure a propellant-driven piston. The equipment configuration is shown in Fig. 1.1 at various stages of a gas gun launch operation. The rear half of the launcher (the pump tube assembly) is a smooth bore solid propellant gun of ordinary design with its projectile replaced by a weighted compression piston. The piston seals the rear end of the pump tube. The muzzle end of the pump tube assembly is sealed by a massive steel cylinder (the central breech.) This unit contains an extension of the pump tube bore, a conical transition to a smaller bore diameter, and a forward-facing receptacle. The receptacle supports a valve mechanism that seals the rear end of the launch tube until it is opened by high gas pressure. A second gun tube (the launch tube) used for accelerating the final projectile package. Typically, the launch tube has a smaller bore diameter (often near one-third) that of the pump tube. A projectile package to be launched is mounted at the base of the launch tube just ahead of the valve.

The gun is loaded by installing a piston at the base of the pump tube, installing a valve and a projectile at the base of the launch tube, and adding a propellant and ignition charge to the rear chamber. Air is then evacuated from the pump tube and central breech volume between the piston face and the valve. A charge of low-molecular-weight gas (helium) is added to the evacuated volume. This situation is depicted in Fig. 1.1a.

Gas gun firing begins with ignition of the propellant charge. Rapidly increasing gas pressure in the chamber causes the piston to accelerate forward into the gas charge thereby compressing it rapidly. Compression continues until the piston uses up the last of its kinetic energy and lodges itself in the transition cone within the central breech. The duration of the compression stroke is short enough to assure that only a very small portion of the energy added to the gas column by compression is transferred to the pump tube wall via radiation, convection, and/or conduction. Thus, the compression stroke of the piston produces nearly adiabatic compression of the gas column. Sometime after the front piston face enters the central breech, pressure within the gas column rises to the point where the quick-opening valve operates. Gas then flows past the valve and impinges on the projectile to start the launch process. A fairly complex situation now ensues in which the volume of the gas column continues to decrease and its pressure increases by continued forward (but decelerating) motion of the compression piston. At the same time, the volume available to the gas is being increased by rapid acceleration of the projectile along its launch tube. Immediately after valve operation, gas volume decreases and gas pressure rises until pressure levels achieve several multiples of the projectile's release pressure. Then the situation reverses as net gas volume increases due to piston deceleration and projectile acceleration. Peak gas pressure is reached as the instantaneous projectile velocity exceeds nine times instantaneous piston velocity for typical gas gun geometry. Finally, the launch cycle is completed as the piston uses up the remainder of its kinetic energy by lodging itself into the transition taper and the driver gas expands adiabatically behind the projectile until it clears the launch tube muzzle.

Professor Workman's group received a contract from the United States Army Ballistic Research Laboratory (BRL) to explore the light-gas gun concept and built a prototype that was tested carefully [1]. Almost immediately, the gas gun launched light projectiles to velocities near 3.0 km/s (9800 ft/s) even without evacuating the launch tube bore. Peak launch velocities were worked up from this level to nearly 4.0 km/s as experience was gained choosing appropriate gun loading conditions. Choosing ideal loading parameters became relatively complicated because so many means of controlling the launch cycle are available. They include:

- Pump piston mass
- Solid propellant mass and burn rate
- Initial light-gas charge pressure
- Projectile release pressure, and
- Projectile mass.

In addition, the following gun design parameters were found to affect launch velocity directly:

- Chamber volume
- Pump tube length
- Pump tube diameter
- Convergence angle of the taper section
- Launch tube diameter, and
- Launch tube length.

All of the loading factors interact closely with one another to determine projectile launch velocity and peak projectile acceleration levels during the launch cycle. It became obvious during early tests that light-gas guns are capable of producing gas pressure profiles intense enough to shatter projectile packages and to deform (or even rupture) gun components. Thus, developing peak performance capabilities of gas guns required delicate balancing of the five controlling loading parameters. As can be imagined, the listed design parameters also control launch velocity, and several of their interactions with each other became understood only after years of analysis and experimentation. Both computational and experimental processes continue today as gas guns operated by at several facilities are adjusted carefully to achieve progressively higher performance levels.

Professor Workman's early results were spectacular and attracted considerable attention among BRL managers. A decision was made that the two-stage gas gun breakthrough was important enough to subject it to military classification. At the time, (about 1950) classified research was not allowed at academic teaching institutions so the light-gas gun development work at NMIMT was terminated and the equipment on hand was shipped to BRL. Unfortunately, development was not continued at BRL, or elsewhere, and the NMIMT equipment was misplaced. Light-gas gun technology lay dormant for several years until political developments caused its resurrection.

1.2.3 Light-Gas Gun Contributions to Space Activities

During the early 1950s, modern ballistic missiles were chosen as one of three primary delivery mechanisms for strategic nuclear weapons. Vehicle development needs, particularly pertaining to reentry science, spurred renewed interest in laboratory launching of macroprojectiles to hypervelocities. Light-gas gun technology had not yet achieved velocities of interest to such efforts but the theoretical potential of gas guns and their position as the fastest launchers available assured their detailed consideration by government laboratories in the United States and Europe and by a few commercial firms involved in defense activities. These efforts were also supported by the beginning of the civilian space effort that started during the mid 1950s when orbiting of man-made Earth satellites began being treated as a current engineering problem. These activities prompted interest of NASA facilities, especially the Ames Research Center (Now NASA Ames). A large light-gas gun laboratory was, accordingly, set up at NASA Ames. The number of facilities employing light-gas guns for space and missile research grew rapidly during the second half of the 1950s. Several dozen operations in the United States, Canada, and Western Europe became active by 1960.

Interest was triggered for investigating the potential problem of meteoroid impacts upon satellites in orbit. The United States Navy planned to place a satellite in earth orbit during the International Geophysical Year, 1957. Development of effective armor to protect it from meteoroid impacts became a priority project. Somewhat later, (1960) methods for defeating hostile ballistic missiles became an important consideration. Attacking ICBM warheads during mid-course flight and during early reentry with fields of macro-particles was looked upon for several years as a fruitful defense measure. Both of these problems lead to intense research into increasing velocity capabilities of light-gas guns and to studies of hypervelocity impact mechanics (impacts that produced initial stress levels many times greater than those that could be resisted by target or projectile materials strengths.)

1.2.4 Organizations that Supported Light-Gas Gun Development

The earliest organized activity to coordinate individual research efforts, exchange results, and consider means for proceeding in a coherent progression involved the Tri-service Hypervelocity Impact Committee formed by the United States Army, Navy, and Air Force. The principal activity of this committee was sponsoring a series of seven international symposia on hypervelocity impact, the first of which was held at the Rand Corp. in Los Angeles CA during the spring of 1956. Nearly half the papers presented at the first symposium described efforts to launch projectiles to the highest velocities possible. The highest velocities reported were produced by a variety of explosively powered projectors rather than light-gas guns. Explosive launch effectiveness was limited severely because projectiles had to be formed during launch processes, which limited available shapes and materials severely. Also, no means were then available for accurately determining sizes or masses of launched items. Release from these limitations provided by light-gas guns assured continuing interest in the technology. Many papers covered two-stage light-gas gun developments. Steady progress in attaining higher velocities was reported as many gun geometries and firing parameter combinations were investigated. Peak light-gas gun velocities reported at the first Rand Symposium were just over 5.0 km/s (16,400 ft/s). The symposia were held at approximately 18-month intervals until the 7th Symposium during February 1965. During this period, work on light-gas gun development divided naturally between efforts to achieve maximum possible velocities, mostly without regard for projectile materials or shapes launched, and efforts to exploit the technology for supporting exterior and terminal ballistic tasks. Figure 1.2 presents reported maximum velocities achieved in these two areas of work

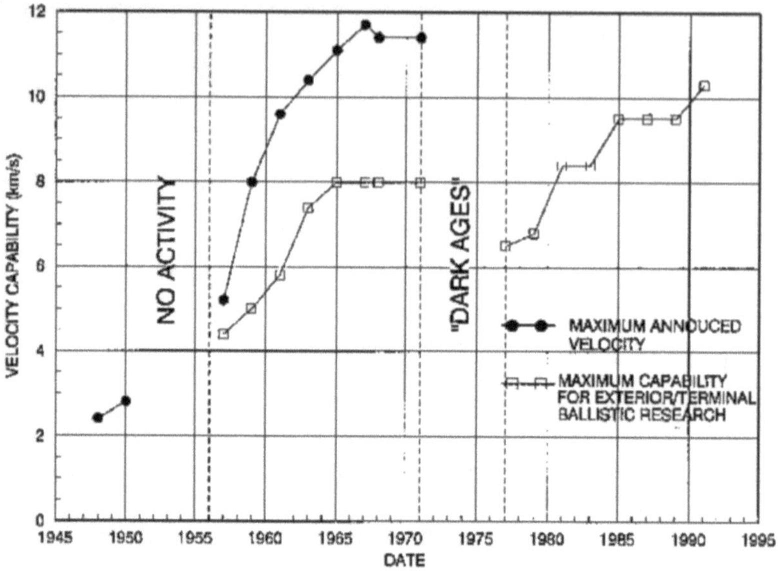

Fig. 1.2. Reported peak light-gas gun velocities vs. time from 1948 to the present

throughout the history of light-gas gun technology. Peak velocities rose eventually to somewhat over 11.0 km/s (36,000 ft/s), but projectiles launched at velocities above 9.0 km/s (29,500 ft/s) were limited to short plastic cylinders of bore diameter that were of limited usefulness for precise quantitative research and development. Velocity limits extended quickly to near 8.0 km/s (26,200 ft/s) for light-gas guns used as research tools (guns that could launch projectiles of multiple shapes made from a variety of materials). The size and complexity of launched projectiles developed rapidly during this period, as progressively larger light-gas guns with controlled peak acceleration levels became available.

A second important development that supported organizing and coordinating gas gun development was the founding of the Aeroballistic Range Association (ARA) in the fall of 1959. The ARA is an association of facilities that operate advanced ballistic research activities. During its early years, each member organization had to operate at least one two-stage light-gas gun. The ARA met at informal symposia every six months where all member organizations were required to present papers on their facility development efforts. The proceedings of these symposia were held confidential to the membership to protect release of trade secrets and to promote frank discussions of potentially embarrassing subjects such as operational accidents and failed projects. The ARA is still an active and vibrant organization that now meets annually. It held its 54th meeting during the fall of 2003. Its membership is still closely controlled but has expanded to encompass other related technology areas and some equipment vendors.

Finally, the light-gas gun field has received substantial support from the Hypervelocity Impact Society (HVIS) formed in 1985 by a group of eminent scientists headed by Dr. Harry Fair, then with the United States Defense Advanced Research Projects Agency (DARPA). The organization is international and has been very active since its inception. It sponsors a series of symposia that started in October 1986 and continue to the present. They are, effectively, continuations of the seven Hypervelocity Impact Symposia. The HVIS also sponsors individual recognition for exceptional contributions to the hypervelocity field with their "Distinguished Scientist" awards that include substantial stipends. They promote academic studies in related fields with student scholarships, student paper competitions, financial support for symposium attendance, visiting lecturers, etc.

1.3 Development of Technologies Needed for Modern Light-Gas Guns

Development of critical light-gas gun components and supporting technologies is described in this section. These developments dominated the occasionally halting progress of light-gas gun technology that has succeed in transporting laboratory curiosities of the late 1940s to important assets needed to support a significant number of current national research and development goals.

Time periods covered here extend from the original concept demonstration in 1948 to the present, so some of the events described earlier must be revisited briefly but from different viewpoints. The same holds for modern and future light-gas gun chronologies discussed at the conclusion of this chapter.

Figure 1.3 presents a sketch of a generic two-stage light-gas gun, identifying the components that are discussed below.

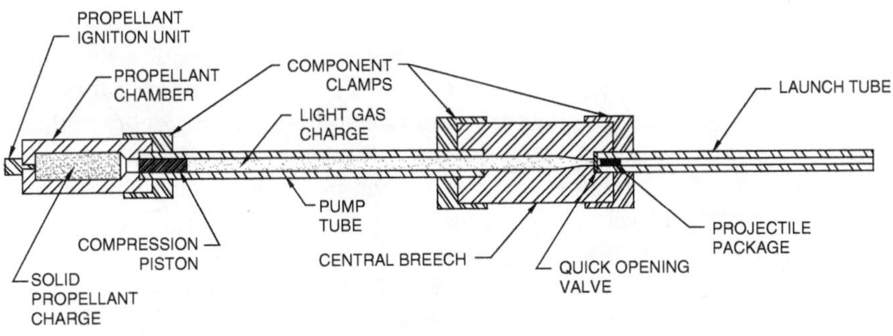

Fig. 1.3. Light-gas gun sketch with names of major components

1.3.1 Sealing Light-Gas Gun Joints

Inspection of early gas gun hardware after firings showed clearly that significant gas leakage occurred at all joints subjected to light-gas charges pressurized during firing cycles. Another factor that supported the assumption that gas leakage affected light-gas gun performance significantly was that launch velocities varied widely (by ±20% or so) for individual firings of particular guns when loading parameters were held constant. Presumably, different shot-to-shot leak rates reduced gun performance to various degrees. At minimum, the joints that leaked badly were those between the pump tube and the central breech and those between the central breech and the launch tube. Typical early gun designs included additional joints within the central breech at both ends of the taper between the pump tube and launch tube diameters. Still more joints, which also leaked, were generally added at both ends of short launch tube extensions that were used to mount the projectile packages to be launched and the quick-opening valves used to release them.

The state-of-the-art for gun joint seals when light-gas gun development began involved using rubber o-rings mounted in grooves on flat faces (perpendicular to the gun axis) and/or on cylindrical surfaces of boss/receptacle joints. Peak gas pressures produced in light-gas charges during gun firing cycles typically exceeded 500 MPa (74,000 $\mathrm{lb/in^2}$), which defeated o-rings by pushing them aside. Face seal surfaces were also subjected to very large forces that tended to separate them. Even small surface movements reduced seal effectiveness to nil. These forces were produced by both gas pressure operating on the surfaces inside the seal and the large recoil forces that arise during the gas gun firing processes.

The first effective steps taken to seal light-gas guns were to reduce the number of joints exposed to highly pressurized light-gas charges to an absolute minimum. The first to go were joints at the ends of tapers between pump and launch tube diameters. The central breech assemblies became single units fabricated from strong and tough steel. Next, the projectile package and the quick-opening valve was moved to the rear of the launch tube, thus eliminating the joints needed to seal launch tube extensions. It is interesting to note that these joints reappeared in modern light-gas gun designs once fully adequate seals came into common use. Finally, the launch tubes were made from single steel forgings to eliminate joints along the bore walls as well as bore surface imperfections produced by joint misalignments.

The second effective effort for joint sealing involved development of seals that could withstand very high gas pressures without leaking. The new seals required both high intrinsic strengths and insensitivity to joint motion (opening/closing and diametrical deformation) produced by gas pressure and recoil forces. Two basic seal designs have come to dominate the field. The first, and most classic, uses a metal ring having the cross section of a rather flat metal wedge to produce the seal as is sketched in Fig. 1.4. Gas pressure acting on one face of the feather edge presses it against the mating seal surface with a

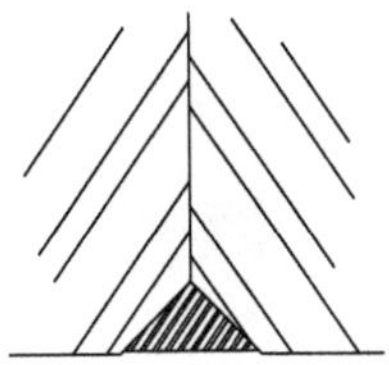

Fig. 1.4. Bridgman type gas seal applied to light-gas gun

force that grows proportionately with the pressure of the gas that is trying to escape. The overall effect of this situation is that the mechanical strength of the seal material becomes progressively less important as the gas pressure increases. These seals can be designed to withstand at least moderate axial and diametrical joint movements during gas pressure application by having the feathered surface engage a mating surface whose radial position varies only slowly with small joint opening motions.

A more radical group of seal designs (that have come to dominate sealing functions in modern light-gas guns) employ conventional rubber o-rings in geometries that allow them to withstand almost infinitely high gas pressures. A fine example of this arrangement is the famous *diamond seal* invented during the 1960s at the General Motors Defense Research Laboratory (GMDRL) and sketched in Fig. 1.5. Diamond seal designs have been copied throughout the industry and they, together with their many variations, are in wide use today. The diamond seal operates between two flat faces that are fitted with opposing o-ring grooves. A metal ring with an outside diameter that nearly fits the outer surfaces of the o-ring grooves is installed between them. The length of the ring is made just less than the sum of the depths of the two grooves. The inner corners of the ring are precisely beveled to form triangular openings within the inner corners of the two o-ring grooves when the seal is assembled. These triangular openings contain a pair of conventional rubber o-rings whose cross sections are squeezed nearly triangular. The cross-section diameter of the o-ring is chosen so that ring areas are near 80% of the areas of the triangular openings. When gas pressure is applied to the joint surfaces inboard of the seal, it forces gas into the grooves at full pressure. This gas engages the two o-rings, forcing them outward toward the narrow openings between the rings and the groove bases. Extremely high gas pressures would extrude the o-ring materials through these gaps and around the outside diameter of the ring to cause seal failure if the seal ring retained its size and shape during pressure application. However, the outward radial force produced on the ring by the gas pressure causes it to expand, thus blocking escape of the o-ring material and assuring maintenance of gas sealing. Diamond seals operate at peak pressures above 800 MPa (117,000 lb/in^2)

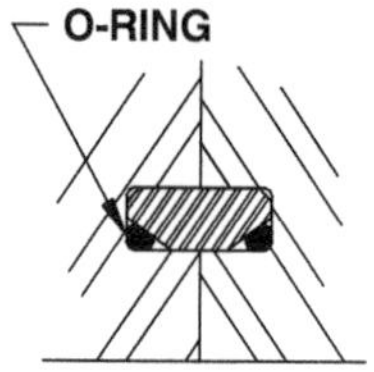

Fig. 1.5. A diamond seal that is typical of many modern light-gas gun seals

through dozens of individual applications without even damaging the rubber o-rings.

A third approach to gas sealing in light-gas guns bears mentioning. Here, the seals are not expected to work perfectly but only to delay significant gas flow for the duration of gas gun firing cycles. One example is the threading of the outside diameter of the launch tube into the down-range end of the central breech. The projectile package and the quick-opening valve are installed before the tube is screwed in place. Before a firing, rudimentary o-ring seals at this joint contain the light-gas charge. These seals fail as high gas pressures develop, but the long and tortuous escape path through the threaded joint assures that negligible amounts of gas escape during firing cycles.

Over a period of thirty years or so (1947–1977) application of modern gas sealing technology succeeded in eliminating gas leakage as an important problem blocking advancement of light-gas gun technology. One result of this success provides striking evidence of how effective modern light-gas guns are in sealing gas within them. During early days, even small gas guns produced ear-splitting roars when they were fired, regardless of atmospheric conditions beyond their muzzles. These blasts were caused by explosive escape of gas charges. Today, even large light-gas guns produce little noise when they are fired into evacuated ranges.

1.3.2 Developing Light-Gas Gun Clamps

A factor closely related to sealing gun components together to prevent gas leakage during firing is the method used to clamp the gun components together. The clamps must hold light-gas guns together during firings and also restrain relative component movement produced by forces associated with high internal gas pressure and forces produced by complex recoil processes associated with gun firing. The gas pressure forces are produced by high-pressure gas operating against opposing surfaces between the gas column(s), and the gas seals. We assume that these surface areas are exposed to fully pressurized gas whether or not any openings between them are expected. We further assume that no gas is allowed to escape past the seals and impinge on these opposing surfaces outboard of the seals. Actually, many gun clamps would fail catastrophically if pressurized gas were allowed to reach outboard surface areas and thereby greatly increase the opening force. Good gas gun design practice dictates that these outboard surfaces be reliably vented to the gun exterior. Obviously, gas pressure forces on flat surfaces are minimized by positioning seals as close as practical to the interior gas columns within the gun components.

Forces produced at seal closure surfaces by gun recoil cannot be either increased or decreased by any seal design features. They arise from acceleration and deceleration profiles of the relatively massive pump pistons and they can become very large. For instance, the recoil force at central seal surfaces can reach 1.32×10^6 N (315,000 lb) during energetic firings of a two-stage light-gas gun with a 50-mm (2.0-inch) diameter pump tube. This force profile appears alternately as an opening and closing force on central gas gun joints. As was mentioned earlier, modern gas seal designs can withstand small movements between gun components but much is to be gained if these movements are minimized. Thus, gas gun clamps need to be both strong and rigid if they are to be effective.

Originally, gun components were clamped together using techniques that produced adequate closing forces and were strong enough to survive firing loads without damage, but they were extremely flexible. Thus, large intercomponent movements were typical during early light-gas gun operations. Steel flanges were screwed onto the outside diameters of components to be connected near the closing ends. These flanges contained matching hole patterns near their outer edges which were aligned with each other and fitted with bolts, washers, and nuts that were tightened to provide clamping force. During gas gun firings, when opening forces were applied, the flanges flexed into conical shapes (like Belleville disc spring washers) allowing partial joint opening. The relatively slender shafts of the holding bolts also stretched under applied axial loads, thereby increasing joint openings. Together, these movements allowed the joints to open to alarming degrees. The situation reached its extreme when flanges and bolts were used routinely to clamp central breeches between pump tubes and launch tubes. Flanges were screwed onto pump tube and launch tube outside diameters near their ends facing one another and long thin bolts were used to provide closure forces for all central gun joints. These clamps proved so flexible that joints between pump tubes and central breeches opened substantially. After each firing, they were pulled closed by elastic recovery of the flanges and bolt shafts with enough velocity to produce component damage. Copper discs were occasionally inserted between critical joints to prevent gun components from being peened over by such repeated impacts.

More rigid seals started being installed between gas gun components during the early 1960s at facilities like GMDRL. Their use spread slowly throughout the world, but examples of flange and bolt joints can still be found on smaller two-stage light-gas guns operating today. Basically, rigidity of modern gas gun clamping is improved markedly by eliminating flanges that can distort out of their plane, and by replacing bolts for providing closure force with thick-walled tubes that surround the clamped components and deform only slightly when subjected to extreme opening forces. A good example of such a coupler is a right-hand/left-hand nut clamp depicted in Fig. 1.6. One component to be clamped is threaded on its outside diameter adjacent to the facing surface with right-hand threads. Similar threads on the other components are of the same pitch but are left-handed. A long hollow nut with female right and left hand machined into each end is started onto each component simultaneously. The nut is then rotated enough to bring the two gun components together to form an extremely rigid coupling.

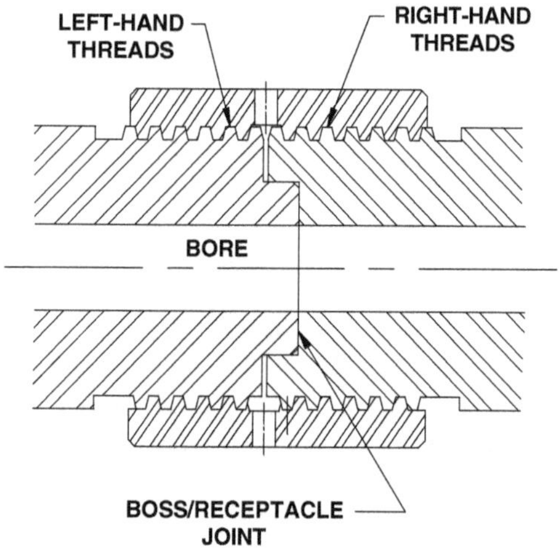

Fig. 1.6. Right-hand/left-hand threaded coupler for joining light-gas gun components

Finally, a jump was made to infinitely rigid couplers. These couplers allow no joint opening at all during light-gas gun firings. Workers at the Ernst Mach Institute (EMI) were among the first to develop gun clamps in which the closing force is produced hydraulically. An example of such a clamp is depicted schematically in Fig. 1.7. These couplers operate by producing joint closing forces that are typically 130% of the sum of maximum gas pressure and gun recoil forces applied to the joint being clamped. Since very high-pressure water pumps are now available at low cost, the sizes of hydraulic enclosures needed to provide such forces could be limited to reasonable values. Under these closing forces, the net force on the joint (closing forces less opening forces) may be counted upon to remain positive (closing) under even the most

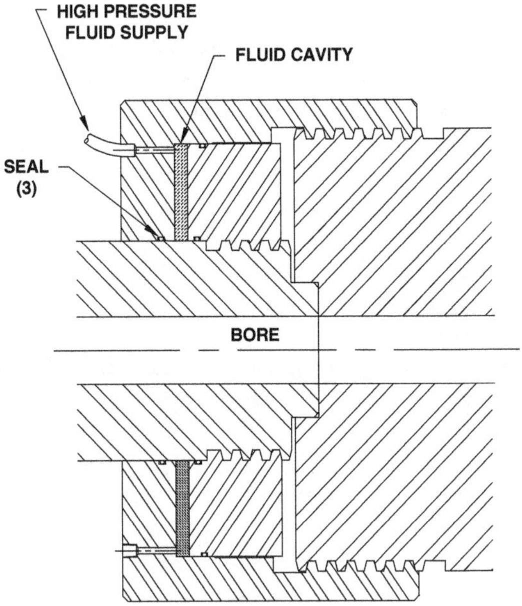

Fig. 1.7. Hydraulic clamp used to join light-gas gun components

extreme joint loads. The joints remain closed as they would if the clamps were infinitely rigid.

Most recently designed light-gas guns use rigid nut clamps of one design or another between pump and launch tube segments. Either rigid mechanical clamps or hydraulic clamps close the joint connecting the propellant chamber and the pump tube. Hydraulic joints are now widely used to close central joints roughly at each end of the central breech. We anticipate that hydraulic clamps will become progressively more common gun design items as time goes on.

1.3.3 Central Breech Development

Central breeches absorb the most intense punishment of any light-gas gun component. During each firing, they must withstand, first, the peak pressure produced by light-gas compression. Second, the transition between pump and launch tube diameters receives the most intense surface heat loading associated with gas gun operation. Gas densities are high at this location and convergent gas flow scours away most of the boundary layer that protects, at least partially, other gun components such as the launch tube bore, which also experience gas flow. Finally, the transition must withstand impact of the compression piston as it comes to an abrupt stop during the last part of its compression stroke. These factors combine to assure that central breeches in all but the most gently fired light-gas guns eventually accumulate enough damage to require their retirement if catastrophic rupture is to be avoided.

Light-gas guns of early design used at the United States Naval Research Laboratory (NRL) and at a few United States Army laboratories were operated with expendable central breeches. Since the cost of each of these breeches had to be amortized over a single gun firing, they were carefully designed to be inexpensive to fabricate. This feature alleviated much apprehension for gun development engineers whose efforts might otherwise be inhibited through fear of destroying an expensive component. Low cost central breeches allowed for the use of compression pistons made from mild steel. These pistons massively distorted transition sections within central breeches when sharp impacts occurred. Light-gas guns at most other facilities used weighted plastic pistons with easily deformable forward sections that produced large stresses when they impacted the transition section, but did not cause central breech failure as a result of a single firing. Today, such pistons are used exclusively, making central breeches semi-capital gas gun equipment (with lifetimes of hundreds or thousands of firings).

Gas dynamicists noted by 1960 that a potential existed to shape transitions between pump and launch tube diameters to minimize gas flow resistance to flows that started subsonic but became sonic as the firing sequence continued. At considerable expense, such surfaces were machined into special transition sections that were tested during energetic light-gas gun firings at several facilities in the United States and Europe. Light projectile packages were fired at velocities as high as possible using these central breeches. No performance advantages could be identified for gas guns using central breeches with tailored transition sections over performance from the same guns using transition sections with more conventional conical designs. Finally, engineers at the Arnold Engineering Development Center (AEDC) conducted an experiment that has effectively eliminated transition-section shape as an important factor for optimizing light-gas gun performance levels. They fabricated and tested a special central breech for one of their smaller light-gas guns (whose capabilities were well researched.) The special unit was made with a flat bottom on the rear bore opening and a coaxial hole of launch tube bore diameter through the remainder of the forging. This transition section could be considered to be a cone with an included angle of 180°. When the special central breech was tested, the resulting light-gas gun performance levels were neither better nor worse than those achieved with central breeches containing standard transition taper angles (30° included angle). Of course, the lifetime was notably short. This design of the special central breech was not adopted at AEDC or elsewhere, but an important point had been made.

Biolettie at NASA Ames, in 1963, proposed a light-gas gun operational concept entitled *accelerated reservoir*. The idea was that a compression piston made from soft plastic would deform at nearly constant material density as it entered the transition section during the later portion of the firing sequence.

Under these conditions, the front piston face must accelerate even though the center of mass of the piston is steadily decelerating due to resistance forces acting on it. Piston face acceleration may be expected to increase light-gas compression rates at just the time when the resultant extra gas pressurization is needed to combat the pressure drop during the latter part of the projectile acceleration profile. Hence, a potential existed for increasing light-gas gun performance. Central breeches were constructed with very long conical transition sections to maximize the accelerated reservoir tendency. These guns operated effectively at very high launch velocities so the accelerated reservoir concept was widely accepted. Later, launch performance levels at and above accelerated reservoir levels were achieved with similar light-gas guns that contained the steeper tapers typical of older central breech designs, and the accelerated reservoir concept slowly disappeared. The author was never sure whether or not the concept might yet be employed to gain future light-gas gun performance advantages.

Heat inputs to the tapered transition section produced by energetic gas gun firings remain an important factor limiting the lifetime of central breeches for large two-stage light-gas guns. The gas flow produces mechanical surface failures called *heat checking*. When all other gasdynamic parameters are held constant, total heat input per unit area to the transition surfaces exposed to energetic light-gas flows rises linearly with gun size so it reaches very substantial values in large guns (generally ones with pump tube diameters of 150 mm, 6 inches, and larger). The heat increases local wall surface temperatures. This temperature increase causes metal expansion, which, in turn, induces large compressive stresses in the affected material since thermal expansion cannot occur. The large stress fields are released through a series of cracks that form in roughly rectangular patterns on the transition wall surface. Fortunately, the cracks extend only characteristic short distances (a few millimeters) into the central breech body and then turn parallel to the transition surface. This is the phenomenon called heat checking. During several gun firings after the cracks first appear, the outlined rectangles start to fall out leaving a surface that looks like a shingled roof with some of the shingles missing. Clearly, although this damage is fatal to the central breech, it appears to present little or no extra danger of catastrophic rupture of the affected central breech. This problem is still vexing because of the obviously large expense of big central breeches that may have their lifetimes shortened considerably by heat checking.

1.3.4 Improving Launch Tube Bores

Early light-gas guns used launch tubes from high performance powder guns (typically smooth-bored ones). Initially, the bores were finished to normal reamed surface smoothness. Bore straightness conformed to capabilities of normal gun drilling practice (0.003–0.005-inch T.I.R. per inch along the bore). The relatively higher projectile acceleration profiles generated by light-gas guns produced failures of launch packages in these launch tube bores. The failures destroyed individual launches and often caused serious cratering of the tube walls. Early launch tube repair practice at many facilities involved removing raised lips from the craters with bore fitting drills or reamers. The tubes were kept in service. Strangely, these repairs worked reasonably well as long as launch velocities remained below, perhaps, 5.0 km/s. Repeated firing failures at higher launch velocities, however, forced reconsideration of launch tube bore wall conditions.

The bore walls were honed initially to smoothness levels of 1–2 μm RMS (25–50 micro-inch) or even better. Bore cleanliness prior to firing also became an extremely important issue. Several laboratories used spring hones and even positive-displacement hones to clean and polish launch tube bores after each firing. Optical and mechanical techniques were used to evaluate bore straightness. During fabrication, special techniques were developed to improve the straightness of gun-drilled holes and bending techniques were used to further straighten tubes after drilling and bore surface preparation.

Joints in gun tubes were found to be inadequate to meet the new higher standards for gun bore integrity. Slight misalignments between bore surfaces at joints produced positive steps that shattered oncoming projectile packages and abrupt changes in direction of bores at even smoothed joints also caused projectile failure. For thirty years (from 1960 to 1990), the first-quality light-gas gun operation required that launch tube joints be eliminated through use of one-piece launch tubes. This restriction limited aspect ratios to those achievable through gun drilling of individual steel rods. Aspect ratios up to about 125 could be achieved readily and values up to 150 or so could be reached in some cases, but only with reduced straightness tolerances. The tubes became extremely expensive, which added to the cost of light-gas gun firing and also made occasional bore damage an extremely serious matter.

Very large diameter tubes with ordinary aspect ratios would prove too long to transport to range sites. Methods were developed to weld such tubes together precisely enough to allow special machining processes to produce excellent bore conditions near and at the original joint sites.

Finally, joint design techniques were developed that allowed tubes to be coupled precisely enough to remove the requirement that no joints be used. Since about 1990, launch tubes with multiple joined sections have become common. Semi-expendable tube segments mounted at the up-range ends of the launch tube assemblies are installed to absorb the pressure deformation, erosion, and general wear and tear that occur during early phases of projectile launches. These segments can be removed and replaced once surface damage becomes irreparable, so to launch tube assemblies can be rejuvenated at a fraction of the cost of complete launch tube replacement.

One example of a modern joint innovation that made launch tube joints possible is the *upset metal boss–receptacle geometry.* Bosses and receptacles have been used to join tubes for many years because they provide good alignment and accept a variety of seal technologies. Normal machine shop practice allows concentricity of the two joined bore walls to be specified within 25 µm (0.001 inch). Of course, tolerances of individual surface are additive for an actual joint. Careful fabrication using modern machine tools can reduce total joint concentricity tolerances to 2.5 µm (0.0001 inch). In addition, clearances must be added to accommodate joint assembly/disassembly. Generally, diametrical clearance of 5.0 µm/cm bore diameter is needed to allow proper assembly of standard boss-receptacle joints. No control is available to determine where bore ends are located within their diametrical clearances. The upset metal joint design concept eliminates the clearance portion of the problem. Basically, the boss of the joint is machined very slightly larger in diameter than that of the receptacle so that the joint must be forced (pressed) together. The press fit is made very light so that conventional gun clamps such as right-hand/left-hand nuts can both close and reopen the joints. The important advantage of upset metal joints is that they eliminate clearance-produced bore misalignments. Obviously, the alignment shift caused by joint movement within the clearance that results from recoil loading also ceases to exist.

Upset metal joint techniques plus other related ones have effectively eliminated serious gas gun operational problems produced by joints. As a result, bore joints have proliferated in modern light-gas gun designs.

1.3.5 Evolution of Quick-Opening Valves

An obviously critical parameter controlling light-gas gun firing cycles is the point in the cycle when the projectile package is exposed to the light-gas charge. An important characteristic of the projectile release valve is the speed with which it operates (from fully closed to fully open). The valve should open quickly enough to ensure complete operation before significant projectile package motion occurs. For small gas guns launching projectiles to hypervelocities, opening times under 10^{-5} s (10 µs) are required to meet this criterion. The valves must also open completely so that no impediment to gas flow remains in the launch tube if the valve is not to limit gas flow efficiency and, hence, light-gas gun capability. Finally, and most importantly, the valve must operate at the proper instant during the firing cycle. The local gas pressure achieving a critical value usually establishes the proper time for projectile launch to occur. This requirement is not as extreme for two-stage gun operation as might be assumed since it usually occurs during a period when gas pressure is rising very rapidly (well before peak light-gas pressure is reached). Thus, an error in release pressure produces only a small change in projectile release time.

The first quick opening valves were shear diaphragms (which are still used with some light-gas guns). This valve is a disc (generally, but not always, metallic) that closes the upstream end of the launch tube immediately behind the projectile mounting location. Conventional o-rings in grooves are used to seal the gas in the pump tube before gun firings. As applied gas pressure increases, shear stress develops across a cylindrical surface through the disc material at the edge of the launch tube bore. A bore-diameter plug is punched to open the valve. This plug accompanies the package as it accelerates along the launch tube.

The most important advantage of shear disc valves over competing devices is their simplicity and predictability of operation. They open very rapidly (within a single microsecond) and expose the entire launch tube diameter to gas flow almost immediately. The most serious disadvantage involves the plug which accompanies the projectile package along the launch tube.

- The plug is nearly bore fitting and its edges occasionally scratch the bore wall as the plug travels the length of the launch tube.
- The plug has mass, which must be accelerated along with that of the launch package.
- The plug accompanies launched projectile packages downrange. It may either disturb the sabot/projectile separation processes or impact the target immediately after projectile impact, thereby obliterating useful data.
- The plug may impact the launch package and destroy it early in the launch cycle. If light-gas seeps around the shear disc and enters the launch tube prior to firing, it can cause the launch package to move forward a short distance. During gun firing, the valve is sheared and the plug is accelerated violently as it approaches and impacts the launch package. Often the launch tube bore wall is cratered by such an event.

Several quick-opening valve mechanisms have been developed to replace shear diaphragms. All of them were designed to prevent launch of valve material along the launch tube during and after projectile release. A notable early design is sketched in Fig. 1.8. It consists of a duct with a rectangular cross section somewhat wider than the launch tube bore diameter and mounted across the up-range end of the launch tube, plus a specially designed metallic slider that can move freely along the track but seals its cross section [2]. The slider, referred to as a *broken beam*, is much too thick to be perforated by the pressurizing light-gas. It contains a blind slot in its up-range face that is exposed to the light-gas charge. Gas pressure within the slot produces forces within the slide that tend to move the right-hand side of the slider to the right and the left-hand side to the left. These forces are resisted by the tensile strength of both the slider material located out-board of each slot end and the material below the bottom of the slot. At a critical gas pressure, the slider fails in tension and its two segments are driven rapidly outward to open a gas flow path along the launch tube bore. The segments proceed nearly to the ends of the retaining slot where they are deformed in one way or another to prevent their reversing direction and re-closing the valve.

These broken-beam valves work very effectively and reliably. As with shear diaphragm valves, critical pressure levels needed to open them can be evaluated easily. They can be configured readily to open at a wide range of gas

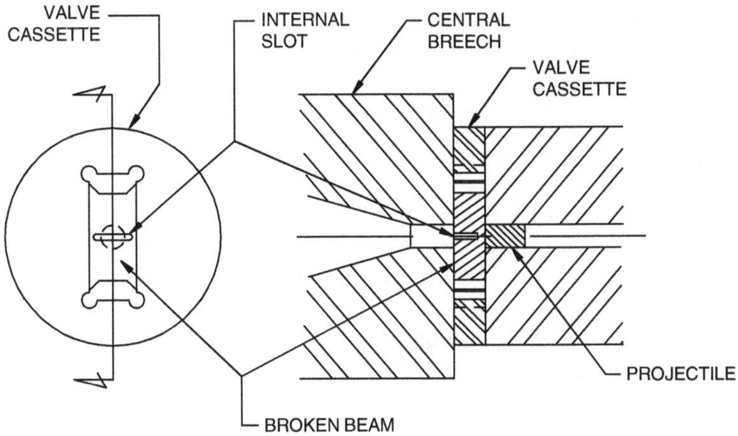

Fig. 1.8. Sketch of a broken beam quick opening valve installed at the base of a light-gas gun launch tube

pressures. Also, they can be configured to operate on command rather than only when critical gas pressure levels are reached. This unique capability is potentially useful for precise control of the time of gun firing.

All of these technologies have been largely supplanted by petaling diaphragms, which have proven to be inexpensive, reliable, and simple to use. These diaphragms are designed to bulge into radiused receivers and then tear open as the applied gas pressure reaches a pre-chosen critical value. The discs contain two or three evenly spaced scored grooves across their diameters, which control the pressure needed to initiate tearing. The formed petals then bend over a rounded retainer surface to open a clear path for gas flow. When they operate properly, no material is projected down-range and no significant resistance is offered to gas flow. The principal disadvantage of petaling diaphragms is complexity of their design [3]. Rupture pressure is established experimentally, even today. Rupture pressure is generally controlled by adjusting the score depth, but each diaphragm thickness chosen can accommodate only a small range of score depths so a variety of petaling diaphragm blank designs is needed if a wide range of burst pressures is needed. If the groove depth chosen is too small, the diaphragm may shear rather than petal or material tearing may not follow the grooves. Diaphragm material is generally projected along the launch tube whenever this class of failure occurs. If the scores are too deep, the petals may not experience enough force initially to open them completely (this factor does not usually plague light-gas gun operation because gas pressures rise rapidly and far immediately after valve operation during two-stage light-gas gun firings). Full-opening failures are more serious for other petaling diaphragm applications such as single-stage light-gas gun operations. With all these problems, petaling diaphragms have still become by far the most popular of the quick-opening valve mechanisms available to light-gas gun designers and operators. The few light-gas guns that do not use petaling diaphragms at present employ shear disc valves.

1.3.6 Choice of Light-Gas for Gun Operation

All early light-gas gun operations were conducted using helium for the light-gas charge. It was chosen as being the only notably light gas which was chemically inert. Helium used in light-gas guns proved effective in boosting gas gun launch capabilities to velocities up to about 4.5 km/s (14,800 ft/s,) but serious problems were encountered when higher velocities were sought. The principal problem seemed to be over-heating of gun surfaces exposed to peak gas pressures (such as the taper within the central breech, the quick opening valve and the up-range portions of the launch tube bore). Evidence of gross material melting was impossible to ignore and hints existed of localized material boiling. Thin layers of steel were "plated" onto ballistic range components exposed to the gun muzzle blast. Obviously, steel was being launched along with projectile packages by energetic light-gas gun firings. At best, the steel gun walls were melted and fine metal droplets were entrained in the gas flow, which caused the total mass launched to climb rapidly above that of the designated package. At worst, steel was vaporized and iron gas mixed with the light-gas charge, which caused the molecular weight of the driver gas to increase well above the 4.0 g/mole value characteristic of helium.

The reason for intense gun heating can be demonstrated at least partially by considering the pressure produced by isentropic compression of an ideal gas. This pressure is given by the equation

$$P = P_0 C^\gamma \,, \tag{1.1}$$

where P is the instantaneous gas pressure, P_0 is the original (loading) gas pressure, C is the gas compression ratio, and γ is the ratio of gas specific heats ($\gamma = 1.67$ for helium.) The temperature of the isentropically compressed perfect gas is given by the equation

$$T = T_0 C^{\gamma-1} \,, \tag{1.2}$$

where T is the instantaneous gas temperature and T_0 is the original gas temperature.

Note that the relatively large valve of γ for helium causes higher gas temperatures than are produced by isentropic compression of alternative diatomic gases such as hydrogen, for which $\gamma = 1.40$. Additionally, the small size of helium atoms and their chemical inertness combine to make helium the best gas there is for transferring heat convectively (the process that controls heating of gun walls exposed to flowing gas).

Most workers in the field concluded that hydrogen should prove more effective than helium for light-gas gun charges. Fear was prevalent, however, that use of hydrogen might cause catastrophic explosions. After much debate, staffs at several laboratories including the staff at NRL and NASA Ames decided that hydrogen could probably be used safely if the gas guns were sealed well and fired into evacuated range tanks so that hot hydrogen never encountered atmospheric oxygen. The range atmosphere contains so little oxygen that hydrogen burning becomes impossible. Some time after each firing, when the launch gas has cooled thoroughly in the range tanks, these tanks could be re-evacuated and the hydrogen expelled through vacuum pumps located some distance from the range and exhaust lines that were vented out-of-doors.

Hydrogen was first tried during the early 1960s and the results were startling. Light-gas gun launch velocities increased immediately and substantially. These results were supported and extended during subsequent gas gun firings at many laboratories around the world. Very few fires or explosions have ever been reported during conduct of several hundred thousand light-gas gun firings using hydrogen. Little serious damage has been done and no range personnel have ever been hurt. These factors have combined to make hydrogen accepted nearly universally as the gas of choice for operating two-stage light-gas guns.

1.3.7 Search for Higher Gas Temperatures

The stunning success of hydrogen as a driver gas for light-gas guns underscored the importance of gas characteristics to gas gun operation. Clearly, the most important gas characteristic of all is the peak speed it can achieve flowing along a launch tube, since this speed provides an ultimate limit for two-stage light-gas gun launch capabilities. A convenient means of ranking such capabilities is to consider the speed of sound in the gas. Sound speed also affects gun operation in another related way. Information traveling at velocities near the local sound speed controls communication between the down-range end of the pump tube where gas pressures are produced and the base of the projectile package already traveling at high speed where package acceleration is produced. If communication along the launch tube is seriously delayed or broken, high gas pressures developed in the pump tube can no longer affect subsequent projectile package acceleration. Guns operating under these conditions are referred to as being *communications limited.*

As is commonly known, sound speed for an ideal gas is specified by the equation

$$a_0 = \sqrt{(\gamma RT)/M}\ , \tag{1.3}$$

where a_0 is the sound speed for an ideal gas, $R = 8.314\,\mathrm{J/(mole{\cdot}K)}$ is the universal gas constant, and M is the molecular weight of the gas ($M_{H_2} = 2.0\,\mathrm{g/mole}$ and $M_{He} = 4.0\,\mathrm{g/mole}$).

The shift from helium to hydrogen increased sound speed significantly (by a factor of 1.29), which is largely responsible for gas gun performance improvements. No gas exists that is lighter than hydrogen so a new approach was needed to seek yet further improvements in light-gas gun performance capabilities. Apparently, the only reasonable choice was to increase the operating temperature of the gas. Researchers already knew that this approach would ultimately be limited when high hydrogen temperatures caused melting and boiling of gun component surfaces exposed to energetic gas flows. Still, a substantial window of opportunity appeared to exist before gas heating limits became serious. Thus, the rush was on to operate light-gas guns at higher gas temperatures in order to achieve higher launch velocities.

The earliest approach investigated was to increase the compression ratio, C, of light-gas charges by making pump tubes longer. Longer pump tubes (loaded to lower initial gas pressures, P_0 in (1.1), produce larger compression ratios when pressure profiles similar to those in gas guns with shorter pump tubes loaded to higher initial pressures are reached. Equation (1.2) shows that doubling the length of a pump tube (thus doubling C) increases the peak gas temperature by a factor of 1.32 (at least for the case of ideal gas charges). Actually, compression ratios rise more quickly as do peak gas temperatures when real gas effects are considered [4]. Guns with pump tubes having aspect ratios above 300 were constructed at NASA Ames and GMDRL. They produced modest performance increases over similar guns with conventional dimensions but nothing to "write home about".

The next gas-heating approach pursued was called the *shock-heated gun.* Velocities of pump tube pistons in conventional light-gas guns are well subsonic with respect to their light-gas charges. Piston velocities were increased to the point where they moved supersonically into their gas charge columns (above 1.0 km/s for helium and 1.3 km/s for hydrogen) [5]. The resulting strong shock waves produced in the gas charge caused gas heating substantially above that anticipated for isentropic (subsonic piston) compression. Temperatures several times those of isentropic compression guns appear to have been developed when shock compression was employed. Unfortunately, very fast gas compression cycles were found to produce final gas pressure profiles that increased far too rapidly for effective light-gas gun launches. Even rugged projectile/sabot packages were often shattered. Shock compression guns could not launch projectile packages successfully even to velocities that were routine for their isentropic counterparts. After thorough investigation at a number of research facilities, shock compression light-gas guns were either converted to isentropic compression units or were retried. To the author's knowledge, no shock compression light-gas guns are currently in operation or are being considered for the future.

Another group of approaches for increasing gas temperature profiles in light-gas guns involved heating the gas in the pump tube before firing the gun. Equation (1.2) shows that increasing the original temperature, T_0 increases the final temperature, T, proportionally. Simply heating the pump tube to increase, T_0, was quickly found to be of limited effectiveness because the gas temperatures must be evaluated on an absolute scale (Kelvin or Rankine) so that the light-gas gun pump tube must be heated by nearly 300°C (540°F) to double the gas temperature.

Pulse-heating the gas within the pump tube of a light-gas gun and then firing it before the gas could cool significantly seemed to be a better idea. An experiment involving a combination of extreme gas compression and gas pulse heating was investigated by Porter at NRL during 1963 [6]. He used a tube with a bore diameter of 75 mm (3 inches) as a primary pump tube to inject hydrogen into the pump tube of a small 20 mm/7.62 mm two-stage light-gas gun. Initial hydrogen pressure below one atmosphere (partial vacuum) was loaded into the composite pump tube assembly before the coordinated firings. When actuated, this apparatus loaded the 20-mm pump tube with hydrogen preheated to 800 K. The small gun was fired immediately after the first compression stroke. Ideal gas sound speed during the small gas gun firing cycle should have increased by nearly a factor of 1.9 over firings initially at room temperature. Instrumentation showed that the system worked nearly exactly as expected from a thermal viewpoint but no performance increases over those from operating the small light-gas gun alone could be achieved.

The author at NRL [7] and Eckerman at AVCO Defense Research Laboratory (AVCO) [8] conducted independent experiments starting in 1962 that involved pulse heating gas charges in light-gas gun pump tubes with intense electrical arcs powered by capacitor energy storage banks. We at NRL were able to demonstrate initial gas temperatures up to 3000 K in the gas gun pump tube (just over ten times room temperature). Just before gun firing, sound speed increases up to a factor of 2.5 above room temperature were expected. In every case, the light-gas gun could be made to achieve launch velocities as great as those achieved when the gun was fired at room temperature, but no performance increase could be achieved.

1.3.8 Importance of Real Gas Effects Upon Light-Gas Gun Operation

During 1963, information came to the attention of the author that showed that sound speeds in real gases are significantly different from those for ideal gases. Specifically, sound speeds rise rapidly with increasing gas density (instead of being independent of gas density, as is the case for ideal gas) [9]. This news was electrifying! Increasing gas temperatures during light-gas gun firing lead to relatively modest increases in ideal gas sound speeds but to substantial reductions in gas density profiles when particular gas pressure levels needed for effective gas gun firings are achieved. A careful appraisal of gas sound speed vs. temperature for real hydrogen at peak pressure levels achieved during high performance light-gas gun firings (near 1.0 GPa, 150,000 lb/in^2) showed that the sound speed was nearly independent of elevated gas temperatures produced by external gas heating [10,11]. Increases in gas sound speed could be achieved by operating the gas gun at lower peak gas temperatures than were typical of those produced during room temperature firings. We at NRL decided to explore cool gas firing regimes. The most straightforward approach to reducing peak gas pressure was to cool the gas in the pump tube before gun firing. Initial temperatures down to 100 K near (liquid nitrogen's temperature) should produce substantial sound speed increases. The most straightforward way to lower initial gas temperatures was to cool the light-gas gun pump tube and let the gas within it come to thermal equilibrium. This direct experiment proved impossible to conduct because the steel of the pump tube and central breech would become brittle when these units were cooled below the critical temperature of the steel. The gun would burst if fired under these conditions. Instead, low initial gas temperatures were simulated indirectly by reducing the original length of the light-gas column, thereby reducing compression ratios. We inquired about how much gas compression is needed to heat real hydrogen, initially at temperatures down to 100 K, to room temperature (295 K)? This "compression" was produced geometrically by introducing diaphragms across the pump tube at appropriate points (of which the most extreme turned out to be the joint between the pump tube and the central breech), adding the hydrogen at room temperature ahead of them and evacuating the pump tube behind them. The piston was launched along the evacuated pump tube up to its customary velocity and was then allowed to crash through the secondary diaphragm and compress the light-gas charge (preloaded to appropriate pressures up to 15.0 MPa, 2200 lb/in^2). The results of these experiments, presented in Fig. 1.9, show that actual gun performance followed the theoretical sound speed quite closely. The peak velocity achieved with the simulated initial gas temperature of 100 K was 9.85 km/s which was, very temporarily, the world velocity record for two-stage light-gas guns.

The conclusions drawn from these results are that very high performance gas guns are communication limited and that extreme performance of light-gas guns appears to be largely independent of compressed gas temperature profiles. This last conclusion has received substantial support from the evolving performance levels of modern energetic light-gas guns. A good example comes from the Ernst Mach Institute. The size and shape of their original ballistic range facility precluded use of light-gas guns with anything but relatively short pump tubes. Thus, operations at high gas temperatures were beyond their capabilities. Their light-gas guns have been fired over the years at progressively higher velocities until their capabilities reached and exceeded those of any other ballistic range facilities in the world.

It is well to ponder this overall situation carefully. Many years of effort and millions of dollars were spent on numerous projects to extend performance levels of high performance two-stage light-gas guns by increasing gas temperature profiles achieved during energetic launches. When such temperature increases were achieved, firing results showed temperature levels to be largely irrelevant. Finally, the physics to explain this situation was understood and

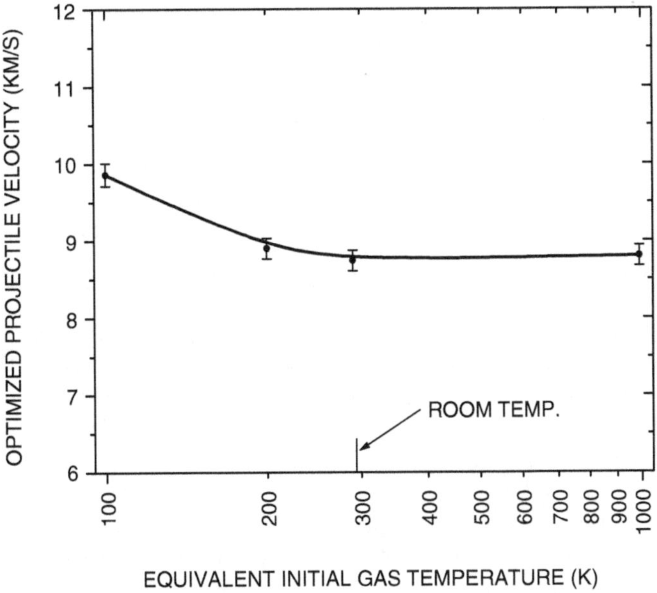

Fig. 1.9. Peak velocity firing results vs. equivalent initial gas temperatures for a high performance light-gas gun

carefully demonstrated. Meanwhile, attention to gun designs and the precise tuning of gun firing parameters was successful in extending performance levels from early capabilities near 2.5 km/s (8200 ft/s) to somewhat over 10.0 km/s (32,800 ft/s).

1.4 Analyses of Light-Gas Gun Operation

Attempts to model two-stage light-gas gun operations mathematically extend back nearly as far as the initial gas gun proposal. Access to an accurate and reliable behavior model is especially important for guiding two-stage gas gun operations because their firing cycles are so complex. In particular, a large number of parameters (both gun design and gun operational ones) are available for tailoring performance to specific needs.

1.4.1 Basic Models of Light-Gas Gun Operation

The earliest models developed used drastic simplifying assumptions to make them mathematically tractable. Their governing equations were almost simple enough to solve in closed form. All gases associated with the firing processes were assumed to be ideal, which flew in the face of general understanding that perfect gas assumptions break down when gas densities rise to high levels, especially when they approach liquid densities (which actually occurs during energetic gas gun firings). Little or no consideration was given to pressure variations produced by piston and projectile motion along the pump and launch tubes. Thus, wave action within gas columns was ignored, and no need was expressed for evaluating maximum gas flow speeds or communication between various gun locations during firing sequences. As might be expected, these models failed to yield accurate performance predictions. Generally, they over-predicted measured launch velocities. However, they were able to reflect with some reliability some important qualitative trends produced by varying gun-loading parameters. For this reason, they were useful for guiding experimental searches for optimum firing parameter combinations to achieve specific operational goals, i.e., they served as interpolation tools for seeking gun performance levels as long as these performances were well within extremes that had already been examined experimentally. The fact that these initial programs were relatively simple to use and ran rapidly on even the slow computers available during the 1960s promoted their extensive use.

The first steps toward developing programs that more realistically modeled gas gun operations involved replacing the ideal gas equation-of-state with equations that accounted for some real gas effects. Equations in closed form such as the famous van der Waals equation were tried, as were some involving series expansions such as the virial equation of state. Some modern semi-empirical equations of state that rely upon tabulated values for key gas-specific parameters were also considered. Expectations that improved descriptions of gas behavior provided by real gas equations would improve code performance failed to occur. The reverse actually occurred, at least initially. Overestimates of launch velocities became gross and the codes seemed to lose their ability to predict performance trends resulting from systematic variations of loading parameters. Later, these problems were partially resolved and the codes gained usefulness as aids for interpolating gun loading parameters to achieve specific performances within proven gun capabilities. These codes are still in use at some laboratories.

1.4.2 Method-of-Characteristics Codes for Modeling Light-Gas Gun Operation

About this time, (the late 1960s) early measurements of gas pressure levels vs. time within pump tubes and along launch tubes of operating light-gas guns became available. These plots showed large and apparently chaotic variations in gas pressure during gun firing sequences, with only general trends being recognizable. Clearly, intense waves propagated in both directions along both pump and launch tubes and their effects dominated gas gun operation. Accurate and reliable analysis codes for two-stage light-gas guns must treat carefully the formation and subsequent movement of these waves if they are to describe gun performance reliably. Workers at the United States Naval Ordnance Laboratory (NOL) formulated the first complete and coherent code for describing light-gas-gun operation in terms of the pressure waves along the gun tubes [12]. The method-of-characteristics formalism was used to describe formation and propagation of such waves. Basically, small pressure disturbances produced at the faces of moving objects (piston and projectile package faces) were evaluated at small time intervals. Propagation of these disturbances along the tubes at local sound speeds was followed mathematically during subsequent time intervals. Particular "characteristic waves" often overtook others generated during earlier intervals to produce larger pressure discontinuities that might or might not evolve into shock waves. For some unaccountable reason, ideal gas sound speeds were employed to evaluate characteristic wave motions in early method-of-characteristics gas gun codes even though sophisticated real gas equations of state for evaluating local gas temperatures and pressure levels were employed.

Results from these codes provided pressure/time variation plots at pressure gauge points along the pump and launch tubes. Measured pressure vs. time plots corresponded qualitatively with computed results. Good quantitative comparisons still eluded investigators. The original method-of-characteristic code also failed to predict gun performance with detailed accuracy, but general reliability was far better than that achieved by any previous code. Later, method-of-characteristic codes developed using the original NOL code as a model have employed realistic sound speed computations. Initially this improvement also rendered these codes notably less reliable than the NOL code for making gun performance predictions, but problems were worked out quickly for codes such as those developed at AEDC and for GMDRL [13]. These codes now give accurate predictions of light-gas gun operations where results are sought that are within the explored gun operating limits. Predictive reliability falls off as progressively more extreme gun performance levels are modeled. These codes and others have proven very useful both for guiding operations of current light-gas guns and even for supporting designs of new ones.

Several investigators, including the author, are of the opinion that some significant physics is still missing from these codes. Errors tend to accumulate during code operations, but balance one-another so that the codes perform well at least over familiar ranges of loading and gun design variables. Improvements tend to upset these balances and degrade overall code performance until the balance can be reestablished. In spite of all these problems, modern method-of-characteristics codes have come into wide use for predicting light-gas gun performance levels and for choosing loading parameters for individual firings when specific operational levels are sought.

1.4.3 Finite Element Codes for Modeling Light-Gas Gun Operation

Meanwhile, a new series of gas gun performance codes has been developed that employ finite-element methods to solve the differential equations governing light-gas gun operation. These codes treat the gun interior as a two dimensional space (positions defined by locations along the tubes and radial positions away from the gun axis, but not azimuthal positions). Boundary layer effects in the gas near metallic surfaces can be treated analytically as can effects of turbulence within gas volumes, especially within the taper section between pump and launch tube diameters. Finally, dynamic mechanical analyses of the light-gas gun structures, also using finite-element formalisms, can be coupled to the gas dynamic programs to produce nominally complete analyses of gas gun operation.

An early version of one of these finite element codes was developed at the Lawrence Livermore National Laboratories (LLNL) [14]. It served well for the design of a large medium velocity light-gas gun but its capability for analyzing operations of conventional gas gun geometries has not yet been evaluated rigorously, especially when the guns are fired energetically in attempts to extend performance limits.

1.5 Recent History of Light-Gas Gun Development

During 1968 a plateau appeared in the performance limits of both ultra-performance light-gas guns and those used as research tools. Increased velocity capabilities were not obtained until the end of depressed activities that occurred between 1970 and 1977. This period saw a general economic slowdown which put pressure on government revenues. Catastrophe struck defense and space research and development activities in the United States. Funding problems quickly spread to Western Europe. Basically, decisions were made in high United States government circles to turn away from government sponsored basic research. The era was epitomized by the Mansfield Amendment to the 1970 United States Defense Appropriation Act. It forbade military sponsorship of research that could not be credibly related to officially sanctioned weapon development projects. The overall philosophy spread to NASA and was reflected in budget cuts for nearly all government agencies that traditionally sponsored non-medical basic research. The effect of this governmental attitude was the rapid shutdown of most research activities that employed light-gas guns. At the depth of the technology recession in 1973 or so, only a few light-gas gun facilities remained in operation and they functioned well below full-scale levels. The attitude toward not sponsoring basic research appeared highly successful to government agencies for the first few years of its application because research costs were held down and technology in the pipeline served current needs admirably. By 1978, however, realization spread that problems associated with reduced research output were accumulating rapidly. Governmental policy began an accelerating shift back toward support for advanced research and development and two-stage light-gas gun activity revived at least tentatively. Revival accelerated rapidly during 1979, as the NATO countries became aware of a rapid military buildup in the USSR that required a response if they were not to experience a definitive shift of relative military power vis-à-vis the Soviet Union. Rearmament activities accelerated somewhat toward the end of the Carter administration in 1979 and much more rapidly during the first Reagan administration (1980–1984). This acceleration continued during the second Reagan administration, especially with the advent of the famous Space Defense Initiative (Star Wars) that aimed at shielding whole continents from ballistic missile attack.

The recent era of light-gas gun development (since 1980) has seen the demise of efforts to maximize launch velocities regardless of the usefulness of the launched projectiles. Rather, emphasis has been placed upon achieving useful experimental results with launched projectiles, i.e., light-gas guns became research tools rather than developmental ends in themselves. A few years were required after the general shut down between 1971 and 1977 for new ballistic research range staffs to achieve launch velocity capabilities already reached by earlier groups that dispersed. General pressure to reach higher velocities emerged slowly after 1980 but only limited progress was made until a breakthrough was achieved at the Sandia National Laboratory (SNL).

The research goals at SNL involve studying the response of materials to extremely high-pressure environments. Flat plates of precisely characterized materials (typically metals) were launched to impact targets mounted adjacent to launch tube muzzles and oriented to produce flat-on impacts. Instrumentation was then used to observe the motion of shock waves resulting from the impact and the subsequent release waves that returned impacted materials to low stress levels. The resulting data could be interpreted to reveal required information about the response of materials to the high-pressure levels achieved. Fortunately, relatively simple projectile shapes (flat plates) sufficed for conducting such experiments. These shapes could be maintained even when projectiles were subjected to extreme acceleration levels (hundreds of mega-gees and above) so the plates could, in principle, be launched by impact-induced shock waves. The breakthrough involved mounting a third stage onto the muzzle of a two-stage light-gas gun. In this third stage, plates of material in which the shock impedance varied through the thickness (typically from low density at the impact face to high density at the back of the plate) was launched against a metal plate of a material like aluminum or titanium that is resistant to fragmentation. Basic isentropic compression physics dictates that the second plate will be projected forward at a velocity that is a multiple of the original impact velocity. Special care was required to prevent melting of the second plate [15]. Launch velocities achieved by this new approach were both immediate and substantial. Velocities approaching 16.0 km/s were achieved as third stage techniques were optimized and improved upon. Further increases of velocity capability are anticipated as a group of related gun muzzle techniques are evaluated in detail. Progress has also been made launching short cylinders with some of these techniques [16]. Chapter 4 of this volume describes these techniques in detail.

Meanwhile, interest in increasing launch velocities of military ordnance re-emerged just after 1980 and remains a priority today. Interest centered initially in the range of 2.5–4.0 km/s [17]. Extensive testing programs were needed to evaluate performance of launched projectiles traveling through the atmosphere and determining their lethality against a variety of targets. Light-gas guns could easily reach needed velocities and were considered carefully for these new tasks. Fortunately, light-gas guns originally designed to launch light projectile packages to very high velocities proved to be effective for launching relatively heavy payloads to lower velocities. The robust designs needed to reach high velocities proved ideal for containing high gas pressures needed for launching massive payloads. Several investigators found that light-gas guns could produce constant peak kinetic energies of their projectile packages if their designs were conventional (with launch tube diameters one-third that of the pump tube) when operated at velocities between 1.5 km/s and 6.0 km/s. This fact allowed the simple equation

$$U_{\mathrm{p}} = \sqrt{\frac{2\,G\,d_{\mathrm{p}}^3}{M_{\mathrm{p}}}} \quad \text{or, equivalently,} \quad M_{\mathrm{p}} = \frac{2\,G\,d_{\mathrm{p}}^3}{U_{\mathrm{p}}^2}\,, \tag{1.4}$$

to be derived. In this equation U_{p} is the maximum attainable launch velocity, M_{p} is the total launch mass, d_{p} is the launch tube diameter, and $G = 40.0\,\mathrm{GPa}$ is an empirical factor adjusted to fit theory to observation. Equation (1.4) has proven useful for making convenient estimates of gun performance limits for light-gas guns as long as velocities below $U_{\mathrm{p}} = 6.5\,\mathrm{km/s}$ are of interest. Experience has shown that predicted performance levels can always be reached (and usually exceeded by small amounts).

Another recent trend involves launching very large payloads to velocities up to about 6.5 km/s (21,300 ft/s). A good example is development of a new very large light-gas gun at AEDC. [18] Their current method-of-characteristics code for evaluating light-gas gun performance was used to evaluate the performance of a light-gas gun with a 355-mm (14-inch) diameter pump tube that is 30.5-m. long. It was found that launches requiring unusually low peak acceleration (below 80,000 g) were possible from launch tubes 83-mm (3.3-inches) in diameter and 30.5-m (100-ft) long at velocities up to 6.4 km/s. In addition, launch tubes with diameters up to 203 mm (8.0 inches) were found useful for launching very large and massive projectiles, many with complex shapes and internal instrumentation. On the basis of these estimates, a decision was made to fabricate the gun, a view of which is presented in Fig. 1.10. To date, this is the largest light-gas gun yet constructed and evaluated experimentally. It has met its design requirements and is expected to provide exceptional performance for many years to come. Much larger gas guns have been proposed (with pump tube diameters up

Fig. 1.10. The largest existing light-gas gun, which is currently operating at AEDC

to 935 mm (37 inches), but detailed design and construction has not been authorized [19].

Finally, a recent trend has developed for powering small- and medium-sized light-gas guns with highly compressed helium rather than with solid propellant (gunpowder). Sensitivity about perceived safety and security problems with chemical propellants is responsible for this development. Except for gun cleanliness, no advantages for powering light-gas guns with compressed gas have yet been identified. Significant disadvantages of using compressed gas to power light-gas guns include increased system cost and relatively intense danger of handling very high-pressure gas (typically between 34 MPa or 5000 lb/in^2 and 102 MPa or 15,000 lb/in^2). Most workers in the field anticipate that this trend will pass and the future guns will return to being powered by solid propellant.

1.6 Future Trends

Predicting future trends in technology is an extremely dangerous sport that exposes most of those attempting it to embarrassment! This consideration leads me to limit my predictions of future light-gas gun developments to only a single area. The demise of two-stage light-gas guns has been predicted and even proclaimed several times in the past. Each time their inherent capabilities and technical "friendliness" have saved them from extinction and have lead to their more or less continuous gain in popularity. They have found use in an ever-widening range of important research and development activities. I feel safe in predicting that light-gas guns will retain their popularity for the foreseeable future. Frankly, I am unsure what new tasks will emerge for them to support as time goes on, but I am convinced that new tasks will arise to make use of their special capabilities.

References

1. E.J. Workman, *Development of New Gun Using Helium Gas for Projectile Acceleration*, Final Report NMSM/RDD/T70, New Mexico School of Mines, June 1952.
2. C.D. Porter, H.F. Swift, and R.H. Fuller; "Summary of NRL Hypervelocity Accelerator Development," Proceedings of the 5th Symposium on Hypervelocity Impact, Denver, October/November 1962.
3. J.J. Rast, *The design of Flat-Scored High-Pressure Diaphragms For use in Shock Tunnels and Gas Guns*, NAVORD Report 6865 (Ballistic Report 29), U.S. Naval Ordnance Laboratory, White Oak, MD, September 1960.
4. H.F. Swift, and J.R. Baker, *Hypervelocity Capability and Impact Effect Program*, Final Report, USNRL Memorandum Report No. 1677 U.S. Naval Research Laboratory, Washington, D.C. March 1966.
5. H.F. Swift, in: *Impact Dynamics* (eds J.A. Zukas, T. Nicholas, H.F. Swift, and D.R. Curran) pp. 215–239, John Wiley Interscience, New York, 1982.
6. H.F. Swift, and J.R. Baker, *Hypervelocity Capabilities and Impact Effects Program, Appendix 1 Discussion of Gas Gun Augmentation*, U.S. Naval Research Laboratory Report No. 1677 Washington D.C., March 1966.
7. H.F. Swift, "Study of Electrically Augmented Gas Guns"; Proceedings of the 7th Hypervelocity Impact Symposium; Volume 1, pp. 61–102 Tampa, FL November 1964.
8. J. Eckerman, and W.L. McKay, "Performance of a Three Stage Arc Heated Light Gas Gun"; Proceedings of the 6th Symposium on Hypervelocity Impact Vol. 1 pp. 247–303 Cleveland, OH August 1963.
9. J.R. Baker, W.H. Getches, and H.F. Swift, *Theoretical Thermodynamic Properties of Gases at High Temperatures and Densities with Numerical Results for Hydrogen*, U.S. Naval Research laboratory Report 6675 Washington, D.C. September 1968.
10. A.E. Seigel, *The Theory of High Speed Guns*, AGARD-O-graph 91, May 1955.
11. J.R. Baker, "Light-Gas Gun Performance: Theoretical Analysis and Experimental Comparison," Proceedings of the 7th Hypervelocity Impact Symposium; Tampa, FL November 1964.
12. J. Hunter, formally with LLNL, Livermore CA, Personal Communication.
13. A.C. Charters and D.K. Sangster, *Fortran Computer Program For Interior Ballistic Analysis of Light-Gas Guns*, GMDRL Report 5643, Goleta, CA July 1973.
14. L.C. Chhabildas, L.M. Kmetyk, W.D. Reinhart, and C.A. Hall, *Int. J. Impact Engng.* **17**, pp. 183–194 (1995).
15. L.C. Chhabildas, T.G. Trucano, W.D. Reinhart, and C.A. Hall, *Chunk Projectile Launch Using the Sandia Hypervelocity Launcher Facility*, Sandia National Laboratories Report No. SAND 94-1273 UC-410, Albuquerque, NM July 1994.

16. H.F. Swift and E.R. Larrison, *Feasibility of Hypervelocity Ordnance (Estimate of the Usefulness and Feasibility of Hypervelocity Weapons in Modern Combat)*, U.S. Army Armament Research and Development Command, Report No. ADOE-400 338, July 1979.
17. F. Lu and I. Marren *Advanced Hypersonic Test Facilities*, Vol. 198, pp. 531–554. A1AA Publication, New York City, 1995.
18. H.F. Swift, "Concept Definition Study for an Extremely Large Aerophysics Range Facility," NASA Contractor Report 4491 NASA Langley Research Laboratories, Langley, VA. 1993.
19. S.O. Baily, A.B.J. Clark, D.A. Hall, and H.F. Swift, "Recent Developments of Light-Gas Guns at the U.S. Naval Research Laboratory," 3rd Symposium on Hypervelocity Impact, Chicago, IL.

2 Scaling and Designing Large-Bore Two-Stage High Velocity Guns

Henri Bernier

2.1 Introduction

A number of different methods can be used to launch a projectile at a high velocity exceeding a few kilometers per second. In preparing this chapter, I have chosen to limit the presentation to a device called a two-stage light-gas gun that is capable of doing this.

The information related here is not always new and may not be completely up-to-date. The use of new materials with new capabilities, and a better knowledge of coefficients such as friction, heat transfer, and head loss, might have modified the final results obtained. Nevertheless, most of the mechanical problems remain unchanged. The general equations of the behavior of the materials and the methods for scaling and designing such a device are the same, or nearly the same, as thirty years ago.

This chapter is divided into four parts: (i) the scaling of a two-stage light-gas gun, (ii) the design of such a gun, (iii) the operation of a two-stage gun and, (iv) a comparison between the numerical simulations and the experimental results.

In this chapter many examples have been chosen from the large two-stage light-gas gun of the Commissariat à l'Énergie Atomique (CEA), studied and built between 1968 and 1974. Unfortunately, this gun has now been dismantled and the laboratory is shut down.

2.2 The Scaling of a Two-Stage Light-Gas Gun

2.2.1 The Single-Stage Gun

The Single-Stage Gun as a Projectile Launcher

For many years, two types of single-stage guns were in general use: powder guns using a standard gunpowder as the propellant, and gas guns using a compressed inert gas as the propellant. A powder gun with a reasonably large bore launches a projectile with a maximum velocity of about 2.0–2.2 km/s. The achievable velocity does not exceed 3 km/s for a low weight projectile. A single-stage light-gas gun using helium or hydrogen as the propellant provides increased projectile velocity.

After a description of this type of single-stage gun, we will examine the reasons for the limitations on this projectile velocity. To increase the projectile velocity, two possibilities are examined: first, increasing the initial reservoir pressure and, second, reducing the decrease in base pressure that occurs as the projectile is launched. These two ideas form the basis of the two-stage light-gas gun.

The single-stage light-gas gun illustrated in Fig. 2.1 can be described in terms of its two main elements:

- a reservoir that is filled with gas under high pressure, and
- a launch tube that guides the projectile.

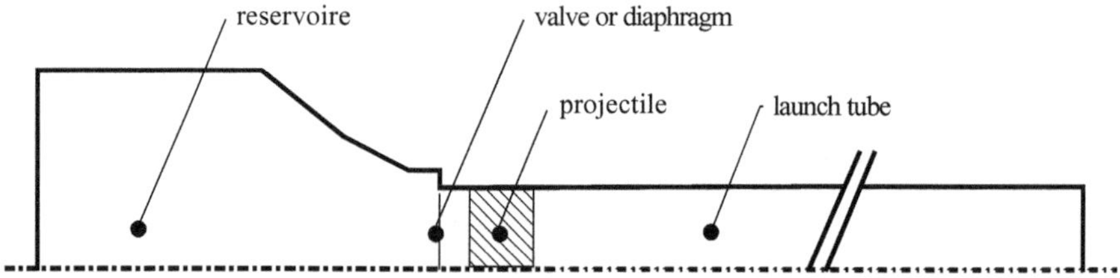

Fig. 2.1. Schematic illustration of a single-stage gas launcher

The downstream part of the launcher is separated from the reservoir by a fast-opening valve. When this valve is open, the pressure of the reservoir rapidly accelerates the projectile from rest. At the moment the valve is opened a rarefaction fan is formed at the base of the projectile and propagates backward to the rear face of the reservoir. As long as the projectile moves forward in the launch tube, its velocity increases but its base pressure decreases. This evolution continues until the base of the projectile reaches the muzzle of the launch tube.

During the motion of the projectile in the launch tube, the internal energy of the driving gas is progressively converted into kinetic energy. A part of this energy resides in the projectile and the remainder resides in the gas. In a first approximation for a given initial internal energy, minimization of the kinetic energy of the driving gas maximizes the kinetic energy of the projectile. The kinetic energy that can be taken by a gas is directly connected to its molecular weight. Therefore, the use of a gas of low molecular weight, such as helium or hydrogen, seems most appropriate. We now consider the advantages that can be obtained by using a low molecular weight driving gas.

The Capabilities and Limitations of the Single-Stage Light-Gas Gun

If the molecular weight of the gas seems an important element, the final velocity of the projectile is, in fact, the direct consequence of the law describing the pressure on the base of the projectile. To get a velocity as high as possible with a launcher, it is important to choose the driving gas for which the pressure decrease at the base of the projectile during its motion in the launch tube is as small as possible. Here we use a simple analytical solution to compare the capabilities of the two light gases, helium and hydrogen. For reference, nitrogen is also considered (Air, a mixture of nitrogen and oxygen, would give almost the same results.)

Using a one dimensional tube, one in which the reservoir has the same diameter as the launch tube, this method has the advantage of providing a complete analytical solution for the flow, assumed to be isentropic, with only one family of characteristic curves as shown in Fig. 2.2 (the length of the reservoir is assumed to be long enough that the rarefaction fan coming from its rear face does not catch the base of the projectile). In this mathematical model, the behavior of the gas is given by a perfect gas equation of state with constant specific heats. Only the main problem is examined. The losses by friction, heat transfer, etc., are not taken into account. This model gives the equation

$$p = p_0 \left[1 - \frac{\gamma - 1}{2\,a_0}\,v\right]^{2\,\gamma/\gamma-1} \tag{2.1}$$

for the pressure decrease at the base of the projectile. In this equation, the values p_0, a_0 and v are, respectively, the values of pressure and sound velocity at the initial time, and the velocity of the projectile; γ is the ratio of the specific heats.

For a launcher with a launch tube of internal cross-sectional area S, of length L, accelerating a projectile of mass M, the equation giving the velocity is

$$\frac{p_0\,S\,L}{M} = \frac{2\,a_0^2}{\gamma+1}\left[1 + \frac{\dfrac{(\gamma+1)\,v}{2\,a_0} - 1}{\left[1 - \dfrac{(\gamma-1)\,v}{2\,a_0}\right]^{\gamma+1/\gamma-1}}\right]. \tag{2.2}$$

This equation gives the length of the launch tube as a function of the velocity of the projectile at its exit (muzzle). The easiest way to solve this equation is to use v as parameter, and calculate the length of launch tube required to achieve this velocity.

From the (2.2), the maximum velocity for the projectile is the value corresponding to a complete release of the driving gas from $p = p_0$ to $p = 0$,

$$\left[1 - \frac{\gamma - 1}{2a_0}\,v\right] = 0 \quad \Rightarrow \quad v_{\text{lim}} = \frac{2a_0}{\gamma - 1}\,. \tag{2.3}$$

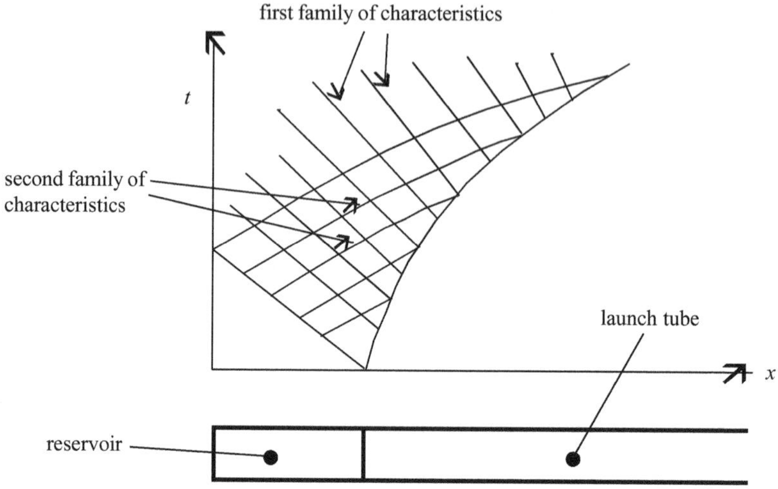

Fig. 2.2. An x–t diagram

Calculations for three gases, with initial conditions chosen for $T = 300\,\mathrm{K}$, are presented in Table 2.1. Obviously, these values are not experimentally achievable but, at least, they show the relative effectiveness of these three gases.

Table 2.1.

Gas	γ	a_0, m/s	$v_{\max}$, m/s
nitrogen	1.400	343	1715
helium	1.663	970	2926
hydrogen	1.406	1258	6197

The distance of the projectile travel, x, is given as a function of its velocity, v, with the same model and with initial conditions p_0 and a_0, is given by the equation

$$x = \frac{M}{Sp_0}\,\frac{2a_0^2}{(\gamma+1)}\left[\frac{\left(\frac{\gamma+1}{2\,a_0}\right)v - 1}{\left(1 - \frac{(\gamma+1)\,v}{2\,a_0}\right)^{\gamma+1/\gamma-1}} + 1\right]. \qquad (2.4)$$

The results given by this equation for a launcher of length L, using a projectile of 100 mm caliber (S = cross-sectional area of this barrel), a mass M of 1 kg, and a reservoir filled to a pressure $p_0 = 50\,\mathrm{MPa}$ (500 bars) are presented on Table 2.2.

Table 2.2.

Launcher length, L		1 m	2 m	3 m	4 m	5 m	6 m	7 m	8 m
nitrogen	p/p_0	0.129	0.080	0.059	0.048	0.040	0.034	0.030	0.027
	V, m/s	434.5	517.8	567.9	603.7	631.5	654.3	673.5	690.0
helium	p/p_0	0.316	0.224	0.178	0.149	0.129	0.114	0.102	0.093
	V, m/s	599.6	753.0	850.8	922.9	980.2	1027.6	1068.1	1103.4
hydrogen	p/p_0	0.449	0.348	0.296	0.255	0.228	0.207	0.190	0.176
	V, m/s	675.6	874.7	1007.5	1108.8	1191.1	1260.6	1321.0	1374.2

One can see that the base pressure decreases very rapidly at the beginning of the launch tube (the first two meters). The acceleration, $\mathcal{A}$, in direct correlation with the base pressure, decreases in the same manner ($\mathcal{A} = pS/M$).

2.2.2 The Two-Stage Light-Gas Gun

To increase the muzzle velocity of a gas gun it is necessary to:

- increase the initial pressure in the reservoir,
- reduce the decrease of the base pressure that occurs during the motion of the projectile in the launch tube.

This is what the two-stage light-gas gun is designed to do.

How to Improve the Capabilities of a Gas Gun

The Effect of Increasing the Initial Pressure in the Reservoir

What results can reasonably be achieved by increasing the initial pressure in the reservoir? To answer this question, we calculated the muzzle velocity of a single-stage light-gas gun with initial conditions obtained by isentropic compression of the driving gas (treated as a perfect gas).

In order to obtain a good comparison, we introduced the same initial compression energy,

$$E_i = E_0 - E_{00} = \frac{1}{\gamma - 1}\left(p_0\, v_0 - p_{00}\, v_{00}\right), \tag{2.5}$$

in each of the gases considered. The initial values chosen for these calculations were: $p_{00} = 5 \times 10^5$ Pa (5 bars) in the first case and 1 MPa (10 bars) in the second case. The reservoir volume, v_{00}, is 500 liters and the initial temperature is 300 K.

For two compression energies E_1 and E_2 chosen to obtain, for hydrogen, the same pressure p_0 at the end of the isentropic compression for the two cases, the results are presented on Table 2.3. The quantity $V_{\rm lim}$ is the theoretical maximum velocity given by (2.3). For nitrogen and hydrogen, the maximum pressure is about 1 GPa (10 kbar), but is 400 MPa (4 kbar) for helium.

Table 2.3.

Case	Gas	p_0, bars	a_{00}, m/s	a_0, m/s	$V_{\rm lim}$, m/s
$P_{00} = 5 \times 10^5$ Pa	nitrogen	10935	343	1029	5145
$E_1 = 5 \times 10^6$ J	helium	3925	970	3662	11049
	hydrogen	10555	1258	3799	18714
$P_{00} = 1 \times 10^6$ Pa	nitrogen	10867	343	931	4655
$E_2 = 7.9625 \times 10^6$ J	helium	4635	970	3297	9947
	hydrogen	10555	1258	3437	16932

To see which gas is the best for accelerating the projectile, the muzzle velocities in m/s for three lengths of the launch tube are presented in Table 2.4.

Table 2.4.

Designation	Gas	$L = 6$ m	$L = 30$ m	$L = 60$ m
	nitrogen	2538	3062	3262
Case 1	helium	3976	5581	6237
	hydrogen	5761	7958	8878
	nitrogen	2357	2823	3000
Case 2	helium	3918	5349	5919
	hydrogen	5458	7443	8267

These results have been obtained for a 100 mm caliber projectile of 0.5 kg mass. Nitrogen is presented as reference, and the 60 m launch tube is certainly a practical limiting value.

Comparing the three gases, the highest muzzle velocity is obtained with hydrogen. The increase of the velocity between 6 m and 30 m is about 500 m/s for nitrogen, about 1500 m/s for helium, and about 2000 m/s for hydrogen. Between 30 m and 60 m, the increase of velocity is about 200 m/s for nitrogen, about 600 m/s for helium, and about 900 m/s for hydrogen. But the most important result of these calculations is that, with the same final pressure, p_0, it is the lowest initial pressure (Case 1) that gives the highest velocity. The reason for this is that the final temperature is higher in the first case than in the second case.

Dynamic compression of the driving gas appears to be a good method for increasing the muzzle velocity. Technically, one uses a tube, called a pump tube, filled with a light-gas. A piston, generally pushed by the combustion of gunpowder, compresses the light-gas. This first part of the total launcher is called the first stage of this gun. When the compression is obtained, the gas has a low volume but is at high pressure. The problem is to use this small reservoir to initiate the motion of the projectile. The solution is to open a valve that connects this end of the pump tube with the launch tube. In many cases, this valve is a diaphragm that ruptures at a predetermined pressure. By this method, the first condition, provision of a reservoir at high pressure, is satisfied.

For the second condition, we have to consider how it would be possible to reduce the pressure decrease that occurs at the base of the projectile during its motion in the launch tube.

Release of the Pressure at the Base of the Projectile

Table 2.1 showed the drastic decrease of the pressure at the base of the projectile, especially at the beginning of its motion.

To reduce this release of pressure, a suggestion is to use the motion of the piston in the pump tube, motion that reduces the volume of the reservoir. The best possibility would be to realize the motion of the piston in such a way that the pressure at the base of the projectile remains almost constant during all the motion of the projectile in the launch tube. This is the very simple basis for a theoretical simulation: to assume that the base pressure is constant throughout the motion of the projectile in the launch tube. This model is called the *constant base pressure model.*

With the assumptions of an isentropic one-dimensional flow, without losses, without heat transfer, and treating the gas as a perfect gas, the mathematical solution is analytic. Such a flow exhibits three principal characteristics:

- the base pressure is constant during all the motion of the projectile in the launch tube,
- the flow velocity behind the projectile is a function only of time, $u = u(t)$, whatever the position, x, of the projectile is, so

$$\frac{\partial u}{\partial x} = 0 , \tag{2.6}$$

- at any given time, all the gas particles in the launch tube experience the same acceleration.

It is easy to show that these assumptions are equivalent, so the choice of any one of them ensures that the other two are satisfied. G. Wilenius [1] chose the second assumption as the basis of his work, J. Curtis [2] chose the third one. This type of study gives the time dependence of the pressure in the reservoir that is required to maintain a constant pressure at the base of the projectile. Knowing this pressure history makes it possible to determine the required motion of the piston, but no one knows how to ensure that the piston actually undergoes this motion. Therefore, one tries to produce a motion approaching this ideal by a combination of three ideas:

- use a pump tube of an internal diameter larger than that of the launch tube (the velocity of the projectile very quickly exceeds that of the piston),
- begin the motion of the projectile before the piston stops moving,
- adjoin a conical shape at the end of the pump tube to reduce the final volume of the light-gas.

With these conditions, the light-gas flows forward and begins pushing the projectile immediately after the diaphragm ruptures. Its pressure decreases in the neighborhood of the projectile but increases near the front face of the piston as it moves forward, and in this way reduces the volume of this moving reservoir. This volume decreases drastically when the piston enters the conical portion of the pump tube. This is frequently called *accelerated reservoir operation.*

The conical shape provides a smooth transition from the pump tube to the and launch tube. In addition, it both forces the extrusion of the front face of the piston and arrests its motion. Figure 2.3 is a schematic illustration of such a gun (the CEA two-stage light-gas gun).

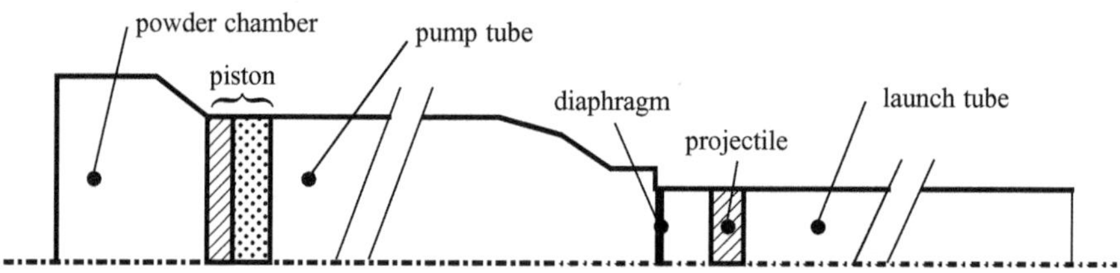

Fig. 2.3. Schematic illustration of the parts of the CEA two-stage light-gas gun

To scale such a gun one must take into account the complexity of all arameters that describe the geometrical characteristics of the elements of a two-stage light-gas gun, and the choice of the best initial conditions to obtain the required projectile motion. Numerical simulation of the behavior of all flows in the two tubes is very useful for this purpose.

General Information on a Two-Stage Light-Gas Gun

Thermodynamic and gasdynamic considerations show that the light gases (those of low molecular weight) are the most attractive gases for launching projectiles at high velocity. This is the reason these launchers are called *light-gas guns.*

The motion of the piston is produced either by the combustion of gunpowder or the release of an inert gas under high pressure.

The operation of a two-stage light-gas gun can be described as follows:

- combustion of the gunpowder (or the release of an inert gas) launches the piston that compresses the light-gas.
- when the gas pressure is high enough, the diaphragm ruptures. The light-gas then enters the launch tube and flows to the base of the projectile.
- under this increasing pressure, sometimes a shock wave, the projectile begins moving and is pushed to the muzzle of the launch tube. During this time, the light-gas pressure decreases.

A general scheme is shown on Fig. 2.4. As this device is made of two guns, the first one pushing the piston and the second one the projectile, it is often called a *two-stage light-gas gun*.

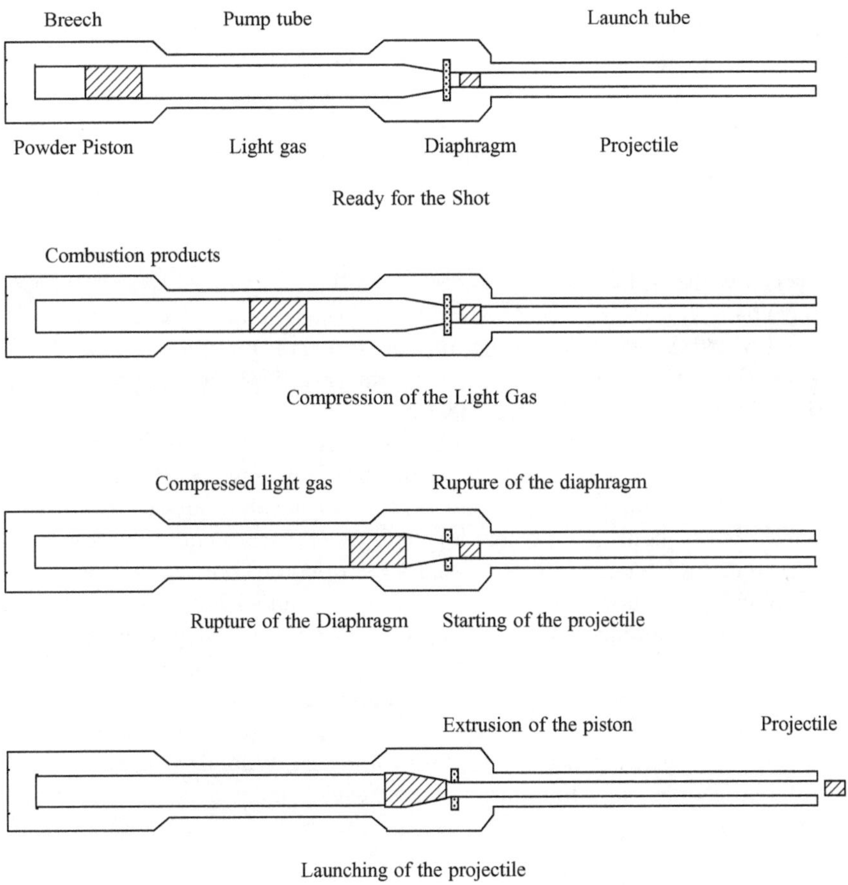

Fig. 2.4. The operation of a two-stage light-gas gun

Some Historical Aspects

We believe that it was at the end of the Second World War that the first two-stage light-gas guns were built. As explained by A.C. Charters [3], the idea of the light-gas with a driving piston was suggested by a group of scientific professors of the New Mexico Institute of Mining and Technology (U.S.A.). W.D. Crozier and W. Hume published the first paper of this group in 1957 [4]. With this concept, several variants of this device were built in many countries, especially in the United States. In France, light-gas guns were built at LRBA in Vernon and ISL in Saint Louis.

At that time, the only method of scaling such a gun was to find a mathematical model to simulate the internal ballistics of the projectile in the launch tube. One of them was the *constant base pressure model.*

In 1962, J.S. Curtis [2] added a cone as a transition just before the diaphragm between the pump tube and the launch tube. The piston is made of two parts: a heavy, rigid rear part and a front part made of plastic material that is easily extrudable. To maintain the pressure at the base of the projectile as constant as possible, the diaphragm is designed to rupture shortly after the piston motion begins. This type of gun is called an *accelerated reservoir gun.*

To define the optimal dimensional characteristics of the different parts of the gun, and the optimal initial conditions for a shot, it was important to simulate, by calculation, the behavior of all parts of such a gun.

At the beginning, with simple assumptions, the system of equations was analytically integrable. Later, rapidly expanding computer capabilities completely changed the approach because they allowed simulation of not only the internal ballistics of the projectile but also the total internal ballistics of the gun (piston motion, compression of the light-gas, diaphragm rupture, release of the light-gas, motion of the projectile, etc.)

R. Piacesi, D.F. Gates, and A.E. Seigel wrote one of the first computer codes in 1963 [5]. Focusing on the main problem, while neglecting friction losses and heat transfer, this code takes into account all the parts (gun powder, piston, light-gas, projectile). The conservation equations for mass, momentum, and energy are written in Lagrangian coordinates for a one-dimensional plane flow (in fact, one and a half dimension). This code, employing finite difference equations, permits analyzing the discontinuities in the flow (shock waves) by the Richtmeyer–von Neumann method. This procedure is used as the basic method of many other computer codes. Other progressively different types of computer codes, using other mathematical methods to solve the general equations of the gasdynamics appeared [6–9]. Many laboratories were therefore able to develop more or less sophisticated codes.

The Code Used by the CEA

As an example, the computer code written and used by the CEA [10] is presented here. This code was first written in 1966–1967, and often improved during the next twenty years.

The Fundamental Equations of Gasdynamics. For one-dimensional nonstationary flow, the general equations of fluid mechanics can be written

$$\frac{\partial F}{\partial t} + \frac{\partial G}{\partial x} = H \qquad (2.7)$$

with

$$F = \begin{bmatrix} \rho \\ \rho u \\ \rho\left(e + \frac{1}{2}u^2\right) \end{bmatrix}, \quad G = \begin{bmatrix} \rho u \\ \rho u^2 + p \\ u\left(\rho e + \frac{1}{2}\rho u^2 + p\right) \end{bmatrix}, \quad \text{and} \quad H = \begin{bmatrix} 0 \\ H_1 \\ H_2 \end{bmatrix}, \qquad (2.8)$$

where ρ, u, e, p, x, and t are, respectively, mass density, material velocity, internal energy per unit mass, pressure, spatial position, and time.

These equations describe the Lagrangian form of the conservation laws of mass, momentum, and energy for all the fluids. The variable H_1 corresponds to the momentum loss (sliding friction losses for the piston and the projectile, frictional force and head loss per unit volume of the fluid), and H_2 represents the loss due to heat transfer.

Thermodynamic Treatment of the Fluids. In these calculations, the burning of the gunpowder and the motion of the driving gas (light-gas) is simulated. The gunpowder is treated as a reactive medium with mixed solid and gas phases. During the burning, the solid phase is treated as incompressible and the gaseous phase is described by the Clausius–Sarrau equation of state,

$$p\,(v - \eta) = n\,R\,T\,. \qquad (2.9)$$

where η is the covolume, n is the mean number of moles in the combustion products, R is the universal gas constant, and T is the current absolute temperature. With this formulation, the internal energy, E, is independent of the volume so that

$$T\frac{\partial p}{\partial T} - p = 0 \;\Rightarrow\; dE = c_v\,dT\,, \qquad (2.10)$$

where c_v is the specific heat at constant volume. We can also write

$$dE = \frac{f}{T}\frac{dT}{\gamma - 1}\,, \qquad (2.11)$$

where f is the specific force of the powder ($f = n\,R\,T$) and γ is the polytropic exponent of expansion of the gaseous combustion products. The internal energy difference between T_0, the explosion temperature of the gunpowder, and T is given by

$$E - E_0 = \frac{f}{\gamma - 1}\left(1 - \frac{T}{T_0}\right) . \tag{2.12}$$

Defining z as the mass fraction of powder that has been burned ($0 < z < 1$), we obtain

$$E - E_0 = \frac{fz}{\gamma - 1}\left(1 - \frac{T}{T_0}\right) . \tag{2.13}$$

The powder, initially solid, becomes gaseous during its combustion. The combustion law is written

$$dz = S\,\Phi\,(z)\,\Theta\,(p)\,dt \quad \text{or} \quad dz = (S/V)\,u\,dt\,, \tag{2.14}$$

where $\Phi(z)$ (or S/V) is a geometrical factor taking into account the shape of the gunpowder granules and $\Theta(p)$ (or u) is the linear combustion velocity of this propellant. This velocity is generally a function of pressure.

The density, ρ, of the solid–gas mixture is given by

$$\frac{1}{\rho} = \frac{1 - z}{\delta} + \frac{z}{\rho_g}\,, \tag{2.15}$$

where ρ_g is the mass density of the gaseous products ($\rho_g = 1/v$), and δ is the mass density of the solid gunpowder.

The pressure, p_f, needed for initiating motion of the piston is obtained by calculation of the burned mass of gunpowder, z_0, at constant volume needed to produce the required pressure:

$$z_0 = \frac{p_f\left(\frac{1}{\Delta} - \frac{1}{\delta}\right)}{f\left[1 + \left(\eta - \frac{1}{\delta}\right)\frac{p_f}{f}\right]}\,, \tag{2.16}$$

where

$$\Delta = \frac{\text{total mass of the powder}}{\text{powder chamber volume}} . \tag{2.17}$$

For the light-gas, the equation of state chosen is

$$\left(p + \frac{a}{v^2}\right)\,[\,v - \alpha\,(v)\,] = n\,RT\,. \tag{2.18}$$

The internal energy is given by

$$dE = c_v\,dT + (a/v^2)\,dv\,. \tag{2.19}$$

With this equation, c_v is a function only of T ($\partial c_v/\partial v = 0$). To take into account the dependence of c_v on temperature, T, the chosen law is $c_v = AT + B$. Three linear ranges are possible.

For helium, $\alpha(v)$ is chosen constant, but for hydrogen, this relation is

$$\alpha\,(v) = \frac{a_1\,v}{v + b_1}\,. \tag{2.20}$$

However, each laboratory generally has its own preferred equations of state of these materials.

Geometrical and Mechanical Simulations. It is generally easy to simulate the geometrical shape of a two-stage light-gas gun in either one and one-half or two dimensions.

In Fig. 2.5 the geometry that was chosen by the CEA is presented. The combustion chamber is a cylindrical tube connected to the pump tube by a conical transition. The downstream end the pump tube is formed by two cones of different angles followed by a short cylindrical section and the diaphragm. The launch tube is cylindrical. The piston is made of two parts. The rear part is treated as an incompressible material. The front part is treated as an elastic–perfectly-plastic material. The projectile is simulated only by its weight. A counter-pressure may be applied on its front face [11].

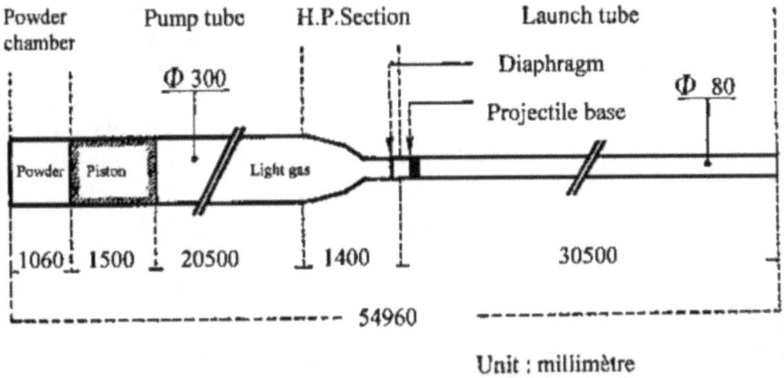

Fig. 2.5. Internal geometry of the CEA two-stage light-gas gun

The piston motion begins when a given initial pressure, corresponding to a certain amount of burned gunpowder, is achieved in the combustion chamber. To stop the piston in the different cones, a friction term is added as counter-pressure on each mesh situated either partially or totally in one of these cones. The head loss associated with the opening time of the diaphragm is simulated by a counter pressure decreasing to zero after a time corresponding to the total opening time of the diaphragm. Experimentally, ingeneral cases, the base pressure increases rapidly at the beginning of the projectile motion. This is followed by a more or less rapid pressure decrease as the projectile moves toward the muzzle.

The constant base pressure model gives the pressure as a simple step function of time, but everyone knows that this pressure history is not achievable in practice. Is this the model that must be followed up? Maybe not; it seems that other possibilities could be used. The computer calculations will easily show whether such ideas are reasonable.

2.3 The Design of a Two-Stage Light-Gas Gun

After the first scaling of the gun, further calculations must be made to determine whether this gun can withstand the stresses associated with all the expected shots. The first calculations concern the stresses correlated with the gas pressures. Other calculations concern the axial stresses associated with the motion of the parts of the gun during the shot.

The scaling allows for the establishment of the internal dimensions of the gun, whereas the mechanical design establishes its external dimensions (thicknesses of the breech, pump tube, high pressure section, and launch tube).

2.3.1 The Stresses Associated with the Internal Pressures

The Tangential and Radial Stresses

The internal ballistics calculated with the digital computer provide the pressure at every point of the gun during the entire firing process. Combining all this information, we can draw the pressure distribution $p(x)$ along the various parts of the gun at any time, t_i. As shown in Fig. 2.6, the drawing of all these curves on the same graph, with the same coordinates, produces an envelope that describes the maximum pressure that the gun must withstand at each point during the shot.

By making further calculations describing all the expected shots for this gun, an envelope of all the maximum pressure curves (one for each shot) can

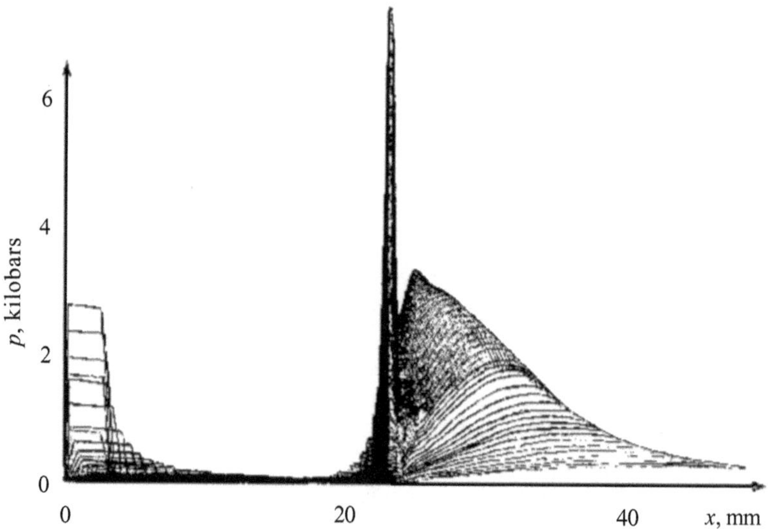

Fig. 2.6. Gas pressure distribution within the gun at various times

be drawn. The final maximum pressure is obtained by adding a safety margin of about 20–30% at each point. This curve will be used for calculating the material strength of all the parts of the gun (see Fig. 2.7).

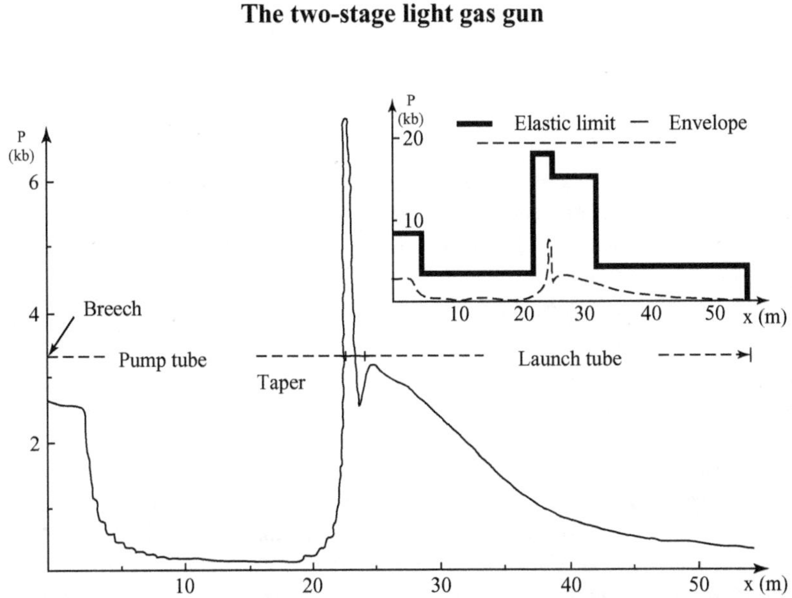

Fig. 2.7. Envelope of the maximum pressure for the CEA two-stage light-gas gun

Except at specific points, most of the metallic parts of the gun (the breech and the important parts of the pump and launch tubes) can be examined by the classical methods of the strength of materials in the range of elastic strains. The values of the radial and tangential stresses are determined for the design values of the internal pressures.

The calculation of these stresses for a thick tube is known as the problem of Lamé. For a tube of internal radius a and external radius b, with an internal pressure, p_a, (the external pressure, p_b, is assumed equal to zero), the radial, σ_r, and tangential, σ_t, stresses are given by

$$\sigma_r = \frac{p\,a^2}{b^2 - a^2}\left(1 - \frac{b^2}{r^2}\right) \quad \text{and} \quad \sigma_t = \frac{p\,a^2}{b^2 - a^2}\left(1 + \frac{b^2}{r^2}\right) . \qquad (2.21)$$

The tangential stress, σ_t, achieves its maximum value,

$$\sigma_t\big|_{r=a} = p\,\frac{b^2 + a^2}{b^2 - a^2} , \qquad (2.22)$$

on the internal skin ($r = a$), and the radial stress on this skin is $\sigma_r = -p$. If the tube is closed, there is an axial stress σ_z given by

$$\sigma_z = \frac{p a^2}{b^2 - a^2} . \qquad (2.23)$$

In applying this equation, it is assumed that the tube is long enough to admit that this stress is uniform over its entire length.

The Special Case of the Taper

The taper is the transition part that connects the pump tube to the launch tube. In this part of the gun the internal pressure may increase to several hundreds of MPa up to 1 GPa (several kilobars, up to ten kilobars). The taper has to withstand pressures higher than the classical elastic limit of a metallic piece, so studies of design options such as frettage (in two or more layers) and autofrettage are needed.

Frettage, shrink-fitting an external cylinder over another tube, produces a state in which σ_t is compressive in the internal tube. Introduction of an increasing internal fluid (or gas) pressure into the assembly first relieves the tangential compression produced in the inner tube by the frettage. A further increase in the internal pressure produces a state of tension in the inner cylinder. The final tensile stress at the inner surface of this cylinder is smaller than the stress that would be produced by the same fluid pressure in a single tube that is initially unstressed. In executing this design, it is essential to control the strength of the external cylinder.

With autofrettage, a high enough internal pressure, called the *autofrettage pressure*, is applied to produce plastic deformation of the internal skin of the tube. When this internal pressure is relieved, elastic tensile stresses are maintained in the external part of the cylinder and compressive stresses are present near its inner surface. If one again introduces an internal pressure that is lower than the autofrettage pressure, the residual tangential compressive stresses are added to the tangential tensile stresses caused by the internal pressure. The tensile stress arising as a result of these two effects is less than it would be if autofrettage were not used, and the material undergoes no plastic deformation if the applied pressure is lower than the autofrettage pressure.

For high pressures, about 1–1.2 GPa (10–12 kilobars), a combination frettage and autofrettage is often used.

For these calculations, the following assumptions are generally made:

- there is no axial stress whatever the internal pressure may be

$$\sigma_z = 0 , \qquad (2.24)$$

- no cross section changes shape whatever the strains may be

$$\frac{\partial \varepsilon_z}{\partial r} = 0 , \qquad (2.25)$$

- the plastic deformation does not cause a change of volume

$$\varepsilon_1' + \varepsilon_2' + \varepsilon_3' = 0 \,, \tag{2.26}$$

- the addition of several equilibrium states is valid if the total of the stresses is again in the elastic state,
- the external pressure is always zero.

Several theories may be used, with either the Tresca or the von Mises yield criterion. With the Tresca criterion, the behavior of the material is given by:

$$\begin{array}{ll} \text{elastic range} & \sigma_\theta - \sigma_r < Y_0 \\ \text{elastic limit} & \sigma_\theta - \sigma_r = Y_0 \\ \text{plastic range} & \varepsilon_\theta - \varepsilon_r = f(\sigma_\theta - \sigma_r) \,, \end{array} \tag{2.27}$$

where Y_0 is the yield strength of the material.

With the von Mises criterion, the following equations are used:

$$\begin{array}{ll} \text{elastic range} & (\sigma_r - \sigma_\theta)^2 + (\sigma_\theta - \sigma_z)^2 + (\sigma_z - \sigma_r)^2 < 2Y_0^2 \\ \text{elastic limit} & (\sigma_r - \sigma_\theta)^2 + (\sigma_\theta - \sigma_z)^2 + (\sigma_z - \sigma_r)^2 = 2Y_0^2 \\ \text{plastic range} & (\sigma_r - \sigma_\theta)^2 + (\sigma_\theta - \sigma_z)^2 + (\sigma_z - \sigma_r)^2 = 2Y_0^2 \end{array} \tag{2.28}$$

In all these calculations, the strength of the material is calculated in a static manner, with the radial and tangential stresses being taken into account. To calculate the axial stresses associated with the different motions of the parts of the gun during a shot (recoil of the breech associated with combustion of the gunpowder, extrusion of the piston, and the force associated with the transition from the pump tube to the launch tube,) it is necessary to make dynamic calculations.

2.3.2 The Dynamic Stresses During the Shot

The first paper on this subject was probably that written by C.M. Leech [12]. With some classical elastic assumptions and introduction of the pressure field $p(x, t)$ given by the results of the internal ballistic calculations, it is possible to determine the time dependence of the axial stress at each point of the gun. The examination of these results confirms, or not, that all the parts of the gun are stressed within the elastic range.

What are the physical assumptions for these calculations? They are generally not explicitly written. The first assumption is that the strains are in the purely elastic range. If a force is responsible of an elastic strain, to add another force, giving again an elastic strain, will give a final strain that is the algebraic sum of the two strains. This assumption is valid if both strains are small enough that the occurrence of plastic behavior is precluded. With this assumption, it is possible to dissociate the calculations of the different stresses. It is important to check, after the calculation of all the stresses (axial, radial, and tangential,) whether this assumption is valid in every part of each piece of the gun that was analyzed.

Another assumption is the validity of the Saint-Venant principle: When a system of pplied forces is changed to another one that is statically equivalent, it is only in the neighborhood of the points of force application that the stresses and strains are changed. For example, this principle allows for the introduction of sudden changes in the diameter of a tube without the stress being significantly altered in regions remote from the transition. The final assumption is that no cross section of the gun changes during the shot.

With these three assumptions, that are valid except at some specific points of the gun, it is possible to simplify the shape of the gun parts and the equations used to calculate the axial stress during a shot.

To simulate the behavior of the material used to build the gun, Hooke's law,

$$\sigma_z = E\,\varepsilon_z\ , \quad \varepsilon_z = \frac{z - z_0}{z_0}\ , \tag{2.29}$$

has often been chosen. In these equations the coefficient E is Young's modulus of elasticity, z_0 is the initial length of an element of the mesh used for the calculation, and z is its length after deformation. The pressure, p, and the density, ρ, are related to the strain by the equations

$$p = -E\,\frac{z - z_0}{z_0} = -E\left(\frac{\rho_0}{\rho} - 1\right)\ . \tag{2.30}$$

With these equations, it is easy to show that the choice of a one-dimensional code (in fact, a one and one-half dimensional code) can give a good value for the velocity of the longitudinal waves in the material. The values obtained for the stresses are valid, but unfortunately volume and density values calculated are incorrect. However, these values are not used for the continuation of the calculations.

By introducing the forces associated with the operation of the gun, the $p(t)$ curves showing the law of combustion of the gunpowder and the force associated with the change of cross section between the pump tube and the launch tube, and by adding the forces introduced by friction, one can calculate the axial stress. Another result of these calculations is a prediction of the motion of the gun during the shot, especially the recoil of the gun.

As an example, the results of a calculation explaining the motion of the gun are presented on Figs. 2.8 and 2.9. The results of some calculations of axial stresses are presented on Fig. 2.10. These calculations show the particular motion of the launch tube. Axial stresses are large, especially in the neighborhood of the rear part of the launch tube. To reduce these stresses, a design called the "sliding launch tube" has been suggested.

To conclude, the combination of these results – the axial, tangential, and radial stresses – confirms that all parts of the gun experience elastic behavior during a shot. These combinations are applied for all of the expected shots.

After many digital calculations of internal ballistics and material strength for all of the expected range of shots, and some iterative calculations, the scaling and mechanical design of a gun can be fully determined.

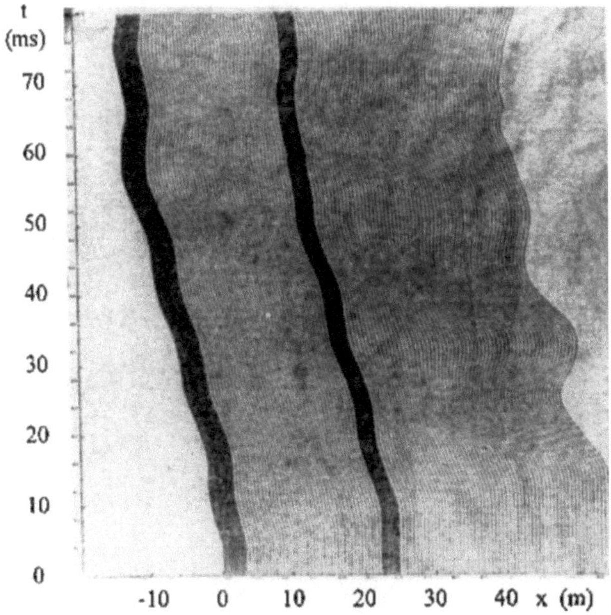

Fig. 2.8. Motion of the gun during a shot

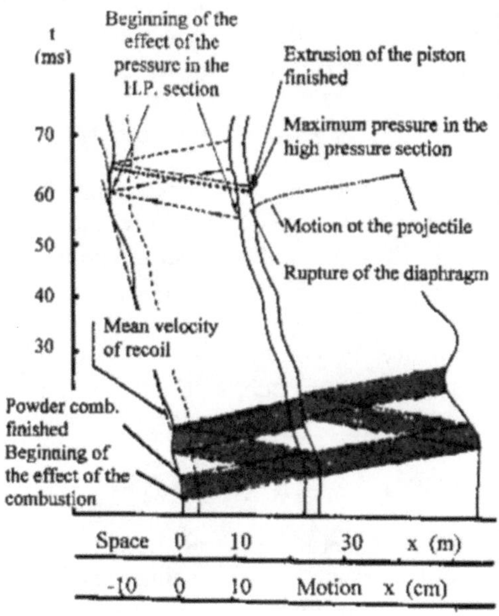

Fig. 2.9. Explanation of the motion of the gun during a shot

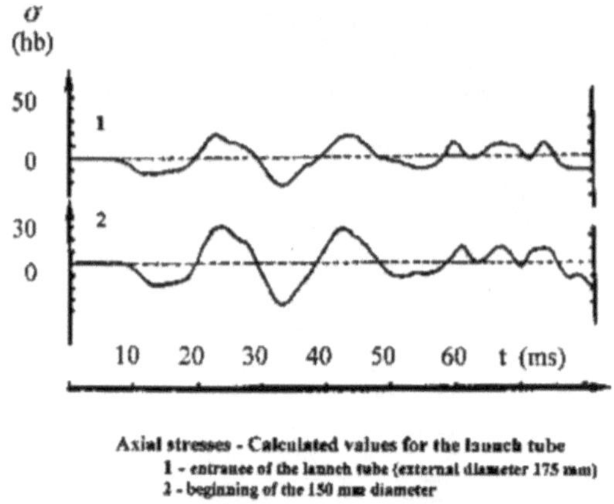

Fig. 2.10. Calculated axial stresses in the CEA two-stage light-gas gun

2.3.3 Some Simple Calculations

For the final design of a gun it is important to calculate its motion, andspecifically its recoil, during a shot. The previous chapter gives a method for making this calculation using a digital computer code, but it can be interesting to obtain a rough result with a simple calculation.

The axial recoil value can be determined with good accuracy by a simple method. The recoil motion can be divided into two phases. The first calculation takes place when the piston has obtained its maximum velocity, V_{p}^*. At this time, it is assumed that the diaphragm has not ruptured and the piston has not yet extruded. The law of conservation of momentum gives

$$Mv_1 = m_1 V_{\mathrm{p}}^* + \lambda_2 m_1' \, V_{\mathrm{p}}^* = (m_1 + \lambda_2 \, m_1' \,) V_{\mathrm{p}}^* \,, \tag{2.31}$$

where M is the total mass of the gun, m_1 is the mass of the piston, m_1' is the mass of the gunpowder, and λ_2 is the ballistic coefficient for the gunpowder (which velocity increases from zero at the rear of the breech to V at the base of the piston; its value is generally chosen equal to one-half of the piston velocity).

The maximum velocity, V_{p}^*, of the piston is determined by application of the internal ballistic computer code, so this equation gives the maximum recoil velocity of the gun, v_1:

$$v_1 = \left(m_1 + \tfrac{1}{2} m_1' \right) V_{\mathrm{p}}^* / M \,. \tag{2.32}$$

The value obtained is generally several meters per second. This result shows that it is difficult to stop the gun during this first phase of the recoil because the recoil velocity is too large to permit relying on friction to arrest the motion of the gun. (This result can be compared with the numerical code calculating the axial stresses and the motion of the gun during and after the shot.)

The second calculation takes place just after exit of the projectile from the launch tube. Assuming that the extrusion of the piston in the taper is finished and the momentum of the piston and gunpowder have been transmitted to the gun, the equation for the total momentum is

$$\mathcal{M}\, v_2 = m_2\, V_2 + \lambda_2\, m_2'\, V_2 = (m_2 + \lambda_2\, m_2')V_2 \tag{2.33}$$

where m_2 and V_2 are the mass and the muzzle velocity of the projectile, m_2' is the mass of the light-gas, and $\mathcal{M} = M + m_1 + m_1'$. Finally, the recoil velocity, v_2, is given by

$$v_2 = \left(m_2 + \tfrac{1}{2}m_2'\right) V_2/\mathcal{M}\ . \tag{2.34}$$

The value obtained is generally 15–25 cm/s. To stop this recoil, it is possible to dissipate the associated energy by friction. The stopping distance, x, using friction only, is given by

$$x = \frac{1}{2\, fg} v_2^2\ , \tag{2.35}$$

where f is the friction coefficient and g is the gravitational acceleration. Using $f = 0.2$, which is a reasonable value, and with a recoil velocity, v_2, of 0.25 m/s, one obtains $x = 0.016$ m, a very small value that permits arresting the gun recoil by friction without use of a shock absorber. Using Δt_1, the duration of the motion of the piston given by the computer code, a rough total recoil is $X = v_1 \Delta t_1 + x$.

2.3.4 The Sliding Launch Tube

We have seen that the recoil of the gun during a shot produces large axial stresses. Study of the motion of the launch tube before exit of the projectile shows longitudinal vibrations and axial stresses of high intensity. To withstand these axial stresses, a thick enough tube is needed. However, a thin-walled launch tube is easy to align accurately because it can be bent with ordinary screws.

The requirement for a thick tube can be reduced by disconnecting the rear part of the gun, the pump tube and taper, from the launch tube. With this solution, illustrated in Fig. 2.11, the motions of the rear part of the gun and of the launch tube are independent.

However, this solution does present some new problems. Before the diaphragm ruptures, a slight volume appears at its front face, between the taper that recoils and the rear face of the launch tube. This volume is not very large, only a few liters, corresponding to an increase of the total launch tube volume of 2–3% (the external diameter and the wall thickness of the launch tube are both small). A slight decrease of the light-gas pressure on the base of the projectile subjects it to a rather smoothly increasing force and lower acceleration than would be produced by an impinging shock wave.

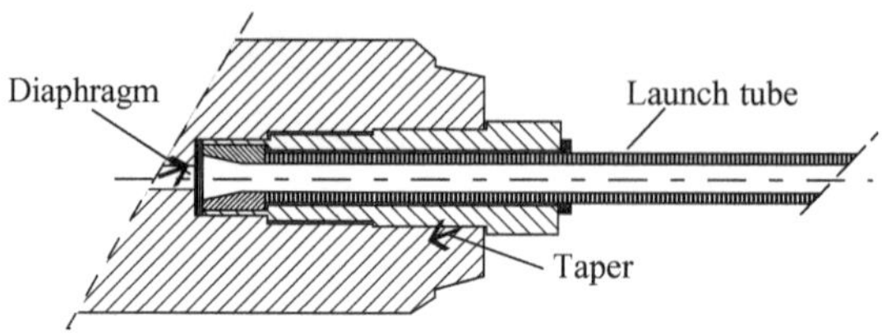

Fig. 2.11. Schematic drawing of the sliding launch tube

A second problem is that the launch tube moves forward, pushed by the gas pressure on its rear cross-section. Some shock absorbers are needed to arrest this motion.

A third problem is the recoil distance of the first stage of the gun. A simple calculation shows that this value does not change significantly. In fact, the only change occurs because the launch tube mass is subtracted from the total mass of the gun (a change that is, in general, less than 10%).

It is interesting to see how the behavior of the gun is changed when a sliding launch tube is introduced. Using the code for calculation of the axial stress, with a slight adaptation, the behavior of different parts of the gun can be simulated. The forces that act on the rear face of the launch tube (derived from the pressure near the front face of the device maintaining the diaphragm) are given by the internal ballistic code. The motion of the gun during a shot is presented in Figs. 2.12 and 2.13.

An examination of the axial stresses in the launch tube shows that they are only about one-half as large as the values calculated in the classical case. The motion of the different parts of the gun and the comparison of the axial stresses between a classical and a sliding launch tube are shown on Fig. 2.14.

Study of the $p(x)$ curves shows that the pressure decreases very rapidly along the first fifty centimeters near the front of the diaphragm. For this reason, it is judicious to put the rear part of the launch tube at this position. The force applied on the cross section of the rear face of the launch tube will be small in proportion to this pressure (the evolution of the pressure at this station is responsible of the motion of the sliding launch tube).

2.4 The Operation of a Two-Stage Light-Gas Gun

2.4.1 The Compression of the Light-Gas

We have seen that it is useful to compress the light-gas before the rupture of the diaphragm. The method to obtain such a result is to push the piston. But how should one push it to get the best result? Thermodynamically, two methods can be considered: isentropic compression and compression by a shock wave. To produce isentropic compression one needs a low piston velocity. To

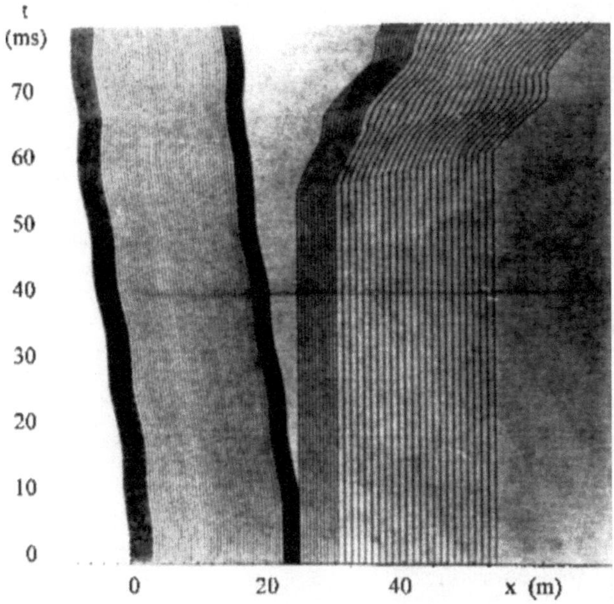

Fig. 2.12. Motion of the CEA gun with a sliding launch tube

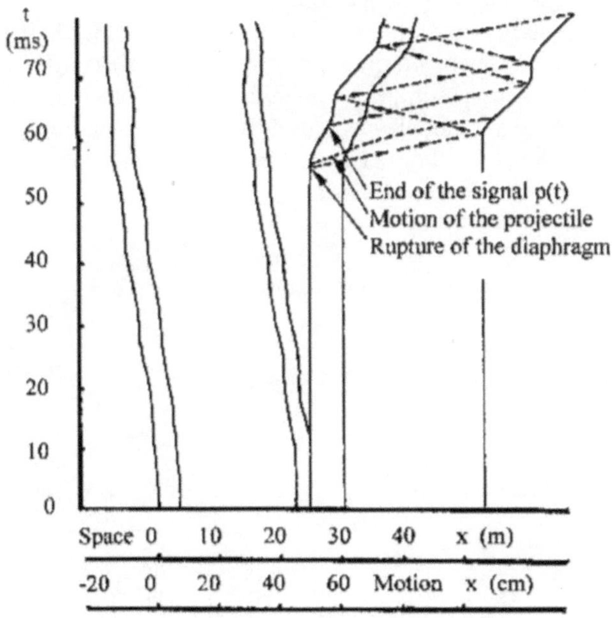

Fig. 2.13. Explanation of the CEA gun with a sliding launch tube

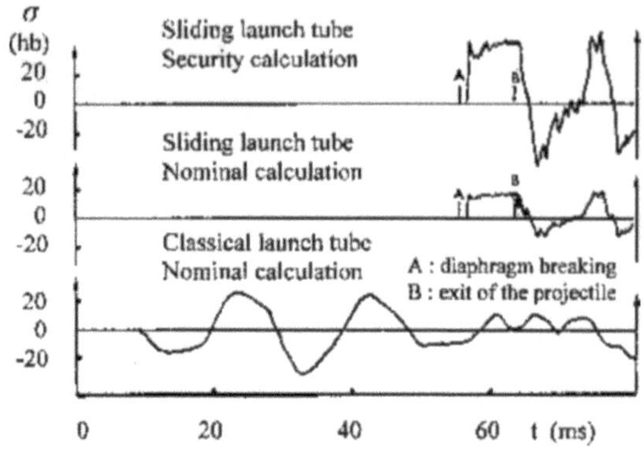

Fig. 2.14. Axial stresses in the launch tube. Comparison of classical and sliding launch tubes (CEA two-stage light-gas gun)

achieve this, the diaphragm must rupture soon after the piston motion begins. The piston and the projectile move together. This solution corresponds to the accelerated reservoir.

But, is this the best method? If one increases the rupture pressure of the diaphragm, the piston compresses the light-gas and, if its velocity is high enough, the compression waves propagating through the gas reflect at the rear face of the diaphragm (before it ruptures) and become shock waves. Figure 2.15 shows the compression waves becoming shock waves. The question now is whether gas heated by shock compression or by isentropic compression is a better propellant, i.e., for the same initial volume, is the final pressure higher for the multi-shock compression or for isentropic compression?

To check this comparison, some one-dimensional calculations have been performed for hydrogen modeled as a perfect gas. From an initial pressure p_i, hydrogen gas is compressed either isentropically or by one shock wave. These

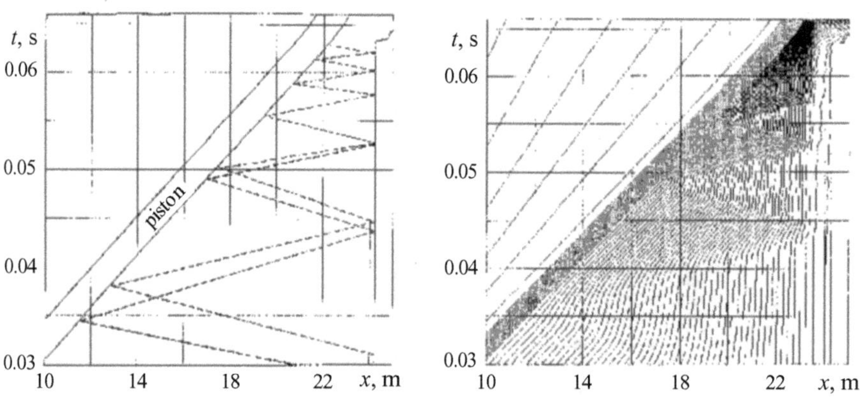

Fig. 2.15. Flow in the light-gas before diaphragm rupture (numerical simulation)

calculations are ideal cases because the compression is neither completely isentropic nor is it produced by only one shock wave. The first result is probably too low and the second one is certainly too high.

The chosen initial pressure is p_i, the final pressure is p_F, the initial temperature is 300 K, the final temperature is T_F, and E_H/E_S is the ratio of the final internal energy achieved by the shock compression, E_H, to the internal energy, E_S, obtained by isentropic compression.

Results for the two cases

- $p_i = 500$ kPa (5 bars), $p_F = 50$ MPa (500 bars)
- $p_i = 1$ MPa (10 bars), $p_F = 200$,MPa (2000 bars) .

that have been studied are given in Table 2.5.

Table 2.5.

	Compression to $p_F = 50$ MPa		Compression to $p_F = 200$MPa	
Compression process	T_F,K	$(E_H/E_S)_F$	T_F, K	$(E_H/E_S)_F$
Isentropic compression	1157	1	1376	1
Shock compression	5324	6.67	10346	10.0

With shock wave compression, the final temperature is about five times and seven times higher, and the increase of the internal energy is about six times and ten times higher than is obtained by isentropic compression.

However, it is not the temperature but the pressure that pushes the base of the projectile. Using the final pressures obtained by compression, with the same assumptions of perfect gas behavior and one-dimensional flow as a single-stage gun with a projectile of 1 kg and a caliber of 100 mm for the launch tube, the release of the gas has been calculated to achieve velocities from 3000 to 6000 m/s. The base pressure, p_B, applied to the projectile at the muzzle, the needed length, L_M, for the launch tube and the time, t_M, to achieve this velocity, are presented in Tables 2.6 and 2.7 and on Fig. 2.16.

Table 2.6.

	Release to achieve a velocity of 3000 m/s			Release to achieve a velocity of 5000 m/s		
$p_F = 50$ MPa	p_B, MPa	L_M, m	t_M, ms	p_B, MPa	L_M, m	t_M, ms
Isentropic compression	7.601	42.39	21.7	1.51	387.95	102.8
Shock compression	25.238	18.26	10.95	15.481	71.21	23.92

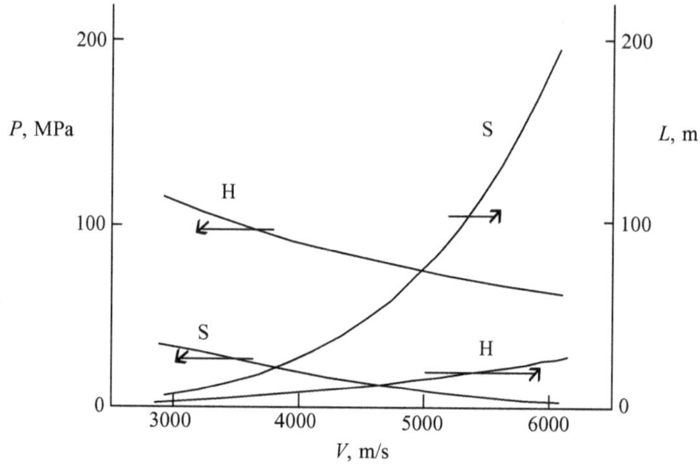

Fig. 2.16. Pressure release and length of the launch tube versus projectile velocity

From the Table 2.6, $p_F = 50\,\text{MPa}$ (500 bars), the pressure at the base of the projectile when it is at the muzzle is about three times larger with shock wave compression than with isentropic compression for a final velocity of 3000 m/s, and about ten times larger for 5000 m/s. The lengths of the launch tube and the internal flight durations required to achieve these velocities are much greater in the isentropic case than in the case with shock wave compression.

From Table 2.7, the results are more interesting because the velocities are greater. The values obtained are more realistic. For example, for a final velocity of 6000 m/s, the needed length of the launch tube is about 26 meters, corresponding to 260 calibers.

Table 2.7.

	$V = 3000\,\text{m/s}$			$V = 4005\,\text{m/s}$		
$p_F = 200\,\text{MPa}$	p_B, MPa	L_M, m	t_M, ms	p_B, MPa	L_M, m	t_M, ms
Isentropic compression	36.9	9.21	4.8	18.8	26.7	9.74
Shock compression	113	4.20	2.56	92.5	8.61	3.81
	$V = 5010\,\text{m/s}$			$V = 6000\,\text{m/s}$		
	p_B, MPa	L_M, m	t_M, ms	p_B, MPa	L_M, m	t_M, ms
Isentropic compression	8.96	72.4	19.7	3.94	193	41.4
Shock compression	75.1	15.5	5.35	60.9	25.8	7.21

From these results, the solution with shock waves seems of great interest. However, the pressure and temperature near the diaphragm may reach very high values (several hundreds of megapascals (several kilobars) and several thousands of Kelvin). For this reason, it is very important to design this part of the gun, called the taper, with great care.

2.4.2 The Motion of the Piston

The existence of several types of gunpowder makes it possible to select the one best suited to produce the needed motion for the piston. Various final motions of the piston can be realized by varying the initial pressure of the light-gas:

- the piston compresses the light-gas in an almost isentropic manner. The diaphragm ruptures soon after the piston motion begins. The velocity of the piston decreases in the conical shape of the pump tube and stops. This is the accelerated reservoir solution.
- the diaphragm ruptures after the light-gas is highly compressed. At the end of its travel the piston compresses the light-gas, usually with shock waves.
  - The light-gas pressure is high, but not high enough to stop the piston before its extrusion in the cone. In the final phase of the compression the gas pressure is too high and the piston rebounds at low velocity. When this happens a strong rarefaction fan forms at the front face of the piston, moves forward into the light-gas, and rapidly decreases the pressure at the base of the projectile. But the piston extrudes later, without counter-pressure.
  - The light-gas pressure is not high enough to prevent the piston stopping in the cone. This is the good behavior sought in the design.

Is it advisable, for the same amount of gunpowder, to use a lightweight piston with a high velocity or a heavy piston with a low velocity for the best transfer of the kinetic energy of the piston to the internal energy of the light-gas? What is the effect of each of these options on the extrusion and stopping of the piston in the conical shape of the taper?

What is the best means of achieving the desired projectile velocity? Certainly, it is a compromise among all the parameters: the weights of the gunpowder and of the piston, the initial light-gas pressure, and the diaphragm burst pressure. The best means of answering this question is to conduct a great many calculations with a computer code.

2.4.3 The Conical Shape for the Piston Extrusion

The first two-stage light-gas guns had a simple change of barrel diameter at the connection of the pump tube to the launch tube. In this case, it was important that the piston not collide with the rear part of the launch tube and damage it. It was necessary to be sure that the piston would rebound from the compressed gas.

After rupture of the diaphragm, when the piston velocity decreases, a release wave moves forward from the front face of the piston, catches up to the base of the projectile and produces a rapid decrease of the applied pressure. However, if the piston rebounds, a stronger rarefaction fan produces a large decrease in the base pressure. To avoid this rebound and the associated decrease of pressure, a conical shape of the front face of the pump tube was designed. The intent in introducing this feature is that the piston will come to rest as it extrudes into the cone.

The choice of a half angle of the order of 15° for the cone seems best suited to stopping the piston. However, forces arising from the extrusion of the piston, combined with the high temperature of the light-gas, damage the tube at the entrance to the cone (see Fig. 2.17). This damage increases with each successive shot and the piston removal becomes increasingly difficult. A method for avoiding this difficulty is to use two cones. The first cone, with a half angle of only a few degrees, decreases the velocity of the piston, and a second one stops it. This biconic shape, has been chosen by the CEA for its two-stage light-gas gun.

Fig. 2.17. Damage to a launch tube

Sometimes the front part of the piston breaks during its extrusion. The broken piece of the extruded part acquires a velocity higher than the rear part of the piston and can move into the launch tube and damage it. The addition of a short cylindrical extension of internal diameter slightly larger than the launch tube diameter solves this problem (see Fig. 2.18.) An extruded piston recovered after a shot is shown in Fig. 2.19.

Fig. 2.18. Launch tube with and without a cylindrical part

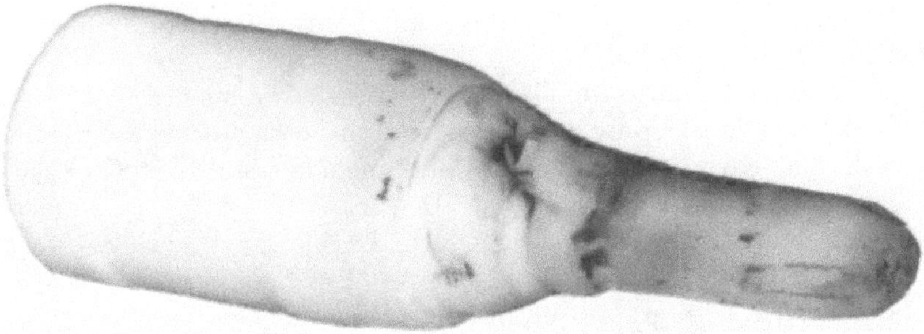

Fig. 2.19. The front part of an extruded 100 mm caliber piston. Some cracks are visible at the rear part of the cylindrically extruded piston

2.4.4 The Diaphragm

In the transition section, to separate the pump tube from the launch tube, many laboratories use a valve in the form of a metal sheet with grooves. When the light-gas pressure on its rear face is high enough, this diaphragm ruptures, generally along the grooves. It opens like a flower with petals.

To scale such a diaphragm, two problems must be solved: the opening at a definite pressure, as accurate as possible and, after opening, the petals must lie against the side of the tube so they do not obstruct the cross-section of the gas flow. Another requirement is that the petals must not break free of the diaphragm.

Each laboratory has its own method. For example, the diaphragm used at the CEA for its large two-stage light-gas gun is as shown in Fig. 2.20.

As the design of a diaphragm seems very simple, many laboratories made only few experiments and, on the basis of these few experiments, generally estimate that the results obtained are satisfactory enough to proceed with

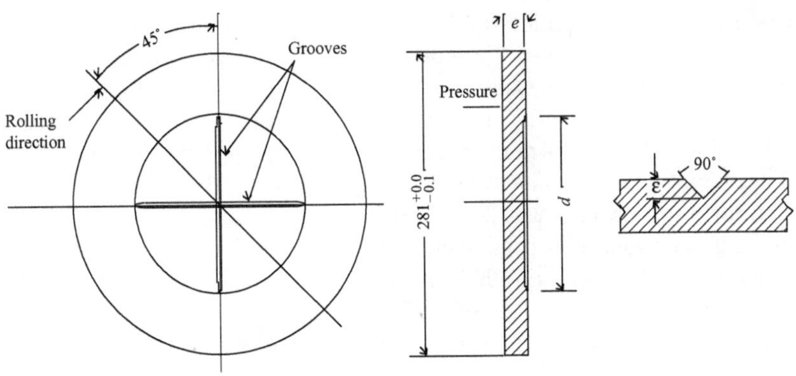

Fig. 2.20. Drawing of the CEA diaphragm

the design of the diaphragm. For this reason, only a few reports have been published.

For the pressure range from several 10^7 Pa (several hundreds of bars) up to 10^8 Pa (one kilobar), only metallic diaphragms with grooves are used. The materials that have been have been tested include stainless steel of several types, aluminum, and copper.

Several solutions have been tested (such as a pile of several sheets of metal for the different breaking pressures), but the most frequently used is a sheet of stainless steel with grooves on one side, sometimes on both sides. Using the same diaphragm, a large range of rupture pressure can be obtained by changing only the depth of the grooves.

For a given diaphragm, for the range $0.05 < \varepsilon/e < 0.4$, the rupture pressure, p, is generally approximated by the semi-empirical relation

$$p = \frac{A}{B} e^{3/2} (\varepsilon/e)^{-1/2} \quad \text{with} \quad B = 1 + 3.75 (\varepsilon/e)^3 , \tag{2.36}$$

where e is the thickness of the working part of the diaphragm and ε is the depth of the groove (obviously this equation is not valid for $\varepsilon = 0$). The coefficient A depends on the diaphragm dimensions, the depth of the groove, and material properties (elastic limit of and the stress intensity factor for a crack propagation in the material) (see [13]). For a given ε/e value, the burst pressure varies like $e^{3/2}$. The radius at the bottom of the groove is an important factor. For the best results, it must be as small as possible [13].

The diaphragm must open without the petals obstructing the gas flow or breaking free and entering the flow. It is important that the rupture pressure be reproducible because this pressure is an important factor affecting the compression of the light-gas. Unfortunately, the variation is generally more than 10%. Depending on the shape of the pressure pulse, the rupture pressure of the diaphragm can be modified by about 30%, with an accuracy of about 10%.

To improve the accuracy of the rupture pressure and its reproducibility, it would be useful to understand the real behavior of the diaphragm. The characteristics of the material are often a function of the sample history. For a diaphragm taken from a sheet of rolled stainless steel, it is important to place the grooves symmetrically relative to the rolling direction. It is likely that a hydraulic pre-deformation may improve the reproducibility.

The method of clamping the diaphragm at the end of the pump tube is also important (see Fig. 2.21). If the diaphragm can slide, the surface that withstands the gas pressure can change significantly. Therefore, it depends on the opening angle of the diaphragm holder. The holder must be configured in such a way that the petals of the diaphragm do not tear or break free, but it is also important that the petals do not obstruct the flow after the diaphragm opens.

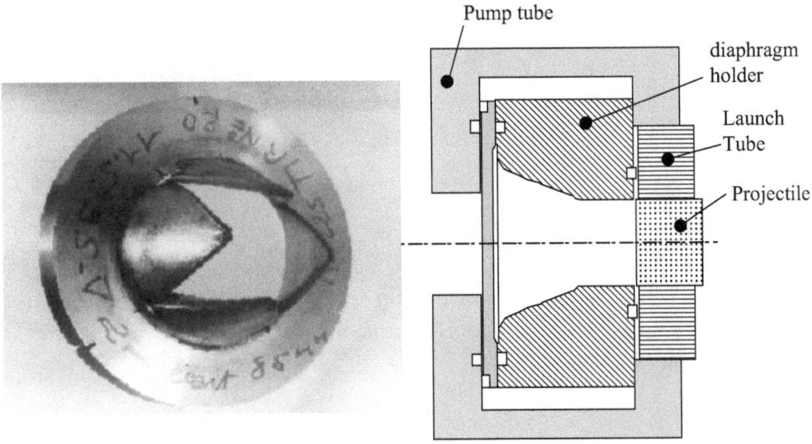

Fig. 2.21. Diaphragm and holder

If the choice of the material from which the diaphragm is made (its characteristic properties and its history) are important factors, the curvature of the holder where the petals lie after opening is also an important factor, as well as whether the shape of the launch tube is of circular or square cross section. Figure 2.22 illustrates exaggerated deformations of diaphragms before breaking. The drawing of the cylindrical diaphragm shows how such a diaphragm could be deformed by pressure. The square diaphragm can be seen as a combination of two cylinders.

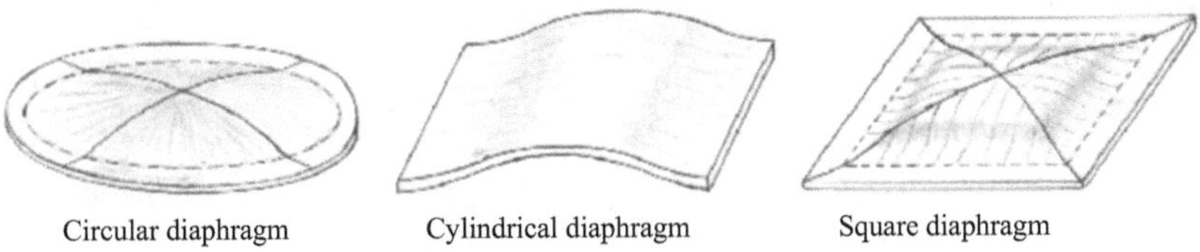

Fig. 2.22. Suggested deformations of diaphragms before rupture

For the square diaphragms a new problem appears. The tearing begins not in the center of the diaphragm, but between the center and the corners of the tube [14]. This problem is difficult to solve. A good possibility is to make the grooves slightly deeper near the center of the diaphragm, in order to reduce the strength of the material in this region. As a beginning of explanation, it seems that the breaking begins at the point of maximum radius of curvature. Theoretical calculations would probably show whether this idea is satisfactory.

Evolution of the Light-Gas Pressure at the Base of the Projectile

The general law of mechanics $F = m\mathcal{A}$, where F is the force applied to a body, m is its mass, and $\mathcal{A}$ is the resulting acceleration, shows that the motion of the projectile is given by

$$pS = M\mathcal{A} = m\frac{dv}{dt} \quad \Rightarrow \quad v = \frac{S}{M}\int p\,dt\,, \tag{2.37}$$

where S is the cross-sectional area of the launch tube, M the mass of the projectile and $p(t)$ the base pressure. The area between the $p(t)$ curve and the t axis, $\int p\,dt$, is a geometrical representation of the final projectile velocity. By the same means, one obtains

$$\int v\,dv = \frac{S}{M}\int pdx \quad \Rightarrow \quad \frac{1}{2}M\,v^2 = S\int pdx\,. \tag{2.38}$$

To obtain the largest achievable velocity for a particular shot, the $p(t)$ graph must be such as to maximize the area, $\int p\,dt$, under the curve. For the same reason, the final kinetic energy of the projectile can be represented by the area, $\int p\,dx$, under the $p(x)$ curve (see Fig. 2.23).

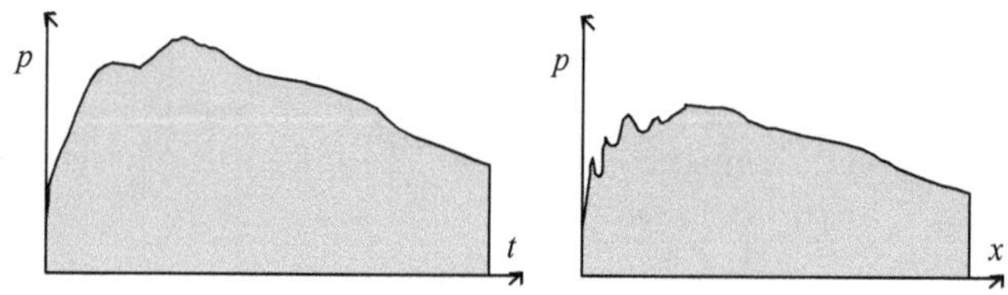

Fig. 2.23. Temporal and spatial graphs of projectile base pressure

Experimentally, in general cases, the base pressure increases rapidly at the beginning of projectile motion. Then the pressure decreases more or less rapidly as the projectile moves toward the muzzle.

The constant base pressure gives a simple rectangular area to the $p(t)$ and $p(x)$ graphs, but everyone knows that this solution is not achievable in practice. Is it this solution that should be followed up? Perhaps not; it seems that other possibilities may be used. The computer calculations will easily show whether such ideas are reasonable.

2.4.5 The Alignment of the Gun – Breech-Pump Tube and Launch Tube

The Breech and the Pump Tube

Before the shot, the alignment of the first stage of the gun is easily obtained by construction. During the shot, all the forces that are responsible for its motion are aligned along the line of sight of the gun, and all the pieces of this tube have this line as axis of revolution.

Therefore, there are no lateral forces of great intensity because the maximum velocity of the piston is generally low (several hundreds of meters per second) and the tube is heavy (several tons). These forces cannot significantly change the recoil direction. Accordingly, it is not necessary to guide the gun during its recoil, but one may leave a small capability of a lateral motion for the breech and the pump tube. The support of this tube can be simply a group of tables that are slightly longer than the calculated recoil. The use of a hydraulic system to reposition the pump tube after a shot allows determination of the friction coefficient needed to calculate the recoil distance. This is the design that has been chosen for the CEA two-stage gun (Fig. 2.24).

Fig. 2.24. The support of the first stage (breech)

The Launch Tube

Precise alignment of the launch tube is a fundamental requirement for a successful shot. Using a thin-walled launch tube, which is easily deformed by simple screws, facilitates its alignment after each shot. If a thick-walled tube is used, the alignment obtained with the first honing is very difficult to restore after each shot (a large device is needed for this). However, one must take into account the axial stresses that occur during the recoil of the gun. Each laboratory has to make a choice about resolving this problem.

2.4.6 The Acceleration of the Projectile

During its motion in the launch tube, the projectile is subjected to a large acceleration, generally several tens of thousands of g. This acceleration induces very drastic stresses within the projectile, and these stresses can destroy its integrity when it begins its free flight.

The stress in the rear part of the projectile can exceed the elastic limit of the material during the first phase of its internal motion although its front face remains elastic.

In the second phase of its internal motion in the launch tube, the base pressure decreases and, after an elastic release, the rear part of the projectile can undergo a plastic release. For this reason, it could have the shape shown in Fig. 2.25 as it exits the launch tube.

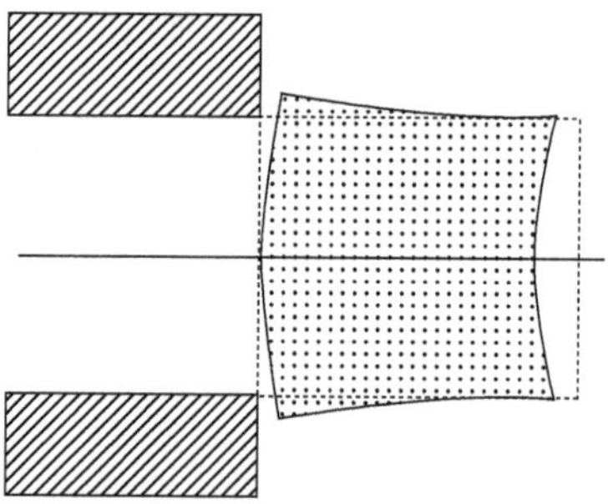

Fig. 2.25. Exaggerated shape of a projectile when its base is located at the muzzle of the launch tube

If the base of the projectile is subjected to a pressure higher than the elastic limit, it will exhibit elastic–plastic behavior at the beginning of its free flight. Some simple calculations [15] show that if the maximum pressure at the base of the projectile is

- lower than Y_0: the behavior of the projectile is elastic during both the increase of the stresses and their release. (Y_0 is the elastic limit of the material, i.e., the quasi-static yield strength).

- between Y_0 and $3Y_0$: the behavior is completely elastic during the projectile motion in the launch tube, but its behavior can be elastic–plastic if the base pressure at the muzzle exceeds Y_0.
- between $3Y_0$ and $6Y_0$: during the increase of the stresses the rear part of the projectile will exhibit elastic–plastic behavior. As long as the projectile is in the launch tube, all the releases are elastic. If the base pressure at the muzzle is lower than Y_0, its behavior will be completely elastic, but if this pressure is higher than Y_0, a part of the projectile has an elastic–plastic behavior.
- higher than $6Y_0$: Elastic–plastic behavior occurs during the increase of the stresses and can occur during release if the stress relief is greater than $6Y_0$.

These calculations are only theoretical ones, because the projectile (model) is generally very complicated. As an example, for a classical plastic material, the usual values are $\rho \sim 1.15\,\mathrm{g/cm^3}, \nu \sim 0.40$, $Y_0 \sim 70\,\mathrm{MPa}$ (700 bars), and $E \sim 1300\,\mathrm{MPa}$ (13000 bars).

The pressure range that seems to ensure the integrity of the projectile is shown in Fig. 2.26.

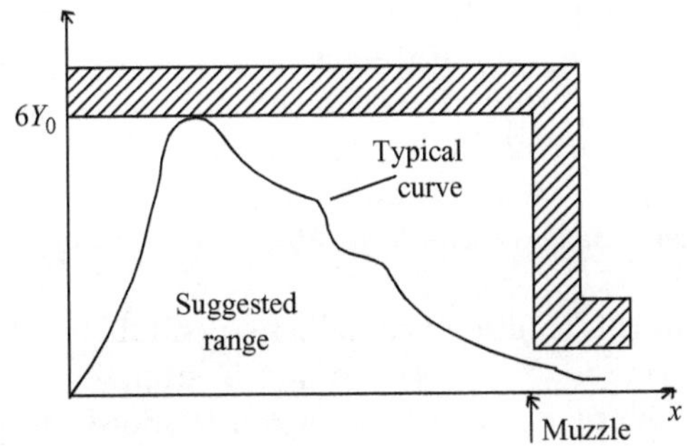

Fig. 2.26. Pressure range in which projectile integrity is preserved

In the case of a projectile made of two (or three) pieces, an impactor (generally a flat piece of metal at the front of the projectile) and a sabot, the interaction between the impactor and the sabot must be considered. The axial stress, σ_i, at the impactor–sabot interface is given by

$$\sigma_i = P_b \frac{S_b}{S_i}\left(1 - \frac{m_s}{m}\right), \tag{2.39}$$

where p_b is the base pressure, m_i is the mass of the impactor, m_s is the mass of the sabot, m is the total mass of the projectile ($m = m_i + m_s$), S_i is the cross-section of the impactor and S_b is the cross-section of the sabot (which is usually the caliber of the launch tube).

If the acceleration of the projectile during its internal flight is too high, and if the stress at the impactor–sabot interface is higher than the maximum stress that the sabot material can withstand, the impactor could become embedded into the sabot.

With the same type of calculation, assuming that the axial stress, σ, is constant all along the sabot and must not take a value higher than σ_{m}, the maximum mass of the impactor is given by the relation

$$m_{\mathrm{i}} = \sigma_m \frac{S_{\mathrm{i}}\, m_{\mathrm{s}}}{p_{\mathrm{b}}\, S_{\mathrm{b}} - \sigma_m\, S_{\mathrm{i}}} \,. \tag{2.40}$$

If the mass of the impactor exceeds this value, it could become embedded in the sabot during the shot.

2.4.7 Stopping the Piston After its Extrusion

After the shot, an important task is to extract the piston from the taper. Different solutions have been designed, depending on the laboratory. Because the piston is heavy, many laboratories use a small gunpowder charge to extract it. Use of a hydraulic device is rarely effective.

When gunpowder is used for the extraction, one needs a certain charge weight to be sure that the piston does not stop in the pump tube. For this reason, the achievable velocity of the piston is not negligible when it arrives at the exit of the pump tube. It is difficult to stop the piston because its weight (generally several hundreds of kilograms) and velocity are large. Several solutions have been tested in different laboratories, but I would like to present the one used in the CEA laboratory for the big two-stage light-gas gun.

The principle of this solution is a ballistic pendulum. The piston impacts in the percussion center of the system and is stopped by several sheets of iron. Most of the kinetic energy of the piston is transformed into rotational energy of the ballistic pendulum. Only few sheets of iron have to be changed to prepare for the next shot (Fig. 2.27).

In our case, the total device weights about 20 tons and after the impact, it rolls on curved rails.

What is interesting about this solution is that it is easy, from the angle of rotation of the pendulum, to know the exit velocity, v_1, of the piston. Assuming that all the momentum of the piston is totally transferred to the ballistic pendulum,

$$m\, v_1 = (m + M) v_2 \,, \tag{2.41}$$

where m is the piston mass, M the ballistic pendulum mass, v_1 is the piston velocity before impact, and v_2 is the velocity of the ballistic pendulum after the impact of the piston.

Assuming that the energy of deformation of the iron sheets is negligible, one can write the energy of the ballistic pendulum

$$1/2(m + M) v_2^2 = (m + M) g h \,, \tag{2.42}$$

Fig. 2.27. The ballistic pendulum

where h is the increase in height of the center of gravity and g the gravitational acceleration ($\sim$10 m/s^2). With R the radius of the rotation circle of the ballistic pendulum, and θ the maximum angle of rotation of the ballistic pendulum during the shot,

$$h = R(1 - \cos\theta)\ . \tag{2.43}$$

The ballistic pendulum velocity is given by

$$v^2 = \sqrt{2\,g\,h} = \sim 4.472\,\sqrt{h} \tag{2.44}$$

and

$$v_1 = [(m + M)/m]\,v_2\ . \tag{2.45}$$

Once the value of h (or θ) is recorded, one can easily calculate the exit velocity, v_1, of the piston.

2.5 The Comparison Between the Numerical Simulation and the Experimental Results

To know the real behavior of a two-stage light-gas gun, and especially to understand its efficiency and to improve its capabilities, a comparison between the numerical simulation and the experimental result is needed. This requires that the gun be equipped with many gauges and measuring devices along its entire length.

In the following section, we present the results of a shot obtained with the CEA two-stage light-gas gun [16]. General views of this gun are presented on Figs. 2.28 and 2.29.

Fig. 2.28. General view of the CEA gun

2.5.1 The Experimental Instrumentation on the Gun

The instrumentation on the CEA two-stage light-gas gun is composed of

- six piezoelectric pressure gauges on the pump tube and taper,
- six strain gauges applied to the launch tube to record axial and angential strains,

Fig. 2.29. General view of the CEA gun

- chronometric gauges near the muzzle of the launch tube to determine the projectile velocity.

The measurements on the powder chamber and the pump tube with the piezoelectric gauges CP1 to CP6 give the pressure history at each gauge location. With these records, a graph of the piston motion in the pump tube can be made. A graph of the motion of the projectile in the launch tube can be made using the records from the strain gauges J1 to J6. These records must be filtered to eliminate the large oscillations.

2.5.2 The Comparison Between the Numerical Study and the Experimental Results

The Pressure History in the Rear Part of the Gun

The different coefficients of the computer code, especially those for the gunpowder, the friction, and the head losses have been chosen to produce a satisfactory agreement with the measurements. The calculations were stopped when the agreement was within the experimental uncertainty (see Fig. 2.30).

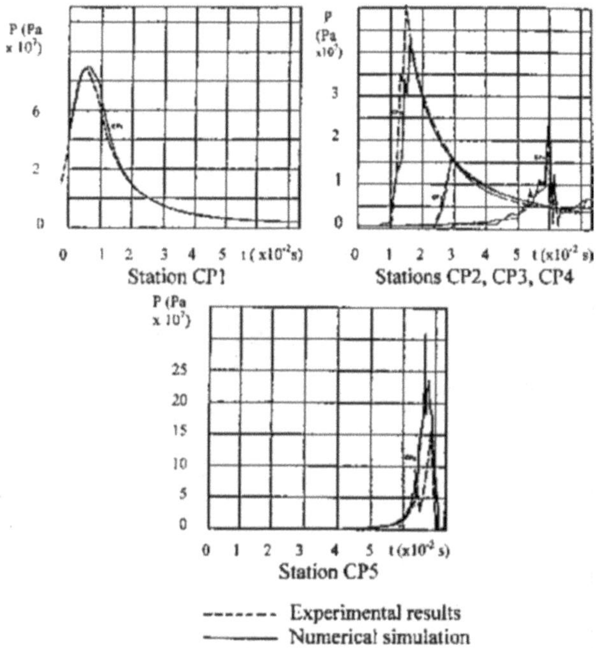

Fig. 2.30. Pressure histories. The dashed curves correspond to experimental results and the solid curves represent the numerical simulation

To take into account the dynamic behavior of the starting pressure of the piston, this pressure has been chosen equal to 60 MPa (600 bars) instead of the 40 MPa (400 bars) that is the mechanical value for its breaking pressure, i.e., the starting pressure.

For the powder chamber (the breech) there is a good agreement between numerical and experimental results (CP1), especially for the maximum pressure (0.11 GPa). The general shape of the pressure history curve is also in good agreement; only the decrease of the pressure shows a slight discrepancy.

For the pressure gauges CP2, CP3 and CP4, the maximum values and general shapes show a relatively good agreement between numerical and experimental results (Fig. 2.30).

For the CP5 gauge, the computed maximum pressure is about 0.3 GPa and the spike is of 500 μs duration. The experimental result does not give this value. The reason could probably be either a mistake in the calculation or the response time of the gauge.

An important result that can be obtained from the records of CP4 and CP5 is detection the reflection of the compression waves from points near the taper in the final phase of the compression of the light-gas. These waves are also obtained by computer alculation.

The Motion of the Piston

From the records of the gauges, a reconstruction of the piston motion is possible. The comparison between the numerical study and the experimental results shown on Fig. 2.31 shows a satisfactory agreement.

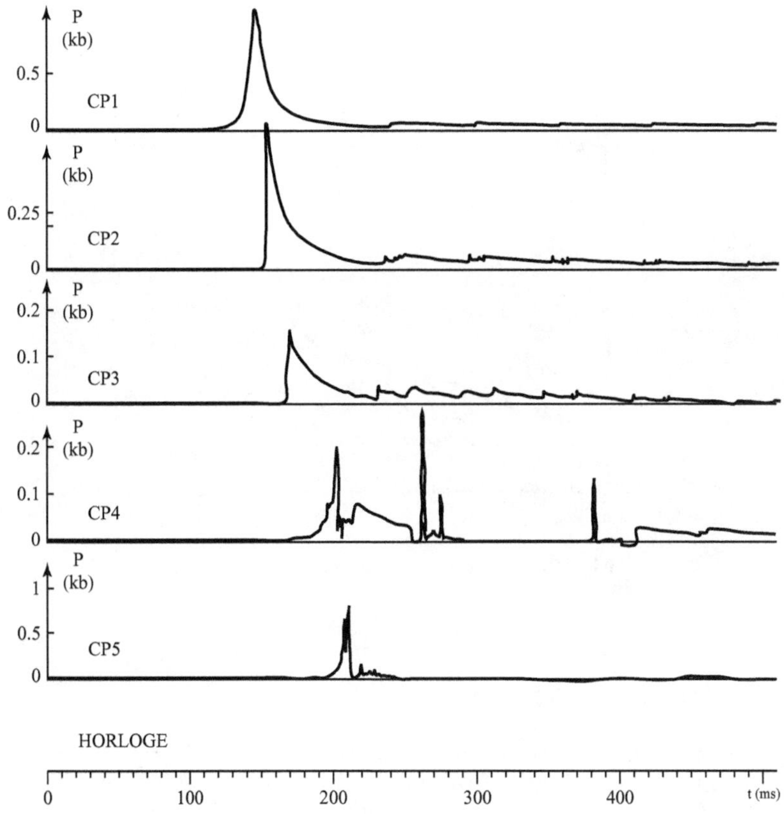

Fig. 2.31. Pressure pulses recorded at different stations on the first stage of the gun during the shot

After the diaphragm ruptures, the code shows that the piston rebounds rather than stopping in the conical shape of the taper. This result has been experimentally verified by two methods. It is possible to detect the rebound from the records CP1 to CP5 (Fig. 2.32). The large extrusion of the piston is also evidence of the rebound. After rebounding, the piston is again pushed by the combustion products and, as no counter-pressure diminishes its velocity, it extrudes deeply into the cone. This result shows that the muzzle velocity could have been made higher than the value obtained by either diminishing the initial pressure of the light-gas (if the launch tube is not too long) or by increasing the weight of gunpowder used.

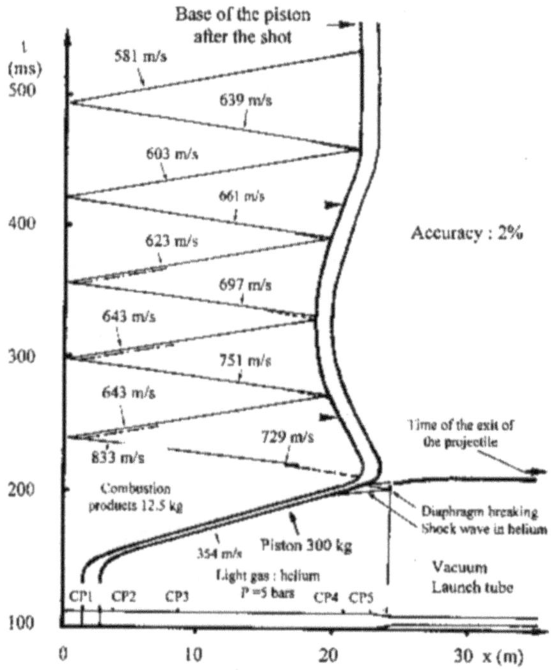

Fig. 2.32. Internal ballistics of the piston

The Motion of the Projectile

By reconstruction from the records of the strain gauges (Figs. 2.33 and 2.34), with the adjustment of the different coefficients of friction, head losses, etc., it is possible to calculate the motion of the projectile in the launch tube with relatively satisfactory accuracy (Figs. 2.35 and 2.36).

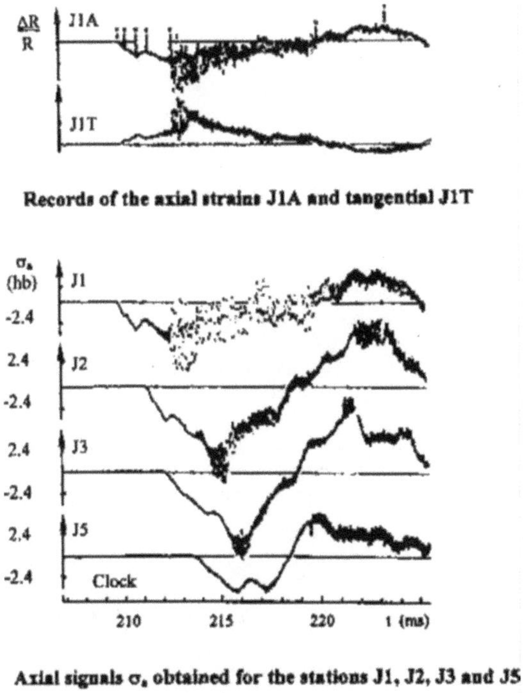

Fig. 2.33. Axial signals σ_a obtained for the stations J1, J2, J3, and J5

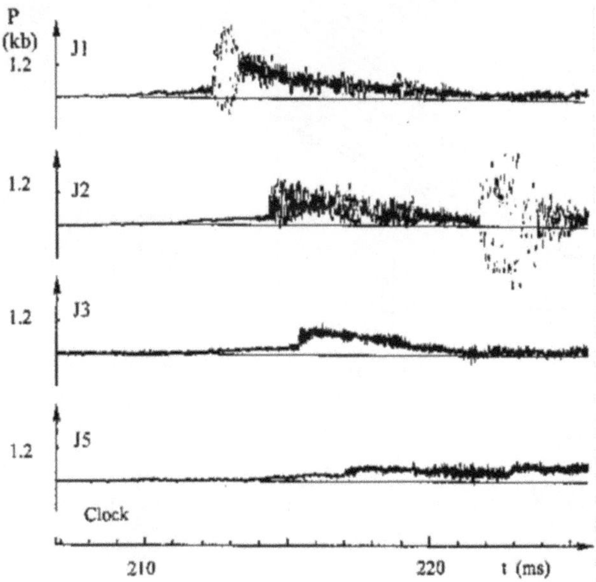

Fig. 2.34. Pressure pulses versus time recorded on the launch tube

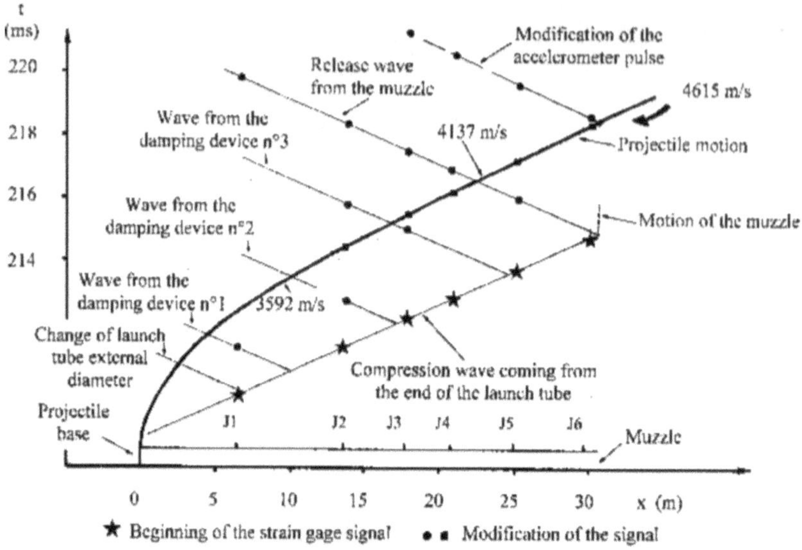

Fig. 2.35. Internal ballistics of the projectile. The symbol ⋆ designates the beginning of the strain gauge signal and the symbols • and ▪ indicate modification of the signal

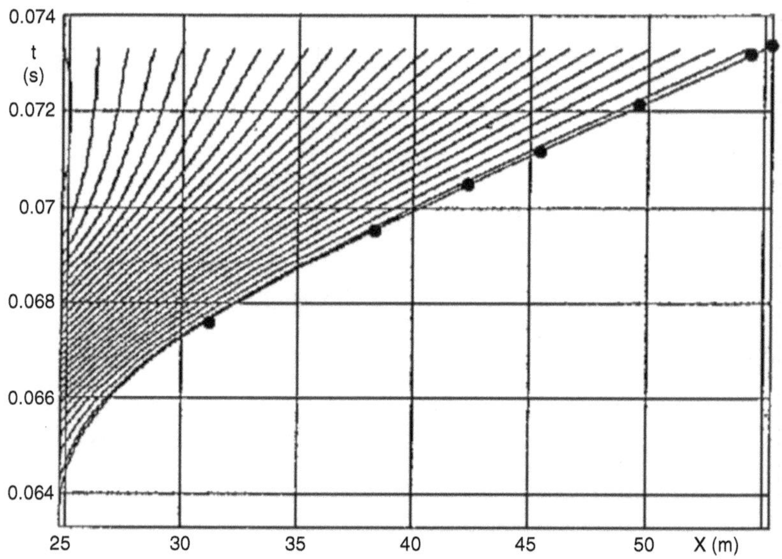

Fig. 2.36. Motion of the projectile. The dots are obtained from the experimental records

The General Simulation of the Experiment

To see a general view of the shot, several graphs show the numerical simulation of an experiment.

The three-dimensional graph of $p(x, t)$ in Fig. 2.37 shows the release of the combustion products, the motion of the piston, the compression of the light-gas, the diaphragm rupture, and the motion of the projectile. The shading, a fourth dimension, shows the temperature.

The three-dimensional graph of $T(x, t)$ in Fig. 2.38 allows for the visualization of the total motions of all the parts of the gun during the shot. In this case, the pressure is shown by the shading.

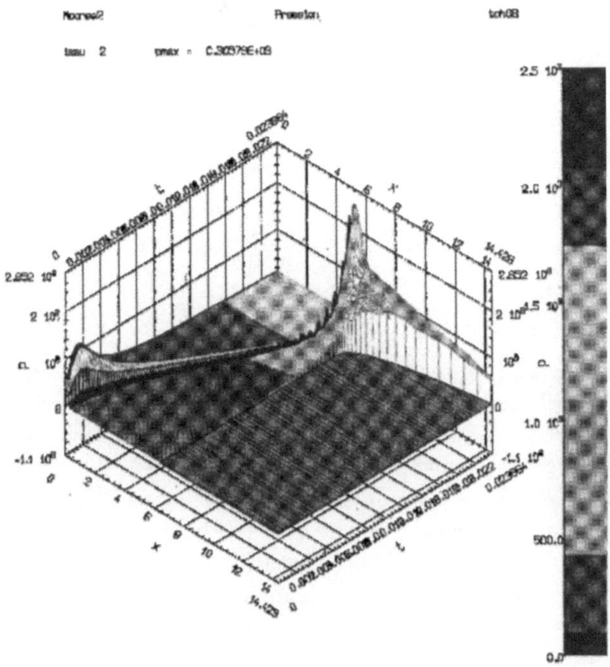

Fig. 2.37. Three-dimensional $p(x, t)$ diagram of a calculated shot

2.6 Conclusion

Scaling and designing a large two-stage light-gas gun for use at high velocities is a complicated challenge. Many calculations are needed. The design of the gun itself is always a compromise between the expected performance and the strength of the metal used to build it (generally special steel for the most highly stressed pieces).

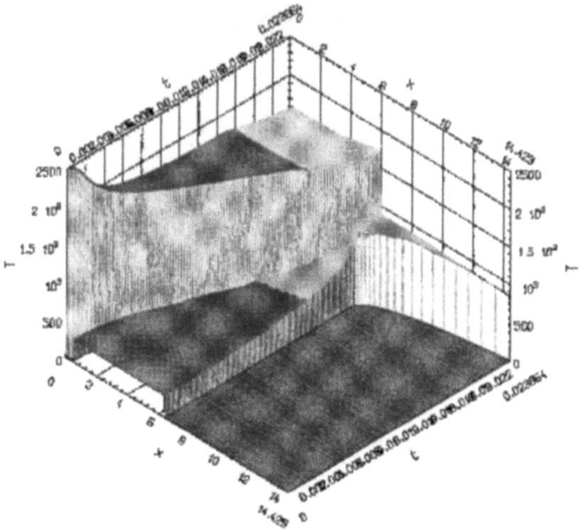

Fig. 2.38. Three-dimensional $T(x,\, t)$ diagram of a calculated shot

From a technical point of view, many problems have to be solved, and many ompromises made. The final result depends on the weakest part of the device.

Here, I tried to present briefly the principle of the computer codes (today many laboratories have developed their own codes), but I especially focused on some of the main difficulties that must be studied and overcome during the conception phase.

For examples, I tried to describe some particular solutions we have chosen for the CEA large two-stage gun.

Many other problems remain to be addressed, (see [17] for some of them) including the fluid connections to the gun, the honing of the launch tube (Do we move the launch tube or the honing machine?), the experimental assembly located in front of the muzzle of the launch tube, etc.

Acknowledgment

I would like to warmly thank all the laboratories, and the members of these laboratories that helped me during my early work on this subject. I would also like to thank all the members of the CEA staff who gave me discretionary powers and carte blanche to scale and design this gun and make the associated studies. I also want to thank all the members of different laboratories, design offices, manufacturers, and the members of my group who helped me study and build this range.

References

I am sorry not to mention many papers that were of great interest during all the work on this project. Those listed comprise only the most important papers, but I think that the references included in each of them would also be of great help to everyone who wants to improve his knowledge of a particular aspect of this subject. For references cited as Aeroballistics R ange Association proceedings, articles can be obtained directly from the authors.

1. G. Wilenius, M. Cloutier, and P.L. Cowan, *A theoretical analysis of a constant base pressure light-gas gun*, CARDE Technical Memorandum 703/62 (1962).
2. J.S Curtis, *An accelerated reservoir light-gas gun*, NASA TN.D-1144 (1962).
3. A.C. Charters, "Development of the high velocity gas-dynamics gun," *Int. J. Impact Eng.* **5**, pp. 181–203 (1987).
4. W.D. Crozier and W. Hume, "High velocity light-gas gun," *J. Appl Phys.* **28**, p. 892 (1957).
5. R. Piacesi, D.F. Gates, and A.E. Seigel, *Computer analysis of a two-stage hypervelocity mode launcher, VIth Symposium on Hypervelocity Impact*, (1963).
6. L.K. Badhwar and J.R.B. Murphy, *A new ballistics computation system for light-gas guns*, Report prepared by Computing Devices of Canada, Ltd., Ottawa, Oct 1965, for C.A.R.D.E. Valcartier.
7. P. Dumouchel and R. Alexander, *A review of internal ballistics and a description of COMDEV computer programs*, Report prepared by Computing Devices of Canada, Ltd., Ottawa, March 1974, for D.R.E.V. Valcartier.
8. R.M. Patin, R.W. Courter, *A one-dimensional simulation model for a two-stage light-gas gun with deformable piston*, A.I.A.A.-86-0046, A.I.A.A. 24th Aerospace Sciences Meeting at Reno, Nevada, January 6–9, 1986.
9. C.P.T. Groth and J.J. Gottlieb, *Numerical study concerning the operation of the D.R.E.V. 250/105 mm light-gas gun for launching high-speed projectiles*, Institute for Aerospace Studies, University of Toronto. for D.R.E.V. Valcartier.
10. H. Bernier, and J. Gambart Moorea, *An internal ballistic code for two-stage light-gas gun study*, 38th A.R.A. Meeting, Tokyo, 1987, (private communication).
11. A.E. Seigel, *The theory of high speed guns*, Agardograph 91 (1965).
12. C.M. Leech, *Axial stresses evaluation in light-gas guns*, CARDE Valcartier, Oct. 1967.
13. H. Bernier, *Flat scored high pressure diaphragms as quick opening valves*, 42th A.R.A. Meeting, Adelaide, 1991, (private communication).
14. Dah Yu Cheng, Robert E. Dannenberg, and Walter E. Stephens, "A novel use of a telescope to photograph metal diaphragm openings," *AIAA Journal* **7(6)**, pp. 1209–1211, (1969).
15. H. Bernier, *Etude de la tenue mécanique d'un projectile soumis à une acceleration importante*, 2ème Congrès Français de Mécanique, Toulouse, Sept 1975.
16. H. Bernier, "Le canon à gaz léger du Commissariat à l'Energie Atomique Exploitation des mesures effectuées sur le canon lors d'un tir," 26th A.R.A. Meeting, San Leandro (1975), (private communication).
17. J.M. Lezaud, *Etude et réalisation d'un canon à gaz léger,* Thèse, Université de Poitiers, April 26, 1976.

3 The Coaxial Plasma Drag Accelerator

E. Igenbergs and M. Rott

3.1 Introduction

The transfer of stored electrical energy to increase the kinetic energy of a mass (which may also be called a projectile) can be characterized by attributes like

- the acceleration process
- the accelerator configuration and principle
- the energy storage and power supply to the accelerator, and
- orders of magnitude describing the process, the accelerator, and the power supply.

Many combinations of these attributes are feasible and have been analyzed theoretically and experimentally. The results are published in journals such as the IEEE Transactions on Magnetics. The present authors do not intend to cover all these very interesting efforts, but do rather restrict the remainder of this chapter to research work which they conducted together with their colleagues and co-workers.

The acceleration process is purely gasdynamic, excluding a small electromagnetic interaction during the acceleration. The acceleration process involves the generation of a high velocity/high density flow of a partially ionized gas (called plasma) using a coaxial plasmadynamic accelerator, and introducing the mass into the flow where it is drag accelerated. The electrical energy is stored in a capacitor bank. This energy supply system must enable the very fast transfer of the stored energy to the accelerator. The expected result of this process is acceleration of particles 0.1–1 mm in diameter to a velocity in the order of 10 km/s. The length of the accelerator should not exceed 1 m and the energy storage system should deliver up to 50 kJ of stored energy within 100 µs.

A model of this type of accelerator is shown in Fig. 3.1, which illustrates a configuration having a plasma generator on the left side and a compressor coil on the right side. The plasma generated is confined by the magnetic field produced by an electrical current in the coil. The plasma density is in the order of 0.01 kg/m^3 and it flows from left to right at a velocity of approximately 50 km/s. The corresponding drag force on a 1 cm diameter sphere would be 10^3 N and the interaction time 20 µs. If the sphere were made from aluminum and not attached the holder, then it would be accelerated to

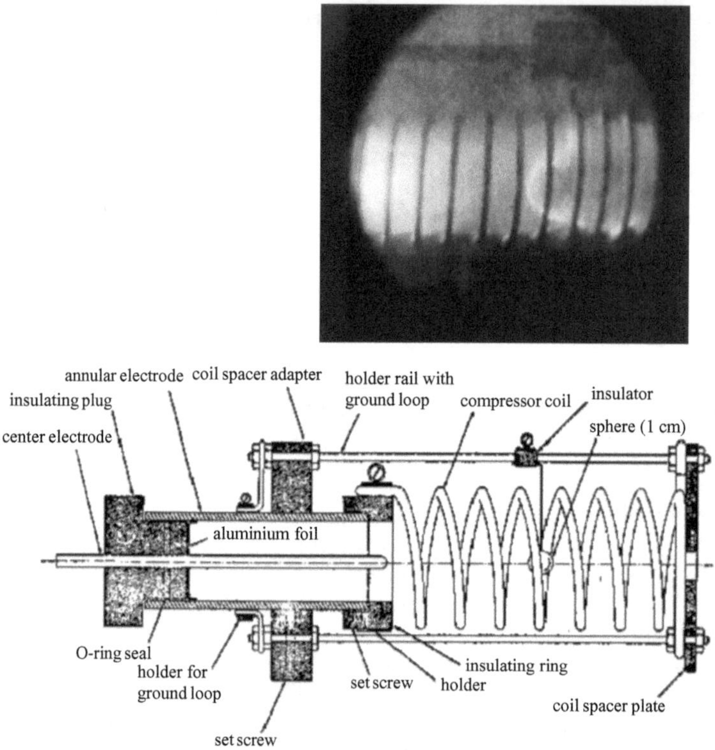

Fig. 3.1. Schematic of plasma accelerator with photograph of plasma flow around a sphere

10–100 m/s. Even a reduction of the size of the sphere from 1 cm to 1 mm diameter would result in an increase of the final velocity to only 100–1000 m/s. This was the situation when the work described in the subsequent sections began.

Starting with the coaxial accelerator, which is the primary plasma generator in all the configurations, we demonstrated in detail why the coaxial accelerator does not provide sufficient acceleration. As the velocity of the plasma flow is high enough, an attempt was made to increase its density. This raised the performance significantly; a velocity of approximately 10 km/s was reached with glass beads of 0.1 mm diameter as projectiles. Measurements and theoretical analysis of the plasma flow in the compressor coil indicated that the maximum dynamic pressure in the direction of the plasma flow occurred inside the coil, near the narrow end. This led to the idea to use a pre-accelerator to inject the projectile into the compressor coil at the optimal time and place. The light-gas gun/coaxial accelerator with a compressor coil configuration accelerated a glass bead with a diameter of 0.6 mm to a velocity between 16 and 20 km/s.

3.2 Test Environment

The experiments described here were conducted at the Space Sciences Laboratory of the NASA George C. Marshall Space Flight Center (MSFC), Huntsville, AL, USA and in the Ernst Stuhlinger Laboratory of the Institute of Astronautics, Technische Universität München (TUM), Germany. Here the test environment at the MSFC, where all this work began, is described. The functionality of the environment at the TUM is the same [2, 3].

3.2.1 Test Range with Plasma Accelerator

The test range is made of a 2.0 m long, 15 cm diameter aluminum tube with a flange at both ends. It has one 38- by 12.5 cm window of 2.5 cm thick Plexiglas for photographing the plasma flow and the particle acceleration. The inner surface of the window is covered with a 3 mm-thick replaceable Plexiglas cover for protection against the hot plasma. Two other windows on top of the range have 3.5 cm-thick aluminum covers with 5.0 cm diameter, vacuum-tight, feed-through plugs. A vacuum pump is connected to the range through a flexible hose. The coaxial plasma accelerator is mounted on a fiberglass plate that closes the front end of the test range. The other end of the test range is closed by an aluminum cover with another feed-through plug and a vacuum gage.

3.2.2 Capacitor Bank and Switch Assembly

The capacitor bank consisted of 72 individual 15 μF capacitors connected in parallel in 4 groups of 18 capacitors each. The experiments were all made with only one-quarter of the bank. The maximum charge voltage was 20 kV and the corresponding stored energy was 54 kJ. Three capacitors connected in parallel by copper bus bars formed a module, and every module was connected to the switch assembly by two coaxial cables. All the cables were the same length. Each quarter of the bank was equipped with four ignitron switches in parallel. These were fired by a trigger assembly containing a set of capacitors and a thyratron to prevent back current to the igniters of the ignitrons. The switch assembly is connected to the coaxial accelerator by six coaxial cables.

The capacitor bank is charged from a high voltage supply which is controlled by an autotransformer and is disconnected before firing. For emergencies a dump circuit is incorporated to discharge any energy stored in the capacitor bank through the charging network. A schematic diagram of the test range, capacitor bank, charging circuit, and switch assembly is shown in Fig. 3.2.

3.2.3 Diagnostics and Particle Velocity Measurement

Photographs of the plasma flow and the accelerated particles were made through the 38- by 12.5 cm window in the range. Two single frame image

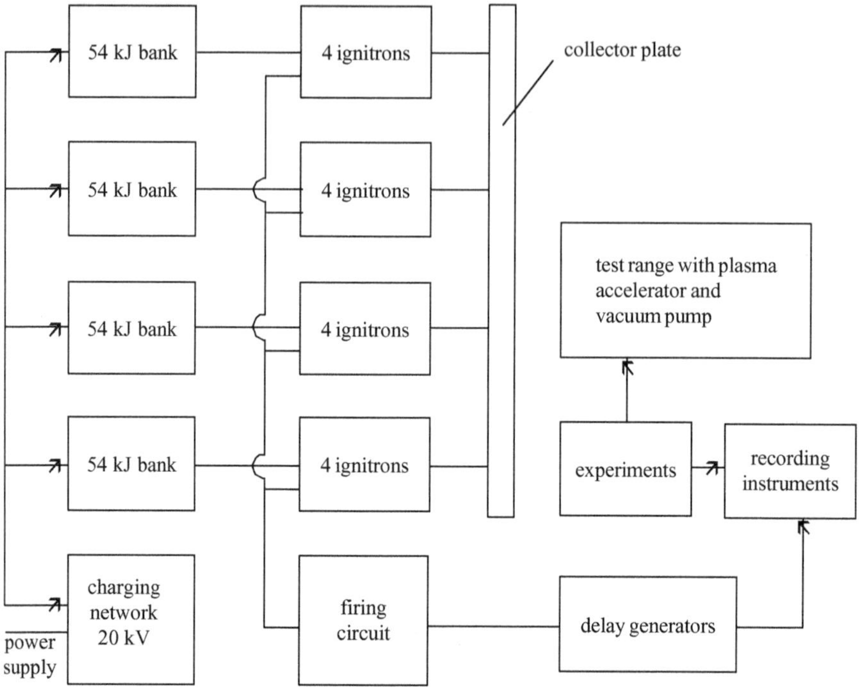

Fig. 3.2. Schematic diagram of the plasma hypervelocity range

converter cameras were used for high-speed photographs. They were triggered by the pulse from the delay generator to the capacitor bank firing circuit. Using the adjustable exposure delay of the cameras, the time interval between the two exposures could be set between 10 and 100 microseconds.

A unit made up of eight individual image converter cameras was used for sequential photographs. In this unit the plasma image is deflected by a prism which is located in the center of a drum-shaped container, to the eight image converter tubes located at 45° intervals. For particle detection in the very bright plasma light, filters were placed between the prism in the center and the image converter tubes. The eight cameras could be triggered at a maximum time interval of 1 µs between successive frame exposures. The minimum exposure time was 10 ns and the maximum exposure time was 100 ns. The trigger pulse for the image converter cameras came from a transient light detector or from the delay generator which also triggered the capacitor bank firing circuit. In both cases, the trigger pulse for the first camera was recorded on an oscilloscope.

Light probes were placed in various positions in front of the plasma range window to measure the plasma front propagation velocity. The output of these detectors was recorded on an oscilloscope. The particle velocity was measured with the eight-camera unit and an impact detector. Using light filters, the particles could be photographed when they left their position at the Mylar film. The time delays between the eight photographs were recorded on time-interval counters. These delay times, together with the recorded trigger pulse for the first camera, bracketed the time when the particles left their position at the front of the compressor coil.

The impact detector placed between 0.2 and 0.4 m downrange had a detector mounted coaxially with the plasma accelerator and compressor coil. An aluminum sheet was held in place by a nut. The accelerated particles perforated this sheet, and a light flash was generated which was deflected by a mirror into a photomultiplier mounted on the top window cover of the range. A hole was punched through one aluminum sheet to determine the origin of the light recorded by the photomultiplier. The plasma light coming through this hole for a discharge with no projectile particles had the same intensity as the signal from a perforation by an accelerated particle. It may, therefore, be concluded that the recorded light flash from an aluminum sheet perforation by an accelerated particle is generated by the plasma, either in the accelerator or in a shock wave in front of the impact detector.

The output of the photomultiplier detector was recorded on the same dual-beam oscilloscope as the trigger pulse for the first camera. The particle flight time, together with the known distance between the compressor coil front end where the photographs were made and the aluminum sheet in the particle impact detector, was used to determine the average particle velocity. The capacitor bank voltage was measured with an inductive pickup coil mounted on top of the positive leads of the coaxial cables leading from the ignitron switch assembly to the coaxial accelerator. The shields of the coaxial cables were stripped from the cables in the switch assembly; the positive leads go through the ignitrons and the shield goes through a ground plate.

3.3 Acceleration of Small Particles

Glass beads with diameters between 0.125 mm and 0.795 mm were used for acceleration experiments. There was no indication of any forces other than gasdynamic forces acting on these beads.

In gas dynamics the force, F, acting on a body can be expressed as the sum of a dynamic pressure term and the product of the pressure gradient, ∇P, and a characteristic length, Δz. These terms are multiplied by a characteristic surface area, S, which is assumed to be the same for both terms:

$$F = \left(C_{\mathrm{D}} \frac{\rho}{2} V^2 + \nabla P \, \Delta z \right) S \,, \tag{3.1}$$

where C_{D} is a constant called the *drag coefficient*. For constant particle size and flow velocity, V, the dynamic pressure term is a function only of the flow density, ρ. Plasma accelerators for neutral particle acceleration were, therefore, designed for high flow densities.

In a magnetogasdynamic flow the pressure gradient term, ∇P, may become large if the plasma density is high enough to permit an immediate transfer of the electromagnetic force into gasdynamic pressure. Maximum gasdynamic acceleration of neutral particles, therefore, is obtained by high density flow and a large negative pressure gradient.

First measurement indicates a plasma density of $\rho = 2.5 \times 10^{-2}\,\mathrm{kgm}^3$. Another estimate can be obtained by dividing the mass of the aluminum foil (the plasma source) inserted into the coaxial accelerator by the volume of the coaxial accelerator of 0.2 m length,

$$\rho_{\mathrm{PLASMA}} = \frac{\text{foil mass}}{\text{accelerator volume}} = 1.06 \times 10^{-1}\mathrm{kg/m}^3 \, .$$

The measured density can be introduced into (3.1). Assuming that there is no contribution from the pressure gradient term, the acceleration of a glass bead may be computed using the dynamic term of (3.1),

$$a = \frac{F_{\mathrm{D}}}{M_{\mathrm{BEAD}}} = \frac{C_{\mathrm{D}}}{2} \frac{\rho_{\mathrm{PLASMA}}}{\rho_{\mathrm{GLASS}}} \frac{3}{4\,R_{\mathrm{BEAD}}} V^2 \, . \tag{3.2}$$

With the drag coefficient $C_{\mathrm{D}} = 2.0$ from the Newtonian approximation for hypersonic flows and the glass density $\rho = 2.5\,\mathrm{g/cm}^3$,

$$a = \frac{3\,V^2}{4\,R_{\mathrm{BEAD}}} \times 10^{-5}\mathrm{m/s}^2 \, . \tag{3.3}$$

A comparison of the results from (3.2) and (3.3) shows that the measured density is approximately one-fourth of the calculated value. This indicates that only a part of the aluminum foil mass is accelerated. High-speed photographs using a movie camera show that plasma columns are accelerated at every half-period of the circuit consisting of the capacitor bank, the cables, and the coaxial accelerator.

3.3.1 Coaxial Accelerator for Particle Acceleration

The electromagnetic force in the coaxial accelerator is generated by the interaction of the radial current density j_r between the center electrode and the annular electrode, and the azimuthal magnetic induction B_Φ around the center electrode. Both are functions of the radial distance r from the center electrode:

$$j_r \sim \frac{1}{r} \quad \text{and} \quad B_\Phi \sim \frac{1}{r} \, . \tag{3.4}$$

The axial electromagnetic force is

$$F_z = j_r\, B_\Phi \sim \frac{1}{r^2} \, , \tag{3.5}$$

and the axial momentum equation becomes

$$\rho\frac{dV_z}{dt} + \frac{\partial P}{\partial z} = F_z \sim \frac{1}{r^2}\,. \tag{3.6}$$

The maximum acceleration and pressure gradient will occur in the vicinity of the center electrode. The plasma leaves a coaxial accelerator with an equally long center and annular electrodes in a conical shape, as shown in Fig. 3.3. During the acceleration, plasma is forced radially toward the wall of the annular electrode. After 10–20 firings, the inner surface of the annular electrode is covered with an aluminum deposit. The maximum acceleration of the particles was observed when they were placed near the center electrode, as shown in Fig. 3.3. A ring-shaped copper holder was inserted into the annular electrode, and the center electrode was shortened. If the ring-shaped holder is in metallic contact with the annular electrode, then the secondary discharge indicated in Fig. 3.4 occurs. This discharge was photographed with the center electrode protruding beyond the annular electrode (Figs. 3.5 and 3.6). The discharge starts when the plasma fills the gap between the center electrode and the ring-shaped holder. The azimuthal magnetic field around the discharge interacts with the axial current density and produces both a radial force and an axial force due to the conical shape of the discharge. A thin Mylar film with glass beads attached to it by a thin coating of vacuum grease was placed across the ring holder. The maximum particle velocity (1–2 km/s with 0.7 mm diameter beads) was achieved with the holder in front of the center electrode inside the annular electrode.

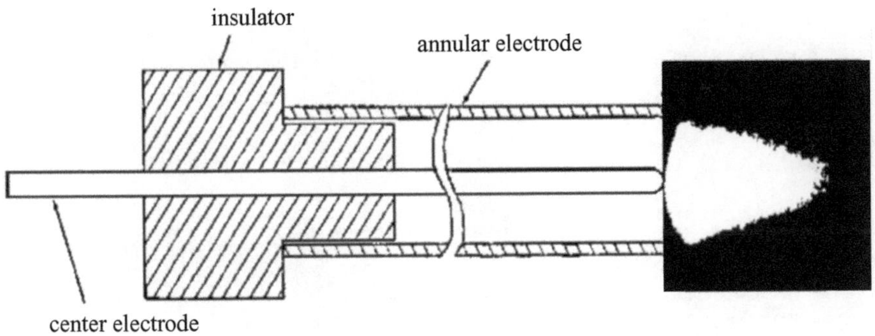

Fig. 3.3. Coaxial plasma accelerator

The center electrode, the insulating plug, and the annular electrode, form the coaxial accelerator. An aluminum foil is placed between the center electrode and the annular electrode before every firing. The positive side of the charged capacitor bank is connected to the center electrode through the switch assembly, and the negative side is connected to the annular electrode.

It should be noted that the configuration with the center electrode longer than the annular electrode (Fig. 3.5) produced a very high velocity luminous front travelling down the range if no Mylar film was placed on the ring holder.

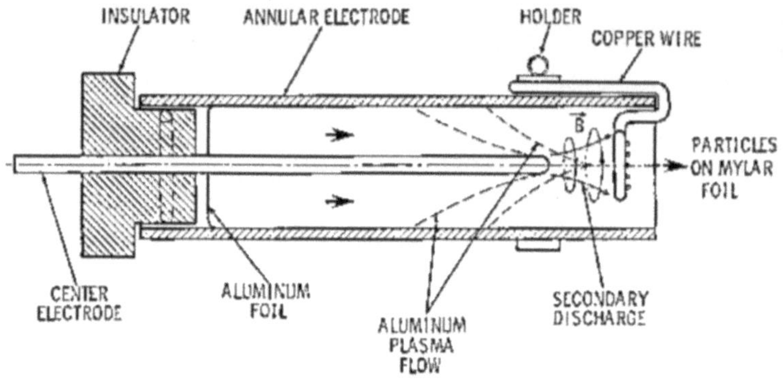

Fig. 3.4. Particle acceleration near the center electrode of coaxial accelerator

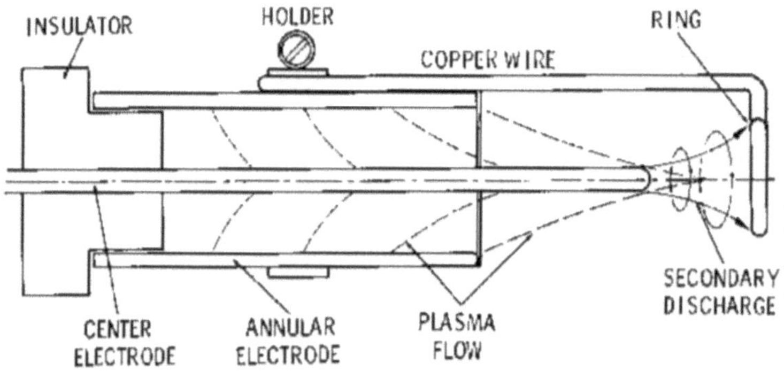

Fig. 3.5. Description of secondary discharge between the center electrode and the copper ring

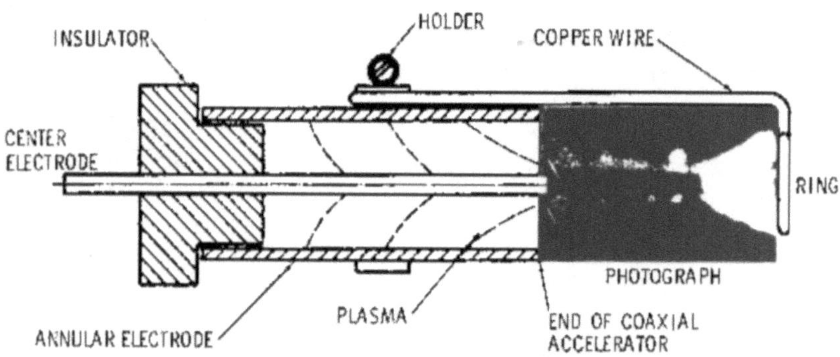

Fig. 3.6. Photograph of secondary discharge

The plasma arrived at the gap between the center electrode and the ring holder with a velocity corresponding to the coaxial accelerator configuration. The secondary discharge provides additional acceleration because of its conical geometry and, thus, acts as a second acceleration stage.

Finally, the coaxial accelerator without the compressor coil produces a fast-moving, low-density plasma. The vaporized and ionized aluminum foil does not stay together, and the plasma leaves the accelerator as a column that is at least as long as the accelerator.

3.3.2 Compressor Coil Configuration [1, 2, 4]

The effectiveness of the system is increased by the addition of a compressor coil at the end of the coaxial accelerator, as described earlier and shown in Fig. 3.7. The emerging plasma initiates a discharge between the center electrode and the copper compressor coil, which is connected to the annular electrode. If the spacing between the coil windings is large enough, the electric current will flow through the copper coil windings and will generate a mostly axial magnetic field inside the coil. This magnetic field changes with time and induces an azimuthal current density in the plasma inside the compressor coil. This current interacts with the magnetic field and results in a radial force on the plasma. While the axial magnetic field is increasing, this radial force acts toward the axis, keeping the plasma inside the compressor coil. The current from the center electrode to the compressor coil has axial and radial components. The azimuthal magnetic field and the radial current density component generate an axial force as in the coaxial accelerator. Therefore, the plasma will be simultaneously compressed and accelerated into the compressor coil. The current density from the center electrode to the compressor coil is controlled by a local equilibrium between the radial force, which tends to push the plasma away from the compressor coil surface, and the current density needed to produce this force.

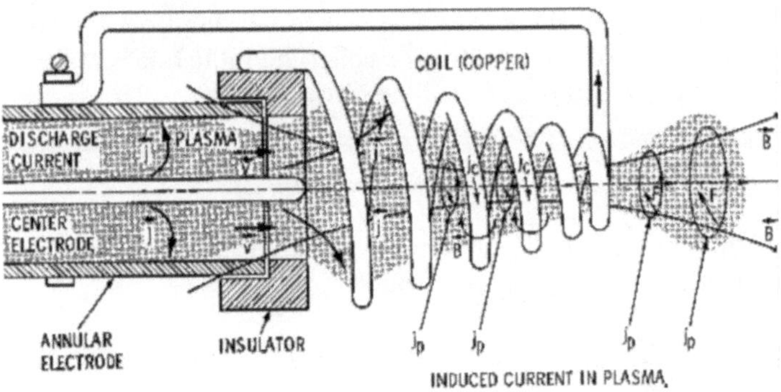

Fig. 3.7. Compressor coil with magnetic field and plasma current density

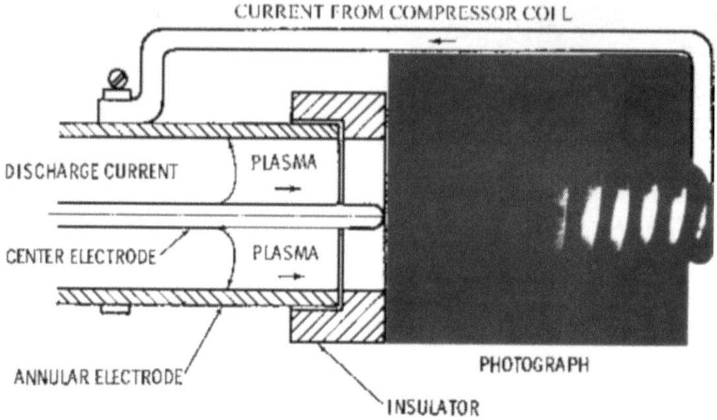

Fig. 3.8. Photograph of plasma flow in the compressor coil during the compression period

There are two ways to connect the compressor coil to the annular electrode potential, and they lead to two different modes of operation. These modes are called the decreasing and increasing inductance modes. In the decreasing inductance mode, the compressor coil is insulated from the annular electrode by an insulator ring Fig. 3.8. The end of the compressor coil is connected to the annular electrode through the holder rail. As the plasma moves into the compressor coil, the current density will assume the same distribution along the compressor coil. The current flow to the wide end of the compressor coil, which is close to the coaxial accelerator end and attached to the insulator ring, will have to traverse all the windings of the compressor coil, whereas the current going to the narrow end of the compressor coil will flow only through the final winding. In an equivalent circuit for this configuration, every current element must be combined with a different inductance, varying from the total inductance L_1 at the wide end to a small inductance at the narrow end. If Δl is a length element of the compressor coil, then the position measured from the wide end of the coil is given by $i\Delta l$, with i varying from zero to n, where $n\Delta l = l$, the length of the compressor coil. The inductance L_i for a current density entering the compressor coil at $i\Delta l$ will be

$$L_i = L_1 - i\,\Delta l \frac{L_1}{n\,\Delta l}\,. \tag{3.7}$$

The total inductance, L, is obtained by adding the different inductances according to the rule for parallel circuits,

$$\frac{1}{L} = \sum_{i=0}^{i=n-1} \frac{1}{L_i} = \sum_{i=0}^{i=n-1} \left[L_1 \left(1 - \frac{i}{n}\right)\right]^{-1},$$

$$\frac{L_1}{L} = \sum_{i=0}^{i=n-1} \left(1 - \frac{i}{n}\right)^{-1}$$

or

$$\frac{L_1}{L} = 1 + \sum_{i=1}^{i=n-1} \left(1 - \frac{i}{n}\right)^{-1} . \tag{3.8}$$

It is assumed that there is no current flowing into the last element of the compressor coil, eliminating the infinite term for $i = n$. This is equivalent to a residual inductance at the end of the compressor coil representing the inductance of the backstrap lead to the annular electrode. Since

$$0 < 1 - \frac{i}{n} < 1 \quad \text{then} \quad \frac{L_1}{L} > 1 ,$$

the initial inductance, L_1, will be larger than the effective inductance. The plasma flow into the compressor coil will, therefore, reduce the inductance in the equivalent circuit from the initial value, L_1, to the value L.

In the increasing inductance mode, the compressor coil is connected directly to the annular electrode. Then the inductance in the equivalent circuit will increase as the plasma moves into the compressor coil.

Both configurations were tried. The decreasing inductance configuration was selected for further investigation after initial experiments indicated that it would be more efficient for the acceleration of glass beads with diameters between 100 and 700 µm. These were attached to a Mylar film and then placed across the hole in the coil space plate.

3.3.3 Theoretical Investigation

Artisimovitch developed the theory of the rail-type accelerator. Other investigators have extended this theory to describe the coaxial accelerator.

Equivalent Circuit

The addition of a compressor coil can be represented by an equivalent circuit which consists of the coaxial accelerator circuit to which is added a mesh with variable plasma resistance and a variable inductance simulating the compressor coil. An additional switch is incorporated in the mesh to simulate the arrival of the plasma in the coil.

In Fig. 3.9 the acceleration of the plasma in the coaxial accelerator is represented by the variable inductance L_2. The parameters L_0 and R_2 are the capacitor bank inductance and resistance, which are constants. The resistance R_1 simulates the arrival of the plasma at the compressor coil and L_1 simulates the compressor coil inductance (see previous discussion of compressor coil configuration). The differential equations for this equivalent circuit with the additional switch closed are

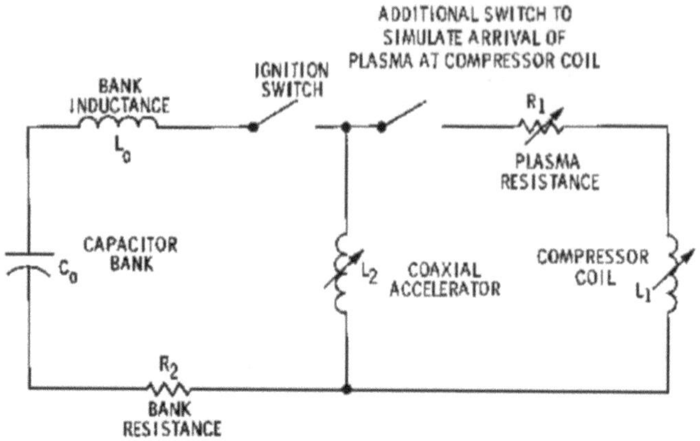

Fig. 3.9. Equivalent circuit for coaxial accelerator with compressor coil

$$U_0 - \frac{1}{C_0}\int_0^t (I_1 + I_2)dt = L_0\frac{d}{dt}(I_1 + I_2) + R_2(I_1 + I_2) + R_1 \cdot I_1 + \frac{d}{dt}(L_1 \cdot I_1) \tag{3.9}$$

and

$$U_0 - \frac{1}{C_0}\int_0^t (I_1 + I_2)dt = L_0\frac{d}{dt}(I_1 + I_2) + R_2(I_1 + I_2) + R_1 \cdot I_1 + \frac{d}{dt}(L_2 \cdot I_2)\ , \tag{3.10}$$

where U_0 is the capacitor bank charging voltage. Equation (3.9) is the current in the compressor coil and (3.10) is the current in the coaxial accelerator.

Magnetogasdynamic Flow

The in the vicinity of the compressor coil axis can be described by the inviscid magnetogasdynamic momentum equation,

$$\rho\frac{d\mathbf{V}}{dt} = -\operatorname{grad} p + \mathbf{j} \times \mathbf{B}\ . \tag{3.11}$$

The magnetogasdynamic force arises from the interaction of the electric current density,

$$\mathbf{j} = \sigma(\mathbf{E} + \mathbf{V} \times \mathbf{B})\ , \tag{3.12}$$

with the magnetic induction $\mathbf{B}$ given by

$$\mathbf{B} = \mu\mathbf{H}\ . \tag{3.13}$$

The components of the vector product $\mathbf{j} \times \mathbf{B}$ are, in cylindrical coordinates,

$$\mathbf{j} \times \mathbf{B} = \begin{pmatrix} j_\Phi B_z - j_z B_\Phi \\ j_z B_r - j_r B_z \\ j_r B_\Phi - j_\Phi B_r \end{pmatrix} . \tag{3.14}$$

The radial component will perform the compression of the plasma. The current density, j_Φ, is

$$j_\Phi = \sigma \left(E_\Phi + V_z \, B_r \right) . \tag{3.15}$$

The first term in (3.14) comes from the change of the magnetic field with time,

$$\dot{\mathbf{H}} = -\frac{1}{\mu} \nabla \times \mathbf{E} . \tag{3.16}$$

The axial field on the compressor coil axis can be written

$$H_z = A(z) \, I(t) \, , \tag{3.17}$$

where $A(z)$ is a function derived from the compressor coil geometry and $I(t)$ is the compressor coil current. By integrating (3.16), the azimuthal electrical field from the time varying magnetic field becomes

$$E_\Phi = -\frac{r}{2} \mu \, A(z) \, \dot{I}(t) . \tag{3.18}$$

The second term of (3.15) is the product of the axial plasma velocity and the radial magnetic field. The latter can be approximated with the assumption that the axial magnetic field does not change in the vicinity of the compressor coil axis with the radius r. Then,

$$\nabla \cdot \mathrm{H} = 0 \tag{3.19}$$

gives an approximation for the radial magnetic field,

$$H_r = -\frac{r}{2} \frac{dA(z)}{dz} I(t) . \tag{3.20}$$

The azimuthal electrical current density is then

$$j_\Phi = -\frac{\sigma \mu}{2} r \left[A(z) \frac{dI(t)}{dt} + v_z \frac{dA(z)}{dz} I(t) \right] . \tag{3.21}$$

The first term of the radial force acting on the plasma is the product of this current density with the axial magnetic inductance,

$$F_r = j_\Phi \, B_z = \frac{\sigma \mu^2}{2} r \, A(z) \, I(t) \left[A(z) \frac{dI(t)}{dt} + v_z \frac{dA(z)}{dz} I(t) \right] . \tag{3.22}$$

As long as F_r is negative, the force acts towards the axis and the plasma is compressed in the coil. The first term is negative during the first quarter period of the compressor coil current when the time derivative of the magnetic induction is positive. The sign of the second term depends upon the direction of the radial magnetic field. This will be negative for approximately the first 80 percent of the coil length. The second term in the radial component of the magnetogasdynamic force in (3.14) will always be negative. It is the interaction between the axial current density from the center electrode to the compressor coil with the azimuthal magnetic field induced by this current density.

The azimuthal component of the magnetogasdynamic force produces an azimuthal plasma velocity which is small near the compressor coil axis because of the symmetry of the configuration. The axial component of the force $F_{z_{\max}}$ cannot be negative if there is to be particle acceleration. The square of the flow velocity in (3.1) indicates that the axial term,

$$F_z = j_r B_\Phi - j_\Phi B_r ,$$

should always be positive. A first approximation would be that these forces are transferred without delay into thermodynamic quantities such as pressure and density. The compressor coil, therefore, increases the density of the plasma flow leaving the copper axial accelerator.

The plasma resistance has little influence upon the characteristics of the equivalent circuit. It may be regarded as part of the impedance Z of the circuit mesh with compressor coil inductance L_1, where a typical example is

$$Z = \sqrt{R_1^2 + \omega^2 L_1^2} , \tag{3.23}$$

with $R_1 = 4.8 \times 10^{-3}$ ohm, $L = 2 \times 10^{-6}$ H, and $\omega = 9.45 \times 10^4\,\mathrm{s}^{-1}$. Then the plasma resistance term is three orders of magnitude smaller than the inductance term.

The increase of the coaxial accelerator inductance L_2 in Step 1 is proportional to the constant

$$K_7 = \frac{U_0^2\, C_0}{2\, m} \frac{b^2\, C_0}{L_0} , \tag{3.24}$$

where b is the inductance gradient. The first factor of (3.24) is the stored energy per unit of plasma mass, m. For a coaxial accelerator with given geometry and constant plasma mass, K_7 changes with the capacitor bank charging voltage U_0.

The theory developed in the preceding section is compared with experiments performed in the Plasma Hypervelocity Range of the Space Sciences Laboratory at the George C. Marshall Space Flight Center.

Plasma Flow

The theory can be correlated with a specific experiment through the elements of the equivalent circuit shown in Fig. 3.9. The inductance of the compressor coil, L_1, and the inductance of the capacitor bank with coaxial cables and switch assembly, L_a, were measured prior to the experiment. Throughout the experiments described here, the capacitor bank and the switch assembly have remained unchanged. Then the only unknown quantities are the plasma resistance, R_1, and the constant K_7, calculated according to (3.24).

The results of the computation of the plasma velocity at the 3.3.4. Experimental Results end of a coaxial accelerator 14 cm long are compared with measured values in Fig. 3.10. For this example, $K_7 = 1.0$ corresponds to a charging voltage of 18 kV. The plasma velocity was measured with two light probes located 20 cm apart, one at the end of the coaxial accelerator and the other behind the compressor coil. The order of magnitude of the velocity from theory and experiment is the same. The velocity increases with larger values of U_0, the charging voltage. The light probes could only measure the propagation velocity of a luminous front, which may not coincide with the major portion of the plasma mass.

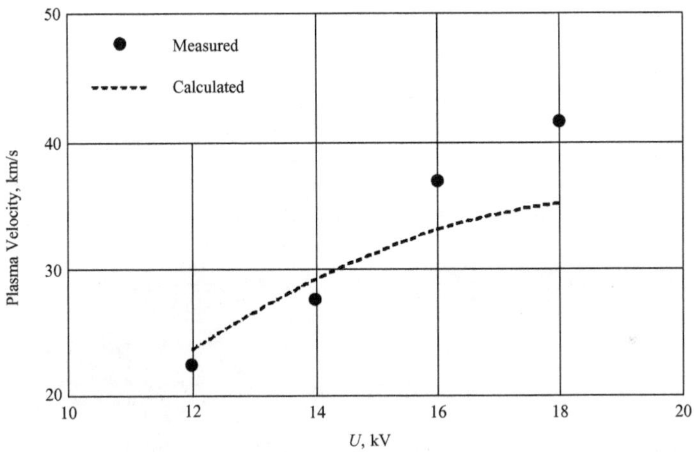

Fig. 3.10. Plasma velocity at the end of the coaxial accelerator

The photographic recording of the oscilloscope trace of the current in the compressor coil as well as the capacitor bank voltage indicates the time when the plasma leaves the coaxial accelerator. The capacitor bank voltage recordings show irregularity when the compressor coil current begins to rise. The arrival time has been plotted in Fig. 3.11 as a function of the capacitor bank charging voltage U_0. For comparison, the recordings of the first light detector used for velocity measurement are also shown.

Capacitor Bank Voltage

The capacitor bank voltage measurement was described in Sect. 3.2.3. The voltage across the capacitor bank is

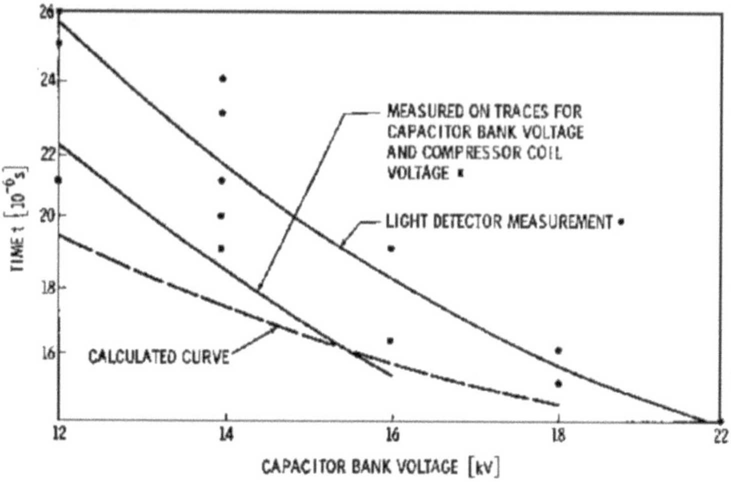

Fig. 3.11. Time of arrival of the plasma at the compressor coil

$$\frac{U}{U_0} = 1 - \int_0^T (X_1 + X_2) dT ,$$

where X_1 and X_2 are the corresponding currents divided by $\omega_0 U_0 C_0$:

$$X_1 = \frac{I_1}{\omega_0 U_0 C_0} , \qquad X_2 = \frac{I_2}{\omega_0 U_0 C_0} ,$$

where

$$\omega_0 = \frac{1}{\sqrt{L_0 C_0}} . \tag{3.25}$$

The recorded curve is compared with the calculated curve in Fig. 3.12. The experimental curve shows a potential drop approximately 30 microseconds after initiation of the discharge. This potential drop has also been observed without the compressor coil, and it coincides approximately with the transition of the coaxial accelerator current through zero. If the coaxial accelerator is still filled with a plasma, then another discharge can be initiated, causing a shift in the potential drop. There is an indication of the same effect after another half-period of the discharge. During these experiments with different compressor coil configurations, high-speed motion pictures were tried. A camera with up to 20,000 frames per second was used so that the exposure time corresponded roughly to one half-period of the discharge. These motion pictures did not give detailed information about the compressor coil flow, but they showed that there were several subsequent accelerations of the plasma cloud. These observations tend to confirm the explanation given for the potential drop observed on the experimental curve. The measured period of the discharge was nearly constant during the experiments, varying by only

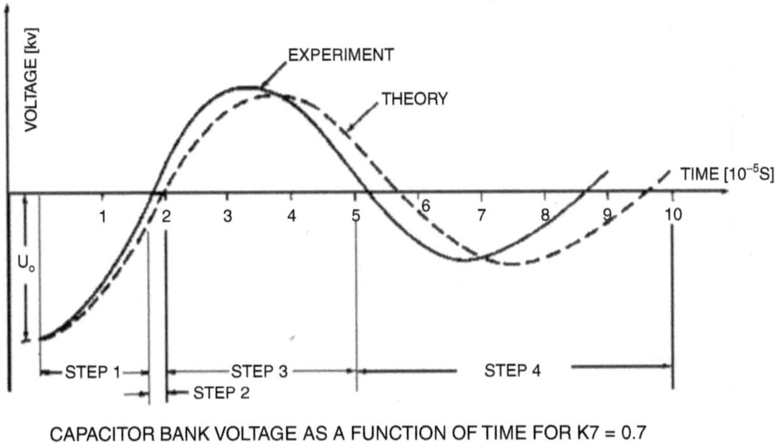

Fig. 3.12. Capacitor bank voltage

2 μs. The integrations of the equivalent circuit equations with different values for the compressor coil inductance showed the same behavior. The frequency of the discharge is, therefore, determined mainly by the capacitor bank parameters C_0 and L_0.

Compressor Coil Current

The magnitude and the change with time of the compressor coil current determine the efficiency of the compressor coil. The experimental curve is compared with the calculated curve in Fig. 3.13. During these experiments, the amplitude of the compressor coil current trace was not calibrated. The absolute value of the compressor coil current can, therefore, only be estimated by comparison with the theory. The theory shows that the magnitude of the compressor coil current depends upon the ratio of the inductances L_0, L_1, and L_2, all of which are well known. Therefore, it seems justifiable to use the correlation between theory and experiment to estimate the magnitude of the compressor coil current:

$$I_1 = X_1 \omega_0 U_0 C_0 \quad \text{or} , \quad X_1 = \frac{I_1}{\omega_0 U_0 C_0} . \tag{3.26}$$

Experimental and Theroretical Data

Data describing the experiment are

Bank capacitance: $C_0 = 2.7 \times 10^{-4}$F
Bank inductance: $L_a = 4.16 \times 10^{-7}$H
Bank resistance: $R_2 = 7.9 \times 10^{-3}\Omega$

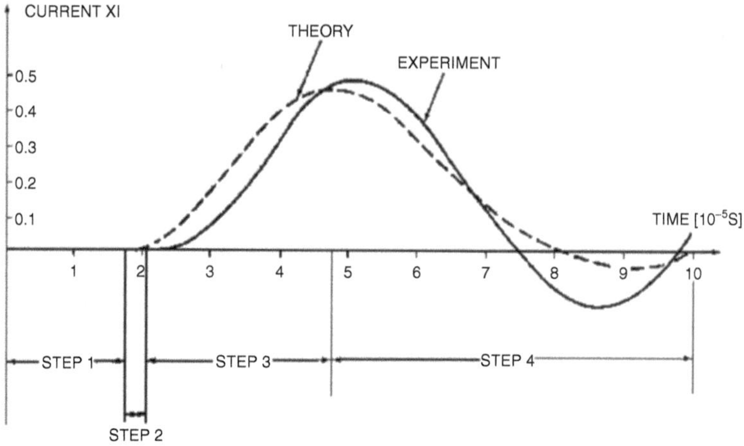

COMPRESSOR COIL CURRENT AS FUNCTION OF TIME FOR K7 = 0.7

Fig. 3.13. Compressor coil current

Coaxial accelerator inductance:	$L_2 = 0.72 \times 10^{-6}$H
Compressor coil inductance:	$L_1 = 4.43 \times 10^{-7}$H
Coaxial accelerator inductance per unit length:	$b = 3.58 \times 10^{-7}$H/m it is H/m!!
Coaxial accelerator length (for plasma acceleration):	20 cm
Compressor coil length:	13 cm
Plasma resistance:	$R_1 = 4.8 \times 10^{-3}\Omega$
Aluminum foil mass:	$m = 0.04$ g
Glass particle diameter:	$D = 0.25$ mm
Charging voltage:	$U_a = 16$ kV
Measured particle velocity:	$v_p = 5.3$ km/s.

With the values given above, the maximum value of the compressor coil current becomes

$$I_{1_{max}} = 4.08 \times 10^5 X_1 = 4.96 \times 10^4\,\mathrm{A}\,.$$

The experiment shows that the compression is terminated when the compressor coil current is at its maximum value.

The comparison of theory and experiment shows that the equivalent circuit and the corresponding theory described in the foregoing section simulate the characteristics of the coaxial accelerator with attached compressor coil. One of the principal reasons is that the behavior of the experiment, as well as the equivalent circuit used for the theory, is dominated by the capacitance C_0 and the inductance L_0 of the capacitor bank, coaxial cables, and switch assembly, and the ratio of L_0 to the inductances of the coaxial accelerator, L_2, and the compressor coil, L_1.

Particle Acceleration

The acceleration of the particles placed on a Mylar film stretched across the narrow end of the compressor coil can be used as an indication of the amount of compression of the discharge. Figure 3.14 is a photograph of the beads after they have left the Mylar film. They have an initial diameter of 0.8 mm and a velocity of 4–5 km/s. The velocity measurement method has been described previously. During the experiments the pressure in the range was always kept at a vacuum below 25 μm of mercury, and the capacitor bank charging voltage and the particle initial diameter were varied. Only spherical glass beads were used.

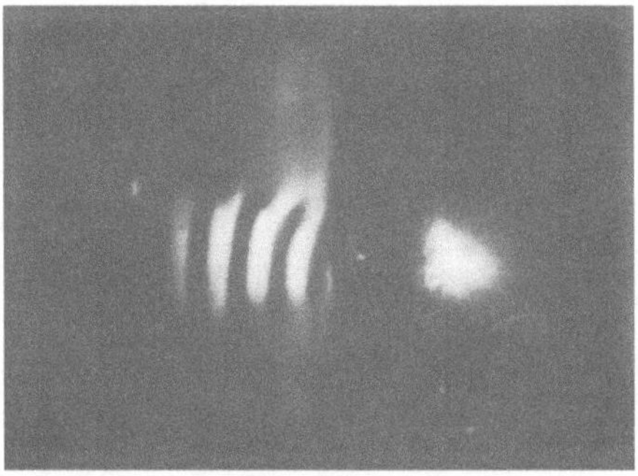

Fig. 3.14. Photograph of accelerated particles in the plasma flow within the compressor coil

Figure 3.15 shows the particle velocity as a function of the capacitor bank charging voltage for three initial particle diameters. The dashed curve in these figures is obtained from the assumption that after the acceleration, kinetic energy of the particles with the mass m_{p} is proportional to the energy stored in the capacitor bank, $C_0U_0/2$, multiplied by the particle cross sectional area, S, and a coefficient C_1 representing the energy transfer efficiency

$$\frac{m_{\mathrm{p}}}{2}v^2 = C_1\, S\, \frac{C_0^2\, U_0^2}{2}\,. \tag{3.27}$$

Then the velocity of a particle with the initial diameter D is given by

$$v = C_2 \frac{U_0}{\sqrt{D}} \quad \text{where} \quad C_2 = \sqrt{\frac{3\, C_1\, C_0}{2\, \rho_{\mathrm{GLASS}}}}\,. \tag{3.28}$$

In the case of the 700 μm diameter particle, the straight line representing this theory could easily be drawn through the experimental results. The theoretical line for particles of another initial diameter will have a slope that is related to the slope of the 700 μm curve by the equation

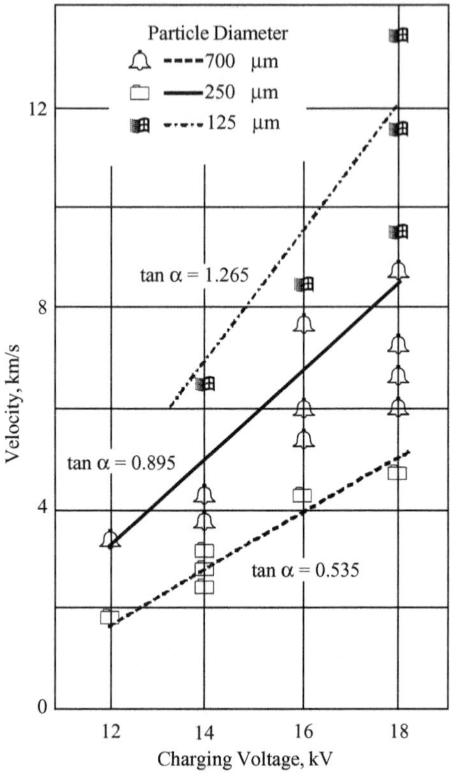

Fig. 3.15. Particle velocity as function of capacitor bank charging voltage

$$\frac{\tan\alpha}{\tan\alpha_{700}} = \left(\frac{700}{D}\right)^{1/2} . \tag{3.29}$$

The slopes of the curves in the figures for the 250 μm and the 125 μm particles were calculated from this equation. The absolute level of these curves was adjusted to the 700 μm experiments. The constant C_2 in (3.28) was determined from the line approximating the 700 μm results. The lines for 250 μm and 125 μm can then be computed with (3.28). In Fig. 3.16 the final velocity is plotted as a function of the particle initial diameter. The theoretical curves are also calculated from (3.28).

The difference between the velocities measured in different experiments with the same initial particle size, which increases with large capacitor bank charging voltage, is very probably due to particle ablation during its interaction with the plasma.

During the experiments which resulted in these data many variations were tried, including

- variation of the length of the center electrode or the coaxial accelerator,

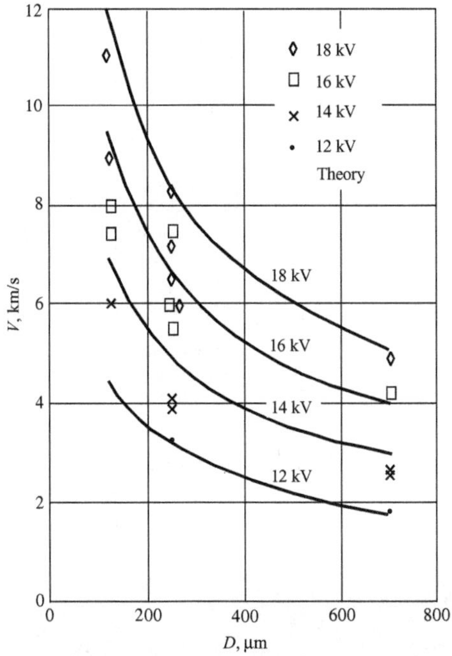

Fig. 3.16. Particle velocity as function of the particle initial diameter

- variation of the length of the compressor coil,
- insulation of the last two front windings of the compressor coil,
- insertion on the glass beads into the compressor coil at a position 1 cm from the narrow end,

together with varying the size of the glass beads and the charging voltage of the capacitor bank. From these observations, it seemed appropriate to consider another method for the injection of the projectiles.

Compressor Coil Diagnostics [5–7]

The measured projectile velocities indicate that pressures in the kbar regime must be generated inside of the compressor coil. The plasma flow pressure inside and outside of the compressor coil was investigated using a coaxial accelerator where, instead of the aluminum foil, gas is injected (Fig. 3.17).

The Pressure Sensor

The current from the center electrode to the compressor coil is of the order of 50 kA, the magnetic induction of the order of 1 Tesla, and the discharge frequency is approximately 15 kHz. The photograph shown in Fig. 3.18 may give an impression of the compressor coil in operation.

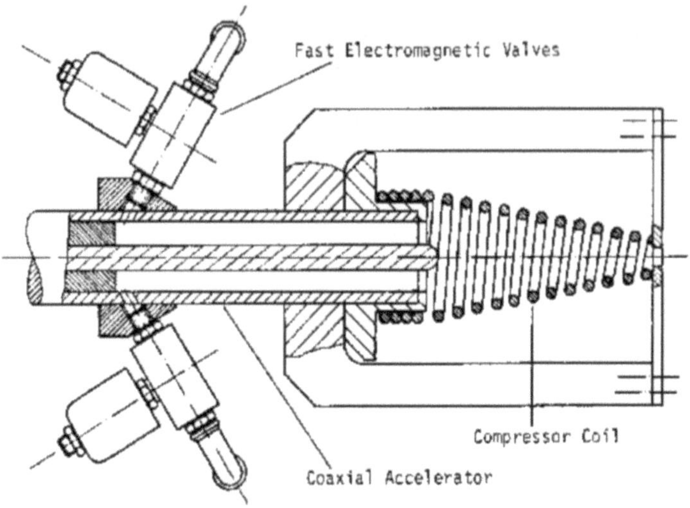

Fig. 3.17. Coaxial accelerator with gas injection

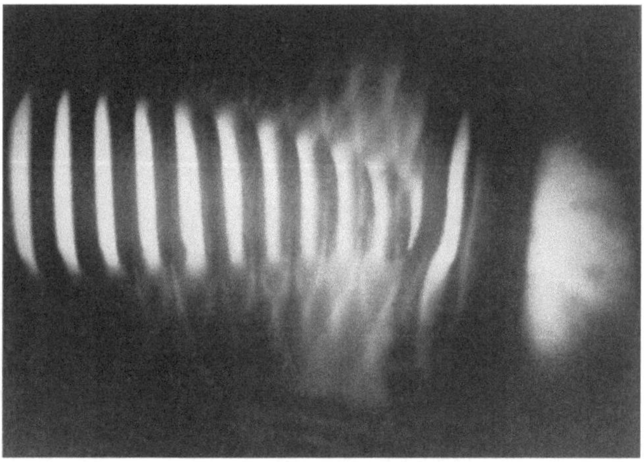

Fig. 3.18. Compressor coil environment

This environment requires a sensor which is not based on an electric principle. The method that was chosen here is the observation of the displacement of a small plate that is exposed to the plasma pressure. This pressure sensing plate is mounted on the front end of a steel tube, as shown in Fig. 3.19. The deformation of the plate by the pressure of the plasma flow is measured by the change of its reflectivity to incident light. A He-Ne laser beam is directed towards its rear surface through an optical fiber inside the steel tube. The reflected light is recorded by a photodiode that is located at the end of another optical fiber outside of the vacuum tank of the plasma accelerator. This is done to reduce the influence of the strong electromagnetic field in the

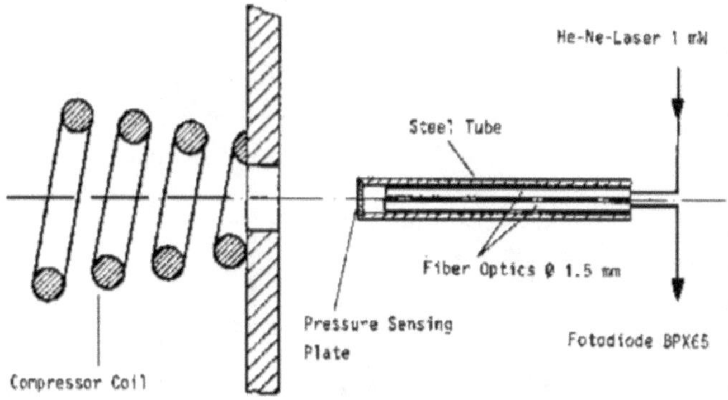

Fig. 3.19. Sensor for the measurement of the transient pressure

accelerator. The change in the output voltage of the photodiode is measured as a function of the displacement of the pressure sensing plate. The theory of shock waves in metals was used to calculate the pressure from the deformation or the displacement of the pressure sensing plate. The accuracy of the method can be estimated from the ratio of the work performed for the deformation of the plate to the work that is required to accelerate the plate. It was found that, in the experiments that are described here, the inaccuracy was less than 5%. It should finally be noted that the pressure that is measured by this method is the total pressure, being the sum of the static and dynamic pressures.

Experimental Results

A typical experimental curve of the voltage output from the photodiode is shown as a function of time in Fig. 3.20.

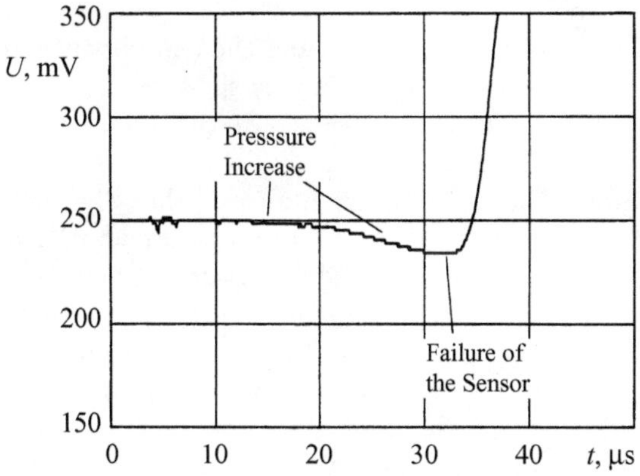

Fig. 3.20. Output of the photodiode in an actual experiment

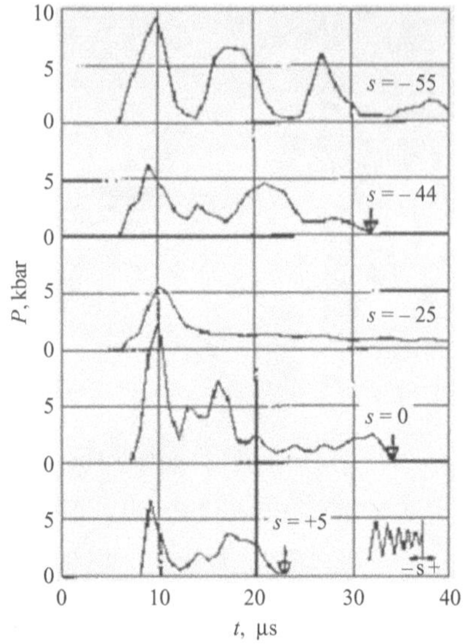

Fig. 3.21. Pressure as a function of time at various locations (arrows indicate failure of the sensor)

The pressure in kbar at various locations inside and outside the compressor coil is shown in Fig. 3.21. Negative values of the coordinate s correspond to locations inside the compressor coil.

In the region where the projectiles are first exposed to the plasma flow, that is, near $s = 0$, a pressure in the order of 5 kbar has been measured during an interval of approximately 10 μs, from 8 to 18 μs, after initiation of the discharge. This pressure will accelerate a 100 μm glass bead (density 2.5 g/cm^3) along a path of 1 cm to a velocity of 7.5 km/s. This is in agreement with the experimental data and present operational characteristics of this type of accelerator.

The current theoretical predictions and the experimental results are compared in Fig. 3.22. The acceleration time is shorter than the duration of the pressure pulse in the compressor coil because the axial extension of the high pressure region is of the order of 1 cm and the mass is very quickly pushed beyond this region. The same figure indicates that the measured pressure buildup is quite different from the theoretical expectations. Two pressure peaks appear at 10 and 20 μs, with a strong decrease being observed between them. This might be a magnetogasdynamic effect. The current from the center electrode of the coaxial accelerator and passing through the compressor coil winding might have been reduced by a short circuit formed between the tip of the center electrode and the narrow end of the compressor coil. The low compression resulted, in turn, in a lower conductivity in the plasma, and the current from the center electrode to the compressor coil windings increased

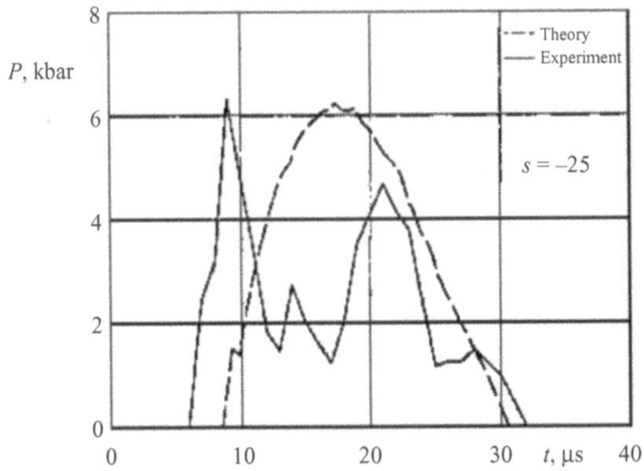

Fig. 3.22. Experimental pressure and theory at 25 mm inside the coil as a function of time

again. At the location $s = -44$ this oscillatory effect occurs three times, as shown at the top of Fig. 3.21. This indicates that the magnetogasdynamic flow inside the compressor coil is indeed non-stationary.

Radial Pressure and Divergence of the Plasma Flow

It has been observed that the projectiles that are exposed to the plasma flow at the end of the compressor coil do not fly axially but rather tend to diverge. Only a small number of particles will be accelerated axially. This can be explained by the fact that the total pressure has a radial component. This radial pressure has been measured with the same sensor as was used to measure the total axial pressure.

Figure 3.23 shows the results of experiments in which the sensor was placed perpendicularly to the axis of the coaxial accelerator and the compressor coil and just beyond the narrow end of the compressor coil at different distances, r, from the axis. The value of the radial component of the total pressure is in the order of 1 kbar.

Figure 3.24 shows the addition of both components and the angle, a, of divergence. The total pressure has its maximum of 10 kbar at approximately 10 μs after the initiation of the discharge. The angle, a, of divergence increases after 20 μs.

The measured axial and radial pressure is in good agreement with the operational characteristics of the plasma accelerator. The pressure sensor seems, therefore, to be well applicable for the measurement of high transient pressures in the environment of a plasma accelerator. During the experiments there was some evidence that more projectiles would hit the target when they were introduced 1 cm inside the compressor coil, at the narrow end.

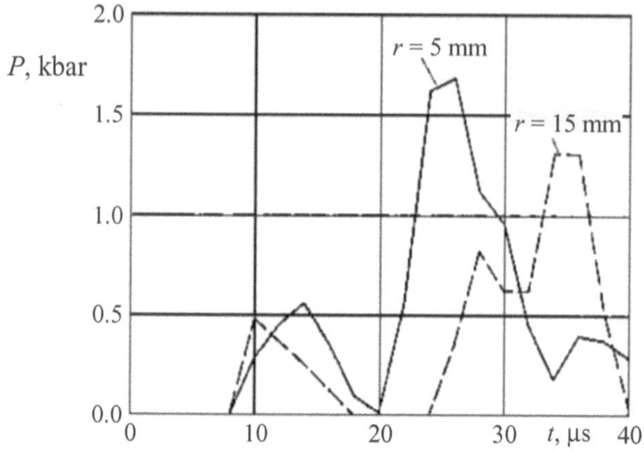

Fig. 3.23. Radial pressure at the end of the compressor coil as a function of time

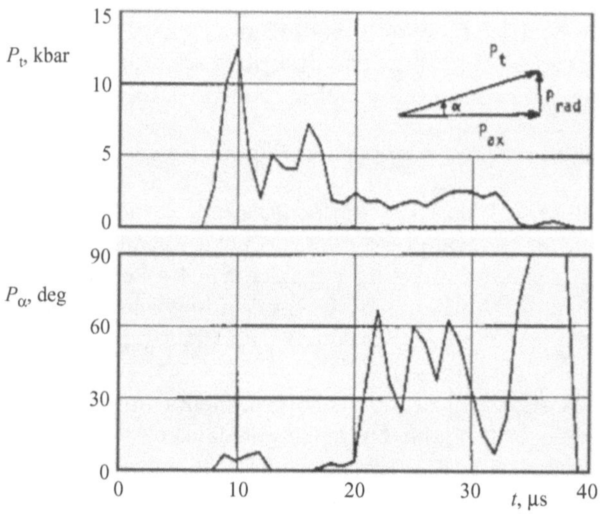

Fig. 3.24. Total pressure at the end of the compressor coil and angle as a function of time

Pressure Measurement with an Active Probe [6]

A series of experiments was conducted to measure the dynamic pressure in front of the muzzle. Kistler Type 6157BA pressure probes were used. The evaluation of these measurements was assisted by measurement of the current in the coil ground loop, the inductance of the system, and by high speed photography. Three different coil shapes were used to evaluate the influence of the coil shape on the compression. A non-conducting coil with a shape identical to one of these coils was tested to determine the contribution of aerodynamic forces to the plasma compression. Additional tests were performed with a non-conducting conical tube to see whether a coil has aerodynamic benefits compared to a closed contour. Finally, the coil was completely omitted. Only the coil holder, attached to the ground loop, remained in the system. This configuration forced a mainly axial discharge from the tip of the center electrode of the coaxial accelerator to the holder, and allowed the investigation of the contribution of the z-pinch to compression. These results are compared to each other and conclusions are drawn on how to improve the coil.

3.4 Effects of Varying the Compressor Coil Configuration

3.4.1 Configurations Investigated

Six compressor coil configurations (Fig. 3.25) were investigated in this series of experiments. Three different copper coils – two conical and one conical–cylindrical form – were used. The first conical coil, referred to as the *large conical coil*, represents the present configuration of the plasma-dynamic accelerator. Due to the high frequency of the discharge, this coil has a rather high inductive impedance, especially in the larger turns. This reduces the amount of current entering the coil, thereby reducing the effect of electromagnetic compression by coil currents. Hence, a coil with fewer turns and smaller diameters, the *small conical coil*, was experimentally investigated. The third coil, the *conical-cylindrical coil*, is introduced to investigate whether a cylindrical extension of the conical part is helpful in obtaining a homogeneous plasma flow parallel to the coil axis.

Fig. 3.25. The six configurations investigated during the experimental series

In addition to the copper coils, a coil made of polycarbonate (PC), a non-conducting plastic material, was employed. This coil had the same shape as the long conical coil. The coil holder plate was also made of PC, and the connection to the coil holder that forms the ground loop of the device was insulated. This minimized the current flow in the system and caused the magnetic fields due to coil currents to be zero. The use of this coil allows for the comparison of the effect of electromagnetic compression with that of mere aerodynamic compression.

Experiments with a PC cone instead of a coil were conducted to investigate whether a coil has superior aerodynamic compression compared to a closed surface. The inner surface of this cone matched the inner contour of the long conical coil. This allowed for the direct comparison of these configurations.

All experiments were performed at an ambient pressure of less than 100 Pa. To stay comparable to previous experiments, the charging voltage was kept at 16 kV, thus providing 39.4 kJ of stored energy.

3.4.2 Experimental Results

High-speed photographs of the plasma flow through the compressor coil show a bright light phenomenon that begins at the tip of the center electrode and extends to the last three or four wire rings. In the first two thirds of the coil it is concentrated on the axis. This phenomenon is most probably caused by an axial discharge. The photographs show the path that the current follows until it enters the coil turns. Most of the current seems to evade the first two thirds of the coil; only the last turns seem to carry notable current levels. This is in agreement with the observation that the inductive impedance of the larger turns of the compressor coil is significantly greater than the plasma resistance. Hence, a z-pinch may have an important influence on plasma compression. So, if one omits the coil completely and retains only the coil holder, keeping it at exactly the same position as it is with the large conical coil in place, the current entering the ground loop takes a mostly axial path through the plasma, thereby causing a z-pinch. Aerodynamic effects are significantly reduced in such a setup. This configuration was also included in the experimental series.

Several experiments have been performed with each of the configurations discussed above. Interesting results can be obtained by comparing the dynamic pressure at the muzzle of the compressor coil.

Kistler Type 6157 BA pressure probes were used. These are piezoelectric pressure gauges that show a linear relation of pressure to signal strength while operated within their design limits. The maximum safe load is 200 MPa. The pressure gauge has to be pre-charged. The output signal was read into an optical transceiver and transmitted to the shielded instrument room.

The experimental setup is shown in Fig. 3.26. The probe is mounted in front and outside of the compressor coil in order to keep the gauge outside

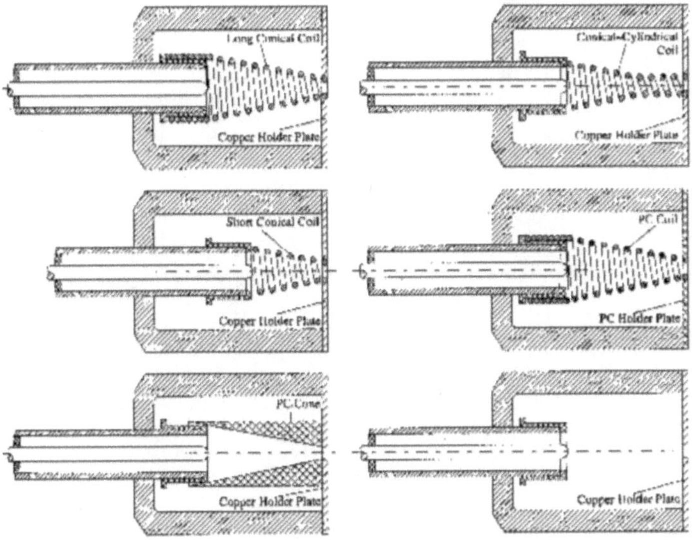

Fig. 3.26. Experimental setup featuring PC coil

of the electromagnetogasdynamic environment within the compressor coil. The probe signal is fed into the instrument room via fiber optics. Inside the shielded room another optical transceiver reconverted the signal. It was then transferred to a Nicolet MultiPro digital oscilloscope. Since the sensor is sensitive to high temperatures and electric discharges, its surface was protected by a thin layer of rubber coupled to the sensor with vacuum grease. Prior to each experiment, the sensor was aligned on the axis of the compressor coil, with the measuring surface being perpendicular to this axis.

Figure 3.27 shows the signal from the Kistler pressure gauge. At the time of the publication, these were not compared with the results shown before on Figs. 3.21 and 3.22, indicating a pressure oscillation with 10 μs between two pressure peaks. The same time difference between two pressure peaks has been measured and recorded on Fig. 3.27. This indicates that these measurements confirm the non-stationary, oscillatory characteristics of the plasma flow in the compressor coil.

The frequency of the coaxial accelerator and the compressor coil is only one tenth the frequency of this internal compressor coil oscillation, and it will be an interesting task to find out why. It can be guessed that the dynamic pressure shown in Fig. 3.28 is, in its order of magnitude, confirmed also by the results shown in Fig. 3.21.

The largest number of experiments concentrated on the long conical coil. A total of 20 data points were taken. The closest distance to the muzzle was 24 mm, the point at which the maximum safe load of the gauge was reached. At this distance dynamic pressures of 75 MPa and 190 MPa were measured, yielding an average value of about 130 MPa. The pressure drops to nearly

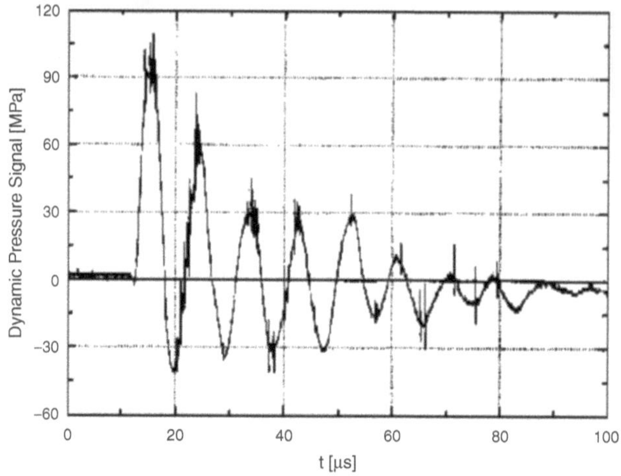

Fig. 3.27. Pressure Signal 32 mm from the muzzle of the long conical coil

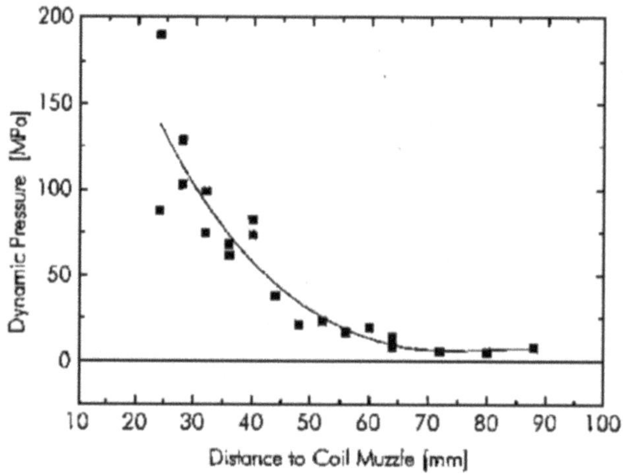

Fig. 3.28. Dynamic pressure in front of the muzzle of the long conical coil

zero within the next 30–40 mm. This means that a particle initially placed in the muzzle of this coil has a maximum acceleration length in the order of 50–60 mm.

Fewer experiments were conducted with the other configurations. Figure 3.29 shows the results obtained. The short conical coil, represented in the plot by open triangles, generally shows twice the dynamic pressure of the long conical coil. One reason for this is that the muzzle of this coil is much closer to the muzzle of the coaxial accelerator, so that less material is lost on the way to the coil muzzle. It is possible that there is also stronger electromagnetic compression involved. The conical–cylindrical coil, represented by

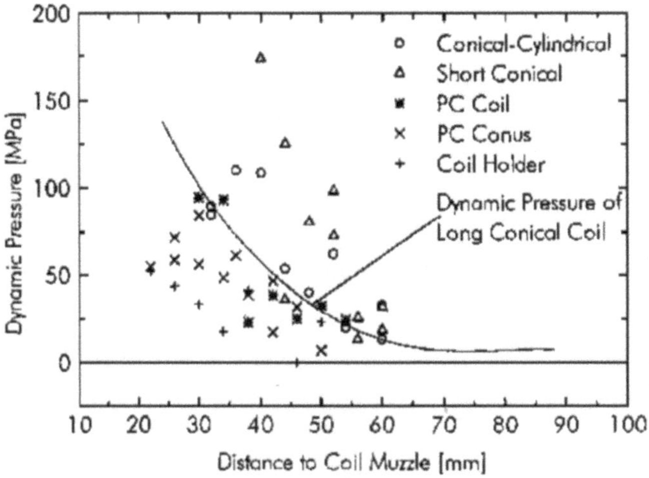

Fig. 3.29. Measured dynamic pressure of alternative shapes in comparison to dynamic pressure in front of the long conical coil

open circles, shows only slightly higher dynamic pressure compared to the long conical coil. The cylindrical extension does not seem to provide the desired effect of forming a homogeneous parallel plasma flow. To validate these measurements, particle acceleration experiments were performed. In these, neither configuration was able to accelerate particles to the same velocity as normally found with the long conical coil. The short conical coil was not able to deliver fast particles with a diameter in the range of the particles loaded. Instead, impacts of very fast, hut also very small particles were detected. Obviously, the glass beads broke apart during acceleration, sustaining the higher dynamic pressure. The conical–cylindrical coil only delivered rather slow particles. High-speed photographs of this configuration show that the plasma flow jams in the transition from conical to cylindrical contour. The pressure gauge only detects the first plasma front that passes the coil uninhibited by this jamming dynamic pressures of over 1000 MPa are possible

- inside the long conical coil, and
- within the coil at a distance of 1/3 of its length, as measured from the muzzle.

Since this pressure decreases rapidly with increasing distance from the coil, it would be reasonable to move the initial particle position to the inside of the coil. This could most easily be done by using a pre-accelerator as an additional first stage, accelerating the particles inside the center electrode. This also allows for the choice of time for the introduction of the particles into the compressor coil by setting the delay between the ignition of this first stage and the plasma dynamic accelerator.

In the current setup of the plasma dynamic accelerator, the most important mechanism of plasma compression is shock-dominated aerodynamic compression caused by the turns of the compressor coil. To enhance the effect of electromagnetic forces due to coil currents, the inductive impedance of the coil has to be reduced. This can be done by choosing coils with fewer turns and smaller turn diameters, or by reducing the discharge frequency. The small conical coil proved that such a configuration can deliver higher dynamic pressures.

Aerodynamically, it is important to use a geometry that allows for venting to prevent the plasma flow from jamming. Also, the distribution of the coil turns is important, because this distribution forms the system of shocks that compresses the plasma while electromagnetic forces are weak. The conical–cylindrical configuration, for example, forms an adverse shock system, leading to inferior performance.

It is obvious that if electromagnetic forces can be established, and if they are large enough to influence the plasma flow, then the effect of such forces is limited mostly to the interior region of the coil. This is also the region where the highest dynamic pressures can be found, extending over greater axial distances than in front of the muzzle. To achieve higher final velocities, it may be sensible to choose the initial particle position to be within the coil. In addition it would be helpful to influence the instant when the particles take this initial position. One way to do this is to use an additional pre-accelerator.

3.4.3 Simulations [7]

The first set of simulations was concentrated on analysis of electromagnetic effects for the long conical coil, including electromagnetic forces and ohmic heating. Here, dynamic pressures at the coil muzzle were calculated and were of the same order of magnitude as measured data. Figure 3.30 shows the calculated axial dynamic pressures at various times, while Fig. 3.31 shows the temperature distribution in the coil at 10 μs. The coil muzzle is at $z = 160$ mm. A test particle, a glass bead with a diameter of 100 μm that was initially placed there, was numerically accelerated to a final velocity of 10.2 km/s – a value that is well within the known performance range of the present accelerator setup. This indicates that shock dominated aerodynamic compression is an important mechanism in the operation of the compressor coil. On the other hand, the pure gasdynamic shock-dominated flow behind the shock or shocks could not exceed the local Mach number, which should make its velocity less than 5 km/s.

Current flow in the wire rings causes compression, but also decelerating forces, like an eddy current brake. Moderate currents, as they occur in the experiment, smooth the formation of shocks. Close to the wire rings, repelling electromagnetic forces form a buffer zone and soften the impact of the fluid. The region of higher dynamic pressure (Fig. 3.30) extends farther out of the coil muzzle, yielding a slightly higher particle velocity of 10.6 km/s.

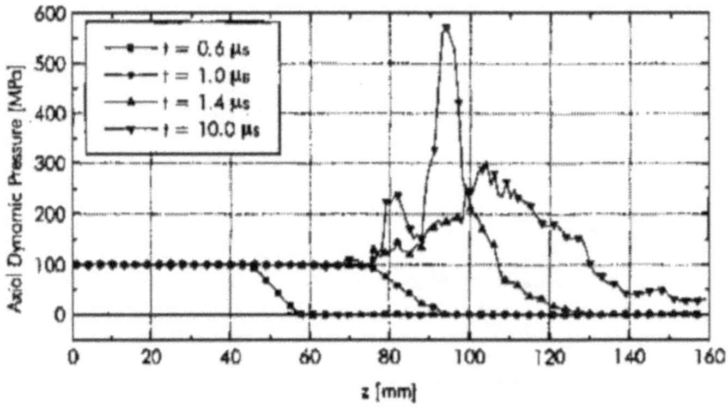

Fig. 3.30. Axial dynamic pressure at $r = 0$ in the long conical coil (neglecting electromagnetic effects) at various times

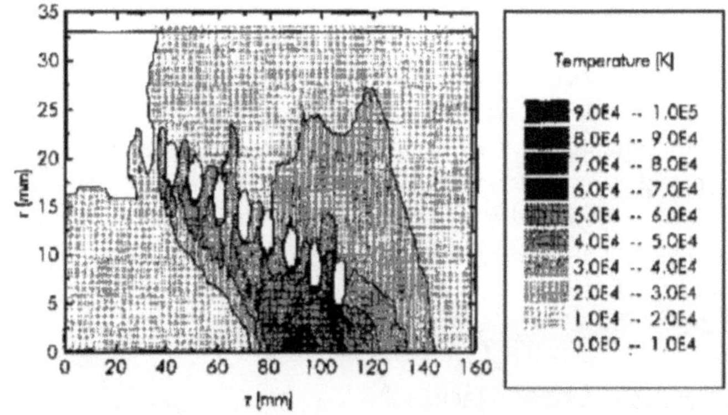

Fig. 3.31. The temperature distribution in the coil at $t = 10\,\mu s$, revealing the structure of the shock system

Figure 3.32 shows a simulation of the axial dynamic pressure, where currents would only flow in the coil windings, not accounting for the z-pinch effect. There is again a strong indication that there will be a high dynamic pressure region 1/3 of the coil length inside the compressor coil, measured from the compressor coil muzzle.

The plasma flow jammed in the transition from the conical to the conical–cylindrical shape, and electromagnetic forces could not compensate for that. This becomes evident when looking at the fluid density at $t = 10\,\mu s$ (Fig. 3.32). Hence, the dynamic pressure at the muzzle decreased so much that the particle velocity dropped below 7 km/s.

In theory, the short conical coil worked much better. Particle velocities of 11.2 km/s aerodynamically, and 10.9 km/s with complete electromagnetics, were calculated. Electromagnetic forces again limit their positive conditioning influence to the interior part of the coil. It was already mentioned that,

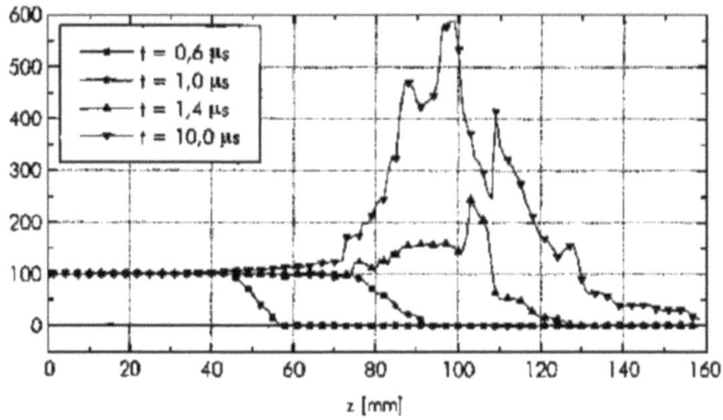

Fig. 3.32. Axial dynamic pressure at $r = 0$ in long conical coil (considering only coil turn currents) at various times

although all other results of the simulations could be experimentally verified, the short conical coil did not deliver the calculated particle velocities in the experiment. Only fast, but very small, particles were detected, indicating that the dynamic pressure was too high and broke the particles apart. This may, of course, also be regarded as indication that the short coil is very effective in introducing magnetogasdynamic forces, perhaps also because of a different internal oscillation frequency.

3.5 Two-Stage Operations

3.5.1 The Coaxial Accelerator with a Compressor Coil and an Electrothermal Accelerator

The Electrothermal Accelerator [8, 9]

Accelerator Design and Electrical Storage System

A schematic of the electrothermal accelerator and its principle is shown in Fig. 3.33. Following the discharge of the capacitor bank by closing of the ignitron switch, a high electrical current vaporizes an aluminum wire in a small chamber and generates a hot plasma by joule heating. It is assumed that the entire plasma is formed from the material of the wire. The expansion of the plasma accelerates the projectile through a barrel. The aluminum wire is enclosed in a cylindrical chamber, and has contact to two electrodes. The axis of the cylinder coincides with the axis of the barrel to minimize the loss due to friction of the plasma flow. The steel barrel and the steel electrode have circular sealing lips, which are pressed into the inner liner (Nylon) of the explosion chamber. This connection is gas tight before the shot and resists

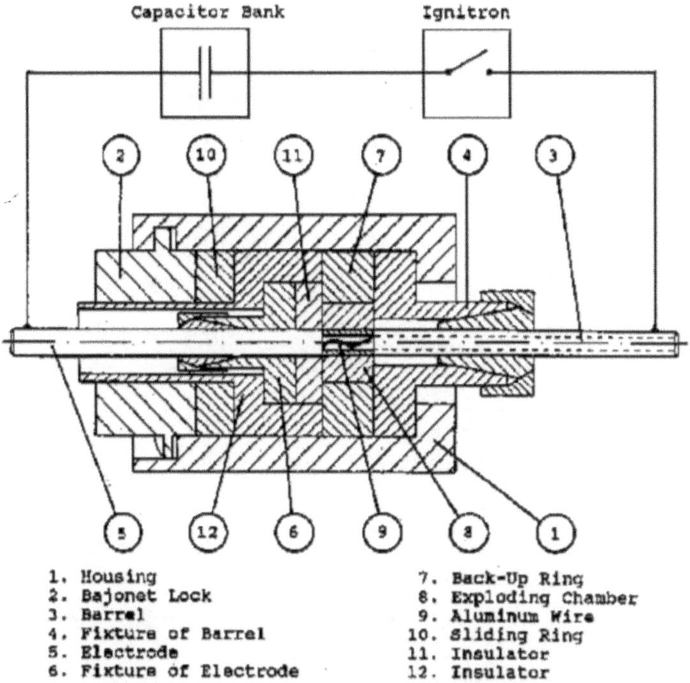

Fig. 3.33. Schematic of the electrothermal accelerator

the high chamber pressure during the experiment. The inner diameter of the chamber is the same as the barrel bore to minimize flow losses.

The following values are representative for an operational accelerator used for the experiments described here:

Length of barrel	140 mm
Inner diameter of barrel	4 mm
Length of exploding chamber	20 mm

To operate the accelerator, a capacitor bank with four modules, each of 14 μF, is used. The switching is done by one GE 7704 ignitron. The following electrical values are representative for the system:

Charging voltage:	up to 12 kV
Capacitance:	56 μF
Inductance:	0.72 mH
Resistance:	24 m Ω
Discharge frequency:	25 kHz.

Experimental Results

Experiments have been performed with the electrothermal accelerator alone to demonstrate its performance and to solve operational problems for the

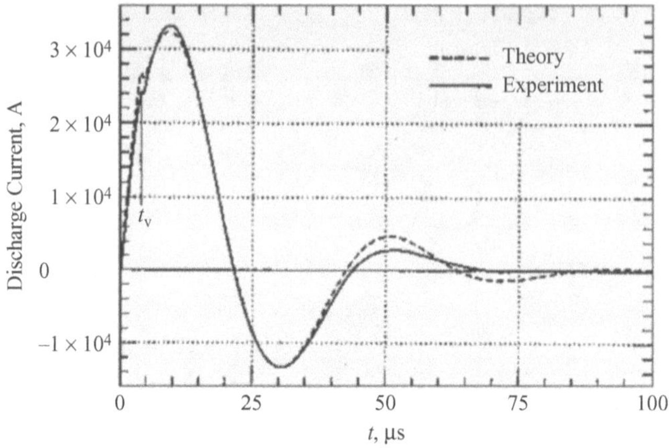

Fig. 3.34. History of the discharge current

proposed two-stage operation. Figure 3.34 shows the measured and calculated discharge current as a function of time. The small notch in the curves at the time t_v indicates the very high resistance in the chamber during evaporation of the aluminum wire.

The performance of the electrothermal accelerator is plotted in Fig. 3.35 as a function of the charging voltage of the capacitor bank. The velocity was inferred from a time-of-flight measurement. The projectile was detected at the muzzle by a light guide, and a piezoelectric sensor gave the impact signal. All experiments have been performed with the 56 µF capacitor bank. A maximum overall efficiency of 7.8% was reached. Meanwhile, a few experiments have been performed with a 352 µF capacitor bank. Velocities of up to 3.6 km/s have been measured for 28 mg particles, but these high velocities are not representative for the operational service of the accelerator. Parts of the insulators were destroyed.

The plasma accelerator is installed in a large vacuum blast tank and is connected to the energy storage system (a capacitor bank consisting of 16 modules with separate ignitron switches) by low inductance cables. The following values are representative for the system

Capacitance:	352 µF
Inductance:	329 nH
Resistance:	4.65 mΩ

The plasma gun itself consists of a coaxial accelerator to which a compressor coil is attached via a ground loop (Fig. 3.36). The discharge of the capacitor bank into the coaxial part of the accelerator generates a plasma between the center electrode and the outer electrode by ionization of a few cm^3 of helium gas that is injected through the outer electrode prior to the

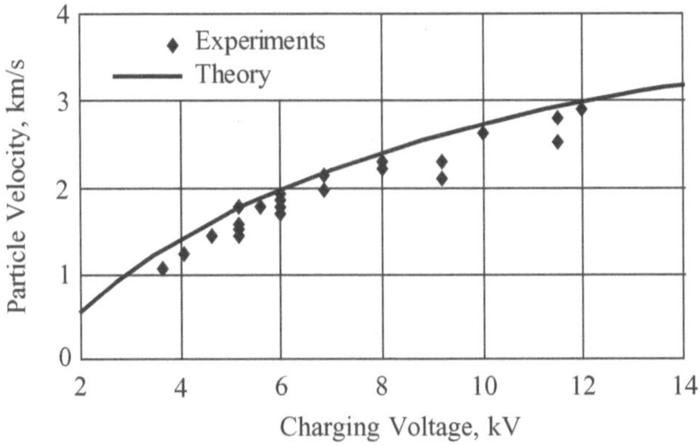

Fig. 3.35. Performance of the electrothermal accelerator for 43 mg projectiles (polyamid cylinders)

discharge. The following dimensions apply to the plasmadynamic accelerator described here

Length of coaxial accelerator:	160 mm
Length of compressor coil:	110 mm
Outer diameter of center electrode:	12 mm
Inner diameter of annular electrode:	33 mm.

For two-stage operation it is necessary to inject smaller particles than the large projectiles that were used for the development of the electrothermal gun. Therefore, many small glass beads, mixed in epoxy and cast into a solid projectile, were accelerated through the barrel. Due to the high accelerating forces, the compound projectile breaks apart, and a loose but compact cloud of single particles leaves the barrel. This property of the compound projectiles was optimized during many experiments.

Figure 3.36 shows the setup of the two-stage facility. The muzzle of the electrothermal accelerator is connected to the center electrode of the coaxial accelerator. For this purpose, the center electrode has to be a tube which is elongated backwards through the cable adapter of the plasma gun. This design leads to a very long injection path of more than one meter. The inner diameter of the center electrode is about 2 mm larger than the barrel bore. With the barrel and the center electrode having electrical contact, the energy storage systems of the two accelerators are connected.

For a successful post-acceleration of the particles through the plasma flow, it is very important to have accurate timing of the two accelerators. About 10 μs after the discharge of the capacitor bank of the plasmadynamic launcher, the plasma front reaches the end of the compressor coil. The main compressed plasma flow is able to accelerate the particles because at lasts about 12 μs. It is obvious that the injected particles have to be within the compressor coil at the time of discharge. The exact timing was realized in

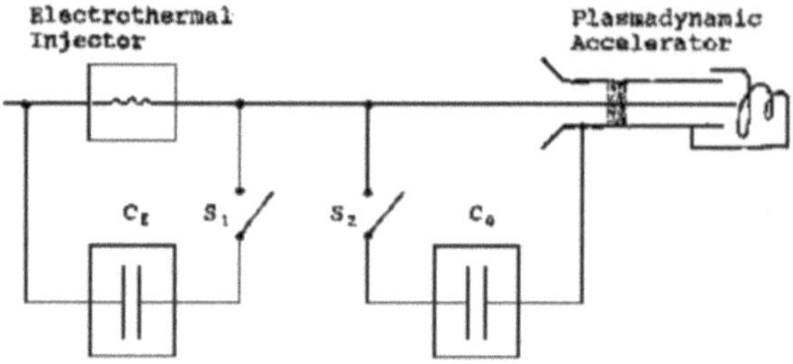

Fig. 3.36. Schematic of the two stage facility

the experiments with a break wire in the compressor coil. A particle out of the injected cloud of glass beads cut a 0.1 mm diameter wire of a battery feed circuit with an LED. A trigger signal then was generated by a fast optoelectronic circuit. This simple device guaranteed very accurate and reliable timing of each launcher.

3.5.2 Experiments with the Two-Stage Facility

To investigate the augmented launch facility described above. The following diagnostics have been used: Rogowski coils measured the transient total current in both systems and the transient current in the compressor coil. The velocity of the plasma flow was obtained by magnetic pickup coils. Cordin single-frame cameras were used to take photographs of the plasma flow in the compressor coil. A time-integrated value of the dynamic pressure of the plasma flow could be derived from the bending angles of thin copper rods that were placed perpendicular to the plasma flow at the end of the compressor coil. This measurement technique is described in detail in [3]. The velocity of the particles was derived from time-of-flight measurements.

The first experiments with the two-stage facility demonstrated a safe and reproducible operation. Both capacitor banks could be discharged in the expected way without mutual interference.

With the first feasibility tests it was not possible to reach the expected velocity increase through the post-acceleration. For most experiments, the final velocity was below the performance of the plasma accelerator alone. The dynamic pressure, $q = \rho/(2V^2)$, dropped to 25% of the nominal value. The total current through the compressor coil increased by a factor of up to two. The measurements of the axial plasma velocity in the compressor coil gave reduced values. From this, it was obvious that the two-stage operation causes a severe malfunction of the compressor coil. In comparison with a nominal shot, the plasma is quenched out of the windings in a typical two-stage experiment. This can be seen in Fig. 3.37.

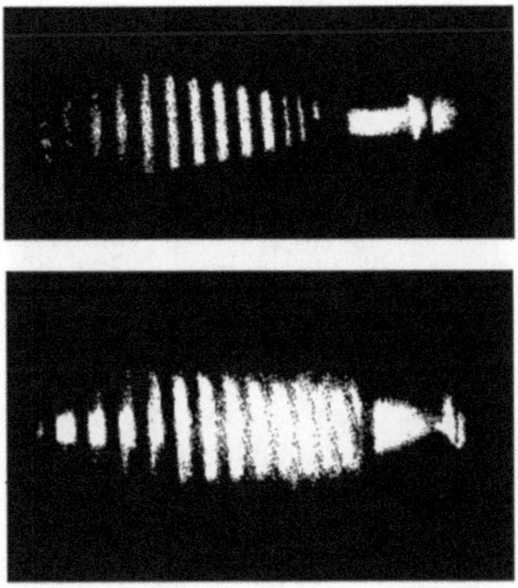

a: without injection 16 ± 1 μs after discharge of capacitor bank

b: with injection 14 ± 1 μs after discharge of capacitor bank

Fig. 3.37. Photographs of the plasma flow in the compressor coil

3.5.3 Theoretical Analysis of the Compressor Coil

More detailed experiments showed that more or less of the accelerating aluminum gas flows into the compressor coil along with the injected particles. At the time of discharge there is a column of aluminum gas or droplets in extension of the center electrode. As aluminum has a low ionization potential, it will be ionized again by the He-plasma flow out of the coaxial accelerator. This effect leads to an increasing conductivity along the centerline of the compressor coil, which is followed by a reduced current through the windings of the coil. The typical conical discharge channel starting at the tip of the center electrode can be recognized. A schematic of the new situation in the compressor coil is shown in Fig. 3.38.

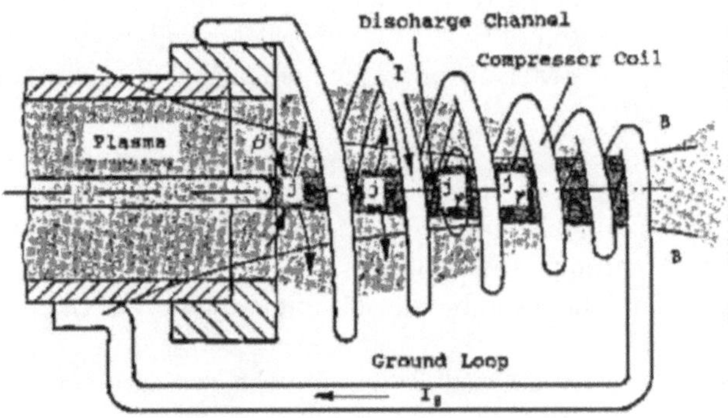

Fig. 3.38. Processes in the compressor coil

Analytically, this can be described with a refined equivalent electrical circuit of the compressor coil, which is depicted in Fig. 3.39. Each winding is represented by its own mesh circuit. The movement of the plasma flow will be simulated by closing the switches S_1 to S_{12} one after another. The plasma resistance R_{AX} is a function of the geometry and the conductivity, q, of the discharge channel. The formulation of all mesh circuits A, B, and C_1 to C_{11} leads to a system of differential equations for the total current in the system and the currents in the windings.

Figure 3.40 shows a comparison of the Rogowski signals of I_s for an experiment with and without electrothermal injection. The capacitor bank is discharged at the time $t = 0$, and the steep rise between $t = 6\,\mu s$ and $8\,\mu s$ represents the motion of the plasma front through the coil.

The high current flow in the case of an injection of the projectiles by the electrothermal accelerator is caused by the plasma generated discharge channel. The same behavior can be simulated in the model with a lower resistance R_{AX} (by the factor 0.5).

The changing conditions in the compressor coil concerning the plasma resistivity, and therefore the current distribution, have a direct influence on the plasma density. The plasma flow can be described by the inviscid magnetogasdynamic momentum equation. Including the z-pinch effect, (3.26) can be rewritten in the form

$$P(z,t) = \underbrace{\sigma \frac{r_s^2}{4} b^2 I \frac{dI}{dt}}_{T_1} + \underbrace{\sigma \frac{r_s^2}{4} V_z b' b I^2}_{T_2} + \underbrace{\mu_0 \frac{r_1^2}{4} j_z^2}_{T_3} , \qquad (3.30)$$

where r_s is the radius of the compressor coil, V_z is the axial plasma velocity through the coil, μ_0 is the permeability of free space, and r_1 is the radius of the discharge channel. The first term T_1 is the term for the tune dependence of the current I, the second term, T_2, is dominated by the inductive change of the current and the plasma velocity, and the third term, T_3, represents the z-pinch effect of the axial current density, j_z. The function $b(z)$ is the gradient of the magnetic induction of the coil. Numerical calculations showed that the term T_1 is about one order of magnitude smaller than the other two terms T_2 and T_3 in (3.30).

Assuming the perfect gas equation, one can calculate the plasma density. The compressing action of the coil, and therefore the dynamic pressure of the plasma flow, is mainly determined by the inductance of the compressor coil, the current through the windings, and the axial plasma velocity. All these quantities are influenced by the aluminum blow-by of the electrothermal injector.

Figure 3.41 gives calculated values for the pressure terms T_2 and T_3 according to (3.30) for various divergence angles, β, of the discharge channel as functions of the total current through the compressor coil. The calculations were made at a location 20 mm inside the narrow end of the compressor coil.

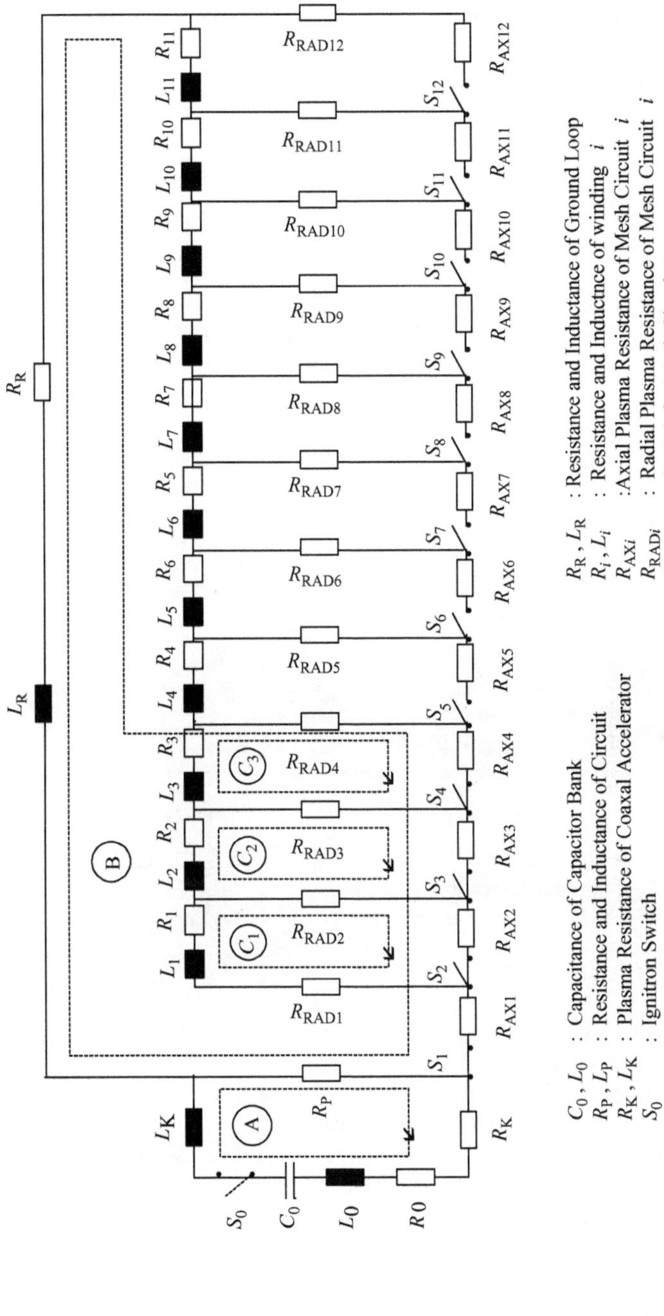

Fig. 3.39. Equivalent electrical circuit of the compressor coil and the coaxial accelerator

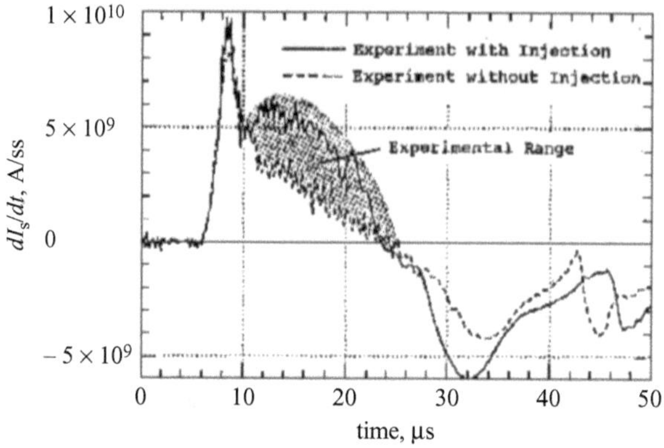

Fig. 3.40. Rogowski signal of the current through the ground loop for an experiment with and without injection of the electrothermal gun

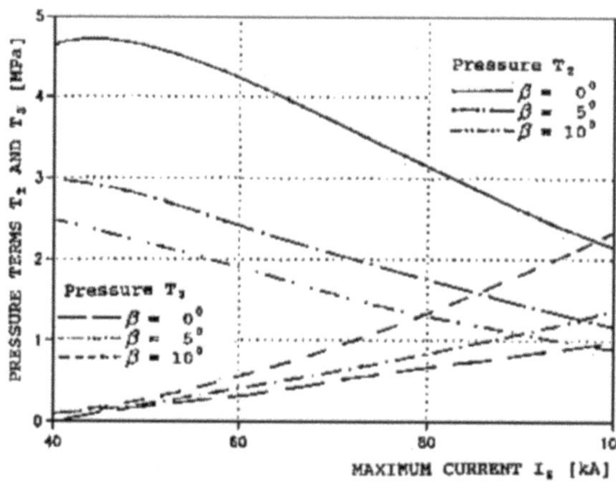

Fig. 3.41. Plot of the calculated pressure terms T_2 and T_3 versus the maximum current I_s through the compressor coil

3.5.4 Sabot Technique

Design Criteria

As result of the first tests, it was concluded that it is absolutely necessary that no aluminum gas from the electrothermal injector enter the compressor coil. Blow-by during the acceleration in the barrel and during the free flight of the particles through the center electrode had to be avoided. These requirements were met by using a special sabot that was completely stopped at the end of the barrel leading to a gas-proof closure of the barrel. The different phases of the acceleration, the separation of the particles from the sabot, and the gas-proof closure are illustrated in Fig. 3.42.

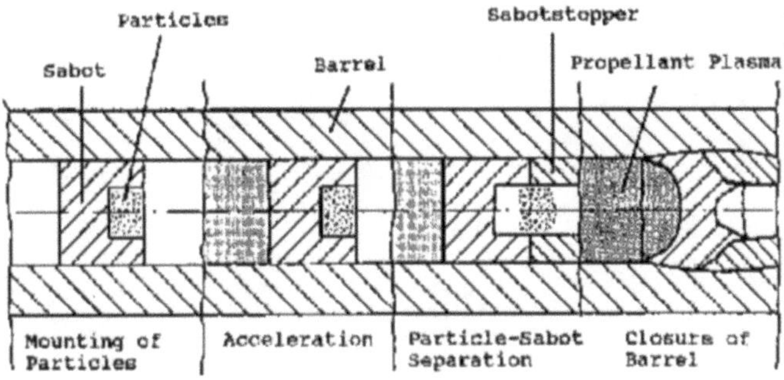

Fig. 3.42. Acceleration of particles in the electrothermal gun with the sabot technique

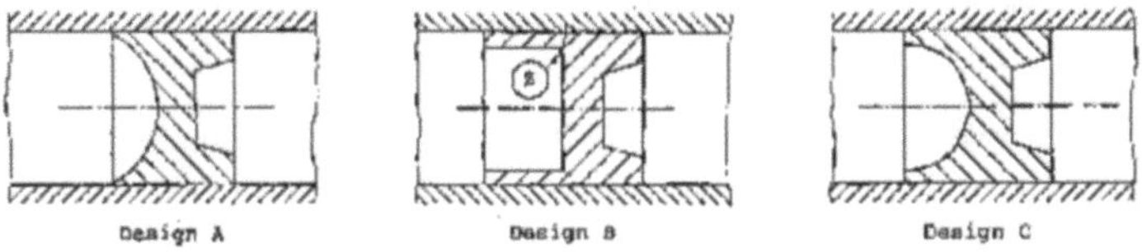

Fig. 3.43. Cross sections of sabots

The chamber for the glass beads in the front side of the sabot was optimized for a good particle/sabot separation. The sabot itself was optimized with regard to minimum mass and a gas-proof motion in the barrel.

In Fig. 3.43 the cross sections of different sabots that were examined in the experiments are depicted. A sabot of design A leaked during acceleration, design B had a high friction between the sabot and the barrel bore, and the sealing lip broke at the location Z during most experiments. Finally a sabot of design C was used. The sabot material was aluminum with a diameter of 5.7 mm. The sealing lip at the back of the sabot was pressed into the polished barrel bore by the pressure of the accelerating aluminum gas.

Experimental Results

With the technique described above, a maximum injection velocity of 1.3 km/s was reached for a 250 mg sabot. Two-stage experiments have been performed with glass beads, ranging in size from 204 μm to 420 μm. The performance of the combined facility is documented in Fig. 3.44 in comparison with the performance of the plasma accelerator alone. With the two-stage operation, an addition of the velocities of the single accelerators was possible. Another important result was a considerable increase in the accuracy of fire, especially for the larger particles.

The severe problems arising from the interaction of the propellant plasma from the electrothermal gun with the He plasma in the compressor coil were solved with a special sabot design. Finally, it was possible to improve the

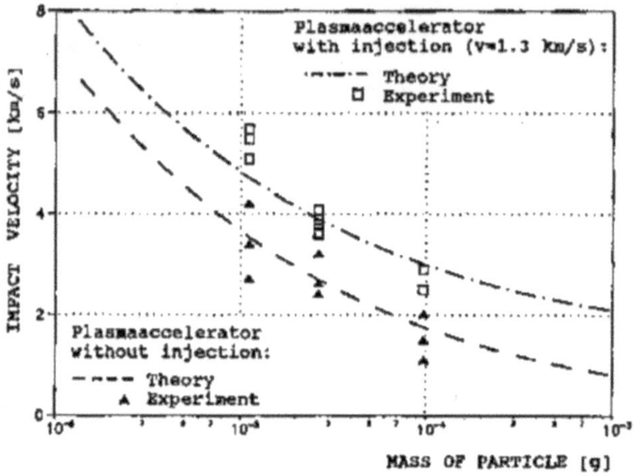

Fig. 3.44. Performance of the two-stage facility in comparison to the plasma accelerator alone

performance of the system. The result was an increase of the particle velocity of up to 45% for 400 µm-diameter glass beads. In addition, the first difficulties led to a better understanding of the processes in the compressor coil. Parameters that had not been considered previously were found to have substantial influence on the operation of the compressor coil. This could be confirmed by a refined theory. On this new theoretical basis, it seems possible that the efficiency of the compressor coil can be improved by injecting a cold insulating gas (e.g., helium) into the center of the coil. This will force more current through the windings, producing a more compressed plasma.

3.5.5 Two-Stage Operation with a Light-Gas Gun First Stage [10, 11]

Description and Operation

In this configuration two completely different systems, a light-gas gun and a plasma accelerator, were combined

- the light-gas gun, which is extensively described in other chapters of this book and powered mostly from chemically stored energy, and
- the electrically powered plasma drag accelerator.

For the combination of these systems, the basic problem of the different time scales of their operation had to be considered first. The time scale of the light-gas gun is in the order of milliseconds, whereas the time scale of the plasma accelerator is in the order of a few microseconds. The solution was to have the plasma system triggered by the light-gas gun system, because the plasma system can easily operate to an accuracy of a millisecond.

The two-stage accelerator has a light-gas gun as the first stage and a coaxial plasma accelerator with compressor coil as the second stage. The light-gas gun is shown in Fig. 3.45. It has a powder charge for initiating the acceleration process, a piston for compressing the light-gas used as the projectile driver, and a pump tube to supply the initial reservoir for the light-gas used. The light-gas reservoir is separated from the barrel and the projectile by a diaphragm.

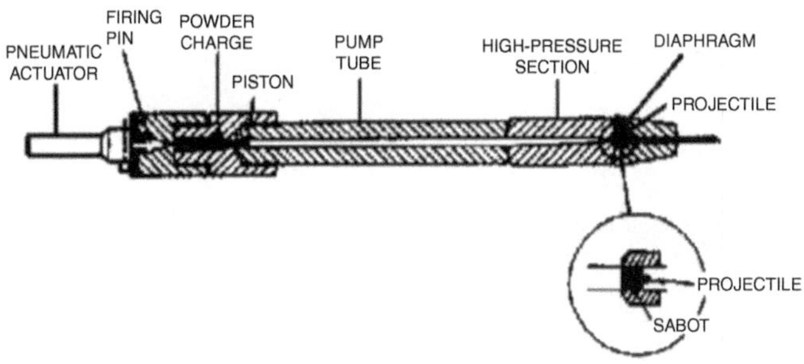

Fig. 3.45. Light-gas gun used in the experiments as first stage

To begin the acceleration process, a pneumatic actuator ignites the powder charge. The powder charge gas expansion accelerates the piston and compresses the gas in the reservoir. The diaphragm will rupture at a known pressure of the compressed light gas. The sabot and the projectile are accelerated in the barrel by the expansion of the compressed light-gas. For maximum performance the pump tube should be filled with hydrogen, but helium was used in these experiments for safety reasons. Glass beads with a diameter of 600 μm were accelerated to an average velocity of 3.8 km/s by the light gas accelerator.

The second stage is a coaxial accelerator with a compressor coil. The example is shown in Fig. 3.46 is identical with the system described previously. Again, the positive side of a charged capacitor bank (270 μF, 20 kV maximum charging voltage) is connected via a discharge switch to the center electrode, which is separated by an insulating plug from the surrounding annular electrode. This center electrode, of course, is now the barrel of the light-gas gun. Prior to the discharge, an aluminum foil is installed connecting the center and annular electrodes. At that time, the aluminum foil operation was the only mode available because the gas injection mode was not yet installed. The annular electrode is connected to the negative side of the capacitor bank. An insulating ring separates the compressor coil from the annular electrode. The holder rail with ground loop connects the front end of the compressor coil to the annular electrode. The front end of the compressor coil is held in place by the coil spacer plate. At the beginning of the discharge the annular electrode is at a negative potential. After the discharge switch is activated, the current

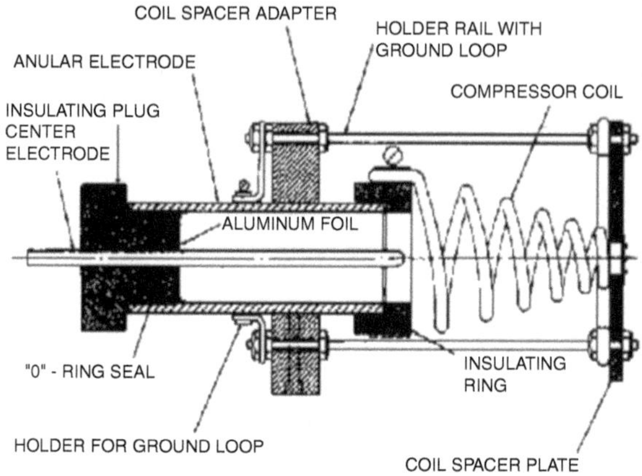

Fig. 3.46. Coaxial plasma accelerator with compressor coil used in the experiments as second stage

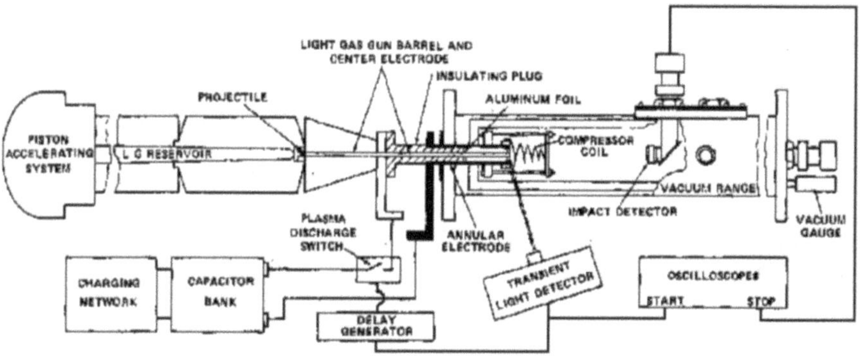

Fig. 3.47. Two-stage accelerator with timing setup

evaporates the aluminum foil and generates an aluminum plasma. The discharge current density from the center electrode to the annular electrode will interact with the azimuthal magnetic field around the center electrode and will accelerate the plasma axially. After the plasma has entered the compressor coil there will also be a discharge between the center electrode tip and the compressor coil. The current in the compressor coil windings generates a magnetic field that interacts with the plasma flow. This results in a compression and an axial acceleration of the plasma. A fast, high-density flow is generated in which projectiles can be accelerated by gasdynamic drag.

The new two-stage accelerator is shown in Fig. 3.47. The light-gas gun barrel is used as a center electrode of the coaxial plasma accelerator. The two-stage operation begins with the loading of the light-gas gun with beads and propellant powder and with the insertion of the aluminum foil into the second stage coaxial plasma accelerator. Then the capacitor bank is charged to the preselected value. The acceleration is initiated by the ignition of the light-gas gun powder charge. The shock wave in front of the beads travelling along the light-gas-gun barrel generates a light flash at the muzzle. This signal is detected by a transient light detector which triggers a delay generator. After a preset delay, this generator triggers the thyratron firing circuit, which, in turn, gives a firing pulse to the ignitrons of the capacitor bank. There is a time interval of 2 μs between the output of the delay generator and the beginning of the capacitor bank discharge.

The successful operation of the two-stage accelerator depends upon the proper timing of the two stages. Each stage must show good reproducibility, and a reliable method is needed to trigger the second stage at a well-defined time during the first stage operation.

The required reproducibility of the first stage depends upon the second stage compressor coil characteristics. From experimental and theoretical investigations of the coaxial plasma accelerator with a compressor coil, the following averaged values were obtained. The time during which the pressure and the density are high enough to provide acceleration is 15 μs. The position from the end of the light-gas-gun barrel along the axis of the compressor coil at which acceleration occurs is 8.5 ± 1.5 cm. With a 4 km/s final velocity of the first stage, the position accuracy becomes ± 6 μs. Reproducibility of the second stage is ± 3 μs. The requirement for timing derived from the position in the compressor coil is the most restrictive one. The light-gas gun first stage must show a reproducibility of $\pm$ 3 μs because the reproducibility of the second stage must be taken into account. A specific loading process of the light-gas gun was developed to obtain the required reproducibility. The projectiles used were 595- to 600 μm diameter glass beads with a density of 2.5 g/cm^3 and a mass of 2.74×10^{-4}g. The beads were inserted into the light-gas-gun barrel between layers of aluminum oxide powder to provide good spread and minimum clumping.

A special effort was made to ensure that each loading was as identical as possible. In addition to the standard procedures used to clean and set up the powder charge, the pump tube, and the high-pressure section, the barrel and the barrel holder were checked carefully. The barrel had to be replaced frequently because it showed damage from the current discharge. Calibration tests of the first stage light-gas gun were performed to determine the time after the muzzle flash at which the beads leave the barrel and their exit velocity. The time was determined with high-speed image converter cameras located in front of two windows in the test range. A transient light detector was focused through one window on the end of the barrel. The muzzle flash triggered the light detector, which, in turn, triggered a delay generator. After a preset delay, an image converter camera is fired and a shadowgraph of the beads emerging from the barrel is taken. These experiments showed that the beads emerged from the barrel at 80.0 ± 2.0 μs after the muzzle flash. The velocity was measured with three photomultiplier tubes mounted along the test range. This standard procedure showed that the beads have an average velocity 3.8 ± 0.2 km/s. These values and their reproducibility are within the limits derived from the second stage characteristics.

Theory of the Two-Stage Acceleration

The performance of the first stage of the light-gas gun was not changed. The velocity of the beads and the time when the beads left the light-gas-gun barrel were kept within the limits described in the preceding section. Therefore, no theoretical analysis of the acceleration of the projectiles in the light-gas gun is necessary, and the experimental values for position and velocity of the beads leaving the first stage became the initial values for the calculation of the acceleration in the second stage.

The theoretical investigation begins with the theory for the equivalent circuit for the capacitor bank with the coaxial accelerator and compressor coil. Then the electromagnetic field inside the compressor coil and the plasma flow are calculated. Finally, a theory for the acceleration of the projectiles by aerodynamic drag forces is used to determine the influence of the timing of the two stages on the final velocity of the projectiles. It is based on the theory developed for the coaxial accelerator with a compressor coil in Sect. 3.3. Equation (3.30),

$$P(z,t) = \frac{\sigma \mu^2 r_0^2 A I}{2}(A\dot{I} + V_z A' I) + \frac{\mu I^2}{4\pi^2 R_e^2}(1 - z/L)^2, \tag{3.31}$$

is also valid in this application. The second term contributes only inside a column of radius $r = R''$ because the axial current density is zero for $r > R''$ and is small in the vicinity of $z = L$.

Theory of the Projectile Acceleration

The projectile in the plasma flow experiences a gasdynamic drag force that is described by (3.1):

$$F = \left[C_D \cdot \frac{\rho}{2}(V_z - v_p)^2 + \nabla P \Delta z\right] S\,. \tag{3.32}$$

The drag coefficient C_D is set constant and equal to 1 using the Newtonian approximation for hypersonic flows. The second term of this equation describes the force resulting from a pressure gradient in the flow where Δz is a characteristic length of the projectile. Both terms are multiplied by the projectile cross sectional area S. The density in the dynamic term is calculated according to (3.11) from (3.24), in which the second term has not been taken into account because numerical calculation showed that it does not contribute significantly to the acceleration of the projectile. The axial derivative of this equation for the pressure $P(z,\, t)$ determines the density and the pressure gradient. The same estimate of the order of magnitude is applied to the left side of the terms of the axial component of the momentum equation for the radial term. The result is similar and shows that the change of the axial plasma velocity V_z is small if the partial derivative with respect to time is neglected. Then a first approximation will give a constant axial velocity V_z. The constant value of V_z is set equal to the velocity of the plasma when it leaves the coaxial accelerator. The resulting equation for the acceleration of the projectile is then

$$\begin{aligned} \frac{dV}{dt} &= C_{\mathrm{D}} \frac{\sigma \mu^2 r_{\mathrm{c}}^2 A I}{4 R T_R} (A \dot{I} + v_z A' I)(V_z - v_{\mathrm{p}})^2 \\ &+ \frac{\sigma \mu^2 r_{\mathrm{c}}^2 \Delta z}{2} \left[2AA'I\dot{I} + V_z I^2 (A'^2 + AA'') \right] \frac{S}{m_{\mathrm{p}}} \, . \end{aligned} \tag{3.33}$$

An analysis of the order of magnitude of the terms of (3.33) shows that the acceleration by the pressure gradient is much smaller than the dynamic term and that the azimuthal current density arises mainly from the interaction between the radial magnetic field and the axial plasma flow. Then only one term remains on the left-hand side for a first approximation:

$$\frac{dV}{dt} = C_{\mathrm{D}} \frac{r^2 \mu^2 \sigma}{4 T_R} I^2 \, V_z \, A \, A' \, (V_z - v_{\mathrm{p}})^2 \frac{A_{\mathrm{p}}}{m_{\mathrm{p}}} \, . \tag{3.34}$$

The position and the time of maximum acceleration of the projectiles can be calculated from the partial derivatives of (3.34) with respect to the axial coordinate and the time. These are then set equal to zero, and the optima are determined. The position for maximum acceleration can be calculated from the derivative with respect to z:

$$(A')^2 + A \, A'' = 0. \tag{3.35}$$

The position is a function of the axial coordinate z only. It can be calculated for a specific compressor coil configuration and current density distribution. The values are indicated in Fig. 3.48 for a compressor coil with a 2 cm initial radius, 0.5 cm end radius, and linear current density distribution. The magnetic field distribution and plasma flow are shown in Fig. 3.49.

The time for maximum acceleration can be determined with the results of the integration of the differential equations for the equivalent circuit or from the experiment. From the differentiation of (3.34) with respect to the time, the condition

$$\frac{dI}{dt} = 0$$

for the maximum is obtained, and corresponds to the first maximum of the compressor coil current.

This analysis shows that for maximum acceleration the projectile should be about 9 cm inside a 12 cm-long compressor coil at the time of the first

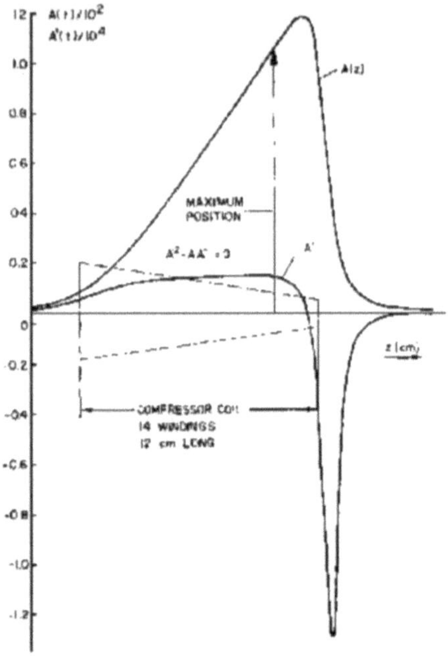

Fig. 3.48. Magnetic field function $A(z)$ and $dA(z)$ for a conical coil with linear current distribution

compressor coil maximum. This is not necessarily the optimum timing for the two stages of the accelerator, but it will probably be close to it and was used for the timing of the first two-stage experiments. These experiments showed the feasibility of the two-stage acceleration.

Theoretical and Experimental Results

In the experiment, the timing between the first and the second stage is adjusted through the delay setting between the light-gas gun muzzle flash and the trigger pulse to the thyratron firing circuit of the second stage. For the theoretical investigation, this delay is converted into the position of the projectile,

$$z = (t - 80.0)\, v_0 \ ,$$

at the time when the integration of the equations describing the discharge of the capacitor bank is started. The coordinate $z_0 = 0$ is at the wide end of the compressor coil which is mounted at the coaxial accelerator end. In the configuration used in the experiments described here, $z_0 = 0$ coincides with the end of the light-gas-gun barrel. A variation of the delay, t, is then equivalent to a variation of the initial position.

Figure 3.50 shows the theoretical two-stage velocity as a function of the delay setting for various values of the capacitor bank charging voltage. The

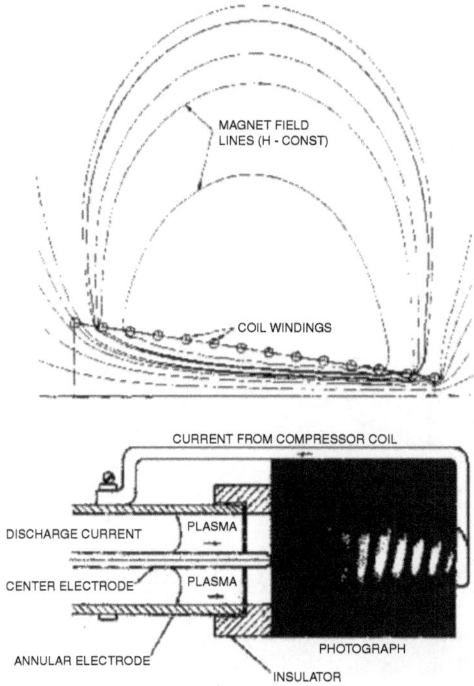

Fig. 3.49. Magnetic field lines and photograph of plasma flow

delay for maximum velocity decreases with increased charging voltage because the plasma velocity increases. The projectile velocity, together with the time of the end of the acceleration and the compressor coil current, is shown in Fig. 3.51. The maximum velocity is obtained when the acceleration ends, approximately at the time of the first compressor coil maximum. The curves in Fig. 3.51 are from the same calculations that were made for the timing of the two-stage experiments. A variation of the delay setting equivalent to the inaccuracy of the operation of the first and the second stages leads to large changes in projectile velocity. From the curves shown in Fig. 3.51, a velocity change up to 4 km/s is obtained for a 5 μs time change. This example shows the importance of the timing of the two stages.

In the experiments, the time interval between the light flash at the muzzle of the light-gas-gun barrel and the impact of the beads at the impact detector was recorded (see Fig. 3.47). This time interval has its maximum value for minimum projectile velocity, i.e., if only the light-gas gun is used. When the plasma gun is added, the time interval decreases and reaches the minimum value for optimum timing or maximum final velocity.

Figure 3.52 shows the theoretical curve and the experimental results. The good agreement of the theoretical and experimental curves indicates that the theoretical model provides adequate timing for the two-stage accelerator. In

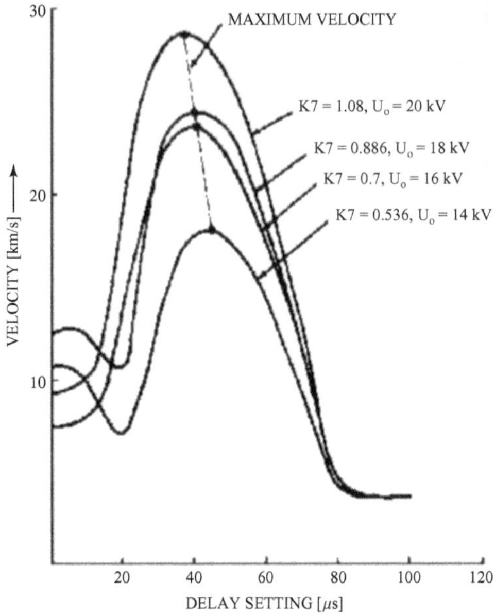

Fig. 3.50. Theoretical velocity as function of delay setting for 14, 16, 18 and 20 kV capacitor bank charging voltage

the same figure, the theoretical velocity is plotted; the maximum theoretical value is 17.9 km/s. Because no other projectile velocity measurement was made, the actual velocity of the projectile is not known. It must, however, be between the maximum theoretical value and a hypothetical minimum velocity of 14.0 km/s, that is computed from the distance between the end of the light-gas-gun barrel and the recorded time interval minus the calibrated delay of 80 µs. This minimum velocity implies that the projectile is accelerated to its final velocity without delay as it leaves the light-gas-gun barrel. It can be assumed that the actual velocity is closer to the theoretical velocity because the experimental and theoretical curves for the time interval between muzzle flash and impact agree well. The curves for the theoretical velocity in Figs. 3.50–3.52 and the curves for the time interval in Fig. 3.52 show a second acceleration. Figure 3.51 shows that this second acceleration occurs at the time of the second maximum of the compressor coil current. This indicates that there are several sequential plasma acceleration and compression phases, as predicted.

Figure 3.53 shows theoretical and experimental results when the charging voltage was increased from 14 kV to 18 kV. The projectile velocity is shown as a function of the charging voltage in Fig. 3.54. The charging voltage was limited by the rating of the capacitor bank and the switch. From the results obtained, it can be assumed that glass beads with a diameter of 600 µm diameter have been accelerated to a velocity near 20 km/sec.

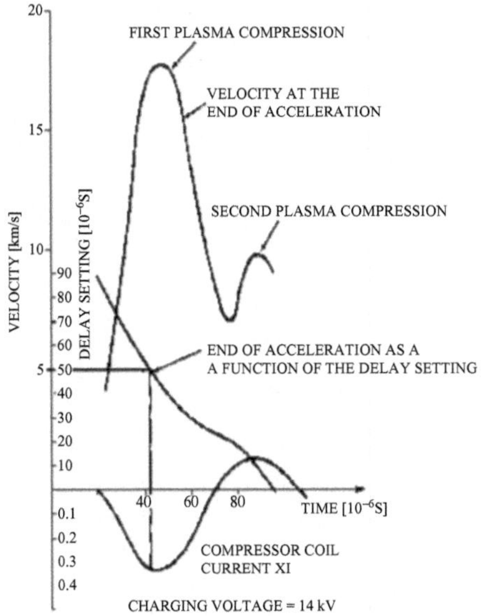

Fig. 3.51. Theoretical curves for velocity, end of acceleration, and compressor coil current

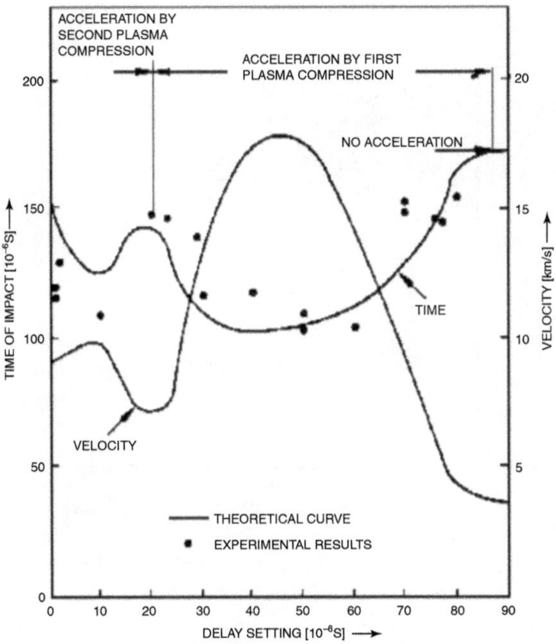

Fig. 3.52. Theoretical and experimental curves for time interval at 14 kV together with the theoretical velocity as a function of the delay setting

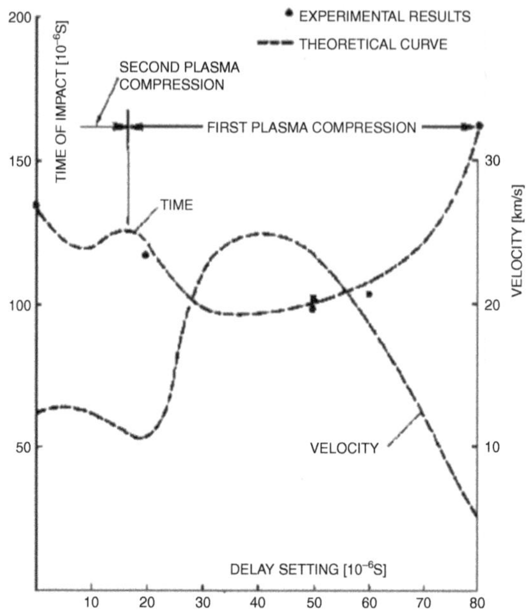

Fig. 3.53. Theoretical and experimental curves for time interval at 18 kV together with the theoretical velocity as a function of the delay setting

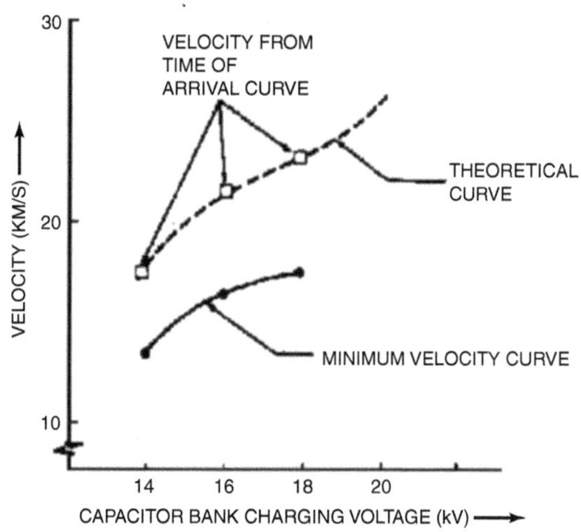

Fig. 3.54. Projectile velocity as a function of capacitor bank voltage for optimum final velocity delay setting

3.6 Concluding Remarks

Both the experiments and the simulations indicate that our understanding of the magnetogasdynamics of the coaxial accelerator with the compressor coil are not good enough, to predict what happens. Instead, we were fortunate to obtain the result described in this report. The magnetogasdynamic flow within the compressor coil is probably only partially ionized, fluid-dynamic/thermodynamic, and electrodynamic effects interact, depending upon the configuration and the corresponding electromagnetic fields.

The functionality does not, as earlier assumed, depend only on the conductivity and therefore the distance between the center electrode and the compressor coil windings, but rather on the inductance and the inductance gradient of the compressor coil. An example for this can be seen in Fig. 3.29, where the short conical coil shows the best performance.

Oscillations with a frequency in the order of 100 kHz were observed, whereas the frequency of the coaxial accelerator with the compressor coil is in the order of 10 kHz. This oscillation within one cycle of the discharge of the main capacitor bank is, of course, a great challenge for an appropriate simulation.

The same is valid for the configuration of coaxial accelerator and compressor coil concerning an optimum electrical current in the windings of the compressor coil in connection with a minimum short current between the tip of the center electrode and the narrow end of the compressor coil or an effective configuration of both. The authors will try to make some additional measurements inside the compressor coil and will also try to measure the electrical current in the coil windings in support of a simulation, which may help us to understand.

All the experimental and theoretical work was done at the NASA MSFC and the LRT of TUM by the authors and their colleagues and co-workers and has been widely published. We learned that the integration of this work that was attempted here taught us more than the addition of the separate publications. It is this expectation that we would like to share with those who read our report and wish, as we do, to continue our work and adventure.

Nomenclature

a = ratio of outer and inner electrodes in coaxial accelerator
A = magnetic field function, amp/m, and area, m^2
b = gradient of inductance, henry/m
$\mathbf{B}$ = magnetic induction = $\mathbf{H}$, gauss
C_D = drag coefficient
C = capicitance, farads
$\mathbf{E}$ = electric field, v/m
F = force, Newtons
$\mathbf{H}$= magnetic field, amp/m,

I = electric current, amp
j = electric current density, amp/m^2
L = inductance, henry
m = mass, kg
P = pressure, N/m^2
r = radius, radial coordinate, meters
R = ohmic resistance (ohm), gas constant, radius
t = time, time coordinate, s
T = temperature, K
U = voltage
v = velocity, km/s
X_1, X_2 = nondimensional current
z = axial coordinate, meter
σ = electrical conductivity, I/ohm m
Φ = azimuthal coordinate
μ = magnetic susceptibility
ρ = density, kg/m^3
κ = polytropic exponent
∇ = gradient operator
Δ = finite difference

Subscripts

0 = time
1 = right-hand equivalent circuit mesh
2 = left-hand equivalent circuit mesh
z = axial
r = radial
Φ = azimuthal
C = compressor coil
d = delay
D = drag
E = center electrode
p = projectile

References

1. E.B. Igenbergs, E.L. Shriver, and D. W. Jex, "Recent developments in micrometeoroid simulation," *Acta Astronautica* **1**, pp. 1337–1355, (1974).
2. E.B. Igenbergs and E.L. Shriver, *Magnetogasdynamic compression of a coaxial plasma accelerator flow for micrometeoroid simulation*, NASA TR R-435, Oct. 1974.
3. E.B. Igenbergs and E.L. Shriver, *Magnetogasdynamic compression of a coaxial plasma accelerator flow for micrometeoroid simulation, J. Appl. Phys.* **44(5)**, pp. 2177–2187, (1973).

4. H. Kuczera, *Theoretische und experimentelle Untersuchungen an einem Plasma- und Teilchenbeschleuniger*, Dissertation, Technische Universität München, 1980.
5. A. Hüdepohl, M. Rott, and E. Igenbergs, "Coaxial Plasma Accelerator with Compressor Coil and Radial Gas Injection," *IEEE Transactions on Magnetics* **25**(1), pp. 232–237, (1989).
6. P. Thomas, *Measurement of the muzzle dynamic pressure of the plasma dynamic accelerator*, presented at the 8th EML Symposium.
7. P. Thomas, "Numerical simulation of the compressor coil of the plasma dynamic acclererator," *IEEE Transactions* **33(1)**, pp. 272–277, (1997).
8. M. Rott, "The LRT/TUM small calibre electrothermal accelerator," *IEEE Transactions on Magnetics* **21(1)**, pp. 597–602, (1993).
9. M. Rott, "The plasma gun augmented electrothermal accelerator," *IEEE Trans. Mag.* **27(1)**, pp. 601–606, (1991).
10. E.B. Igenbergs, B. Cour-Palais, E. Fisher, and O. Stehle, "Plasma flow and fast particles in a hypervelocity accelerator: a colour presentation," *Appl. Opt.* **14(10)**, (1975).
11. E.B. Igenbergs, E.L. Shriver, D.W. and Jex, "New two-stage accelerator for hypervelocity impact simulation," *AIAA Journal* **13(8)**, pp. 1024–1030, (1975).

4 Techniques to Launch Projectile Plates to Very High Velocities

Lalit C. Chhabildas and Marcus D. Knudson

4.1 Introduction

The history of early launch capabilities and the developments of various launcher techniques up to the late nineteen sixties, which includes the developments of explosive lenses [100], explosive liners [109], and smooth-bore guns [6, 22, 25] has been well documented. A detailed account on the history and development of the guns over the last six centuries makes highly recommended reading [22]. Explosive lenses [100] are routinely capable of launching flyer plates to velocities as great as to 6 km/s. Among conventional smooth-bore guns, a two-stage light-gas gun propels projectiles the fastest. Current two-stage light-gas gun technology [6, 22, 25] allows routine launching of 10–20 gm projectiles to a velocity of ~8 km/s, although velocities in excess of 10 km/s have been reported for ten milligram particles [96, 98].

The interest in increased velocity launch capabilities was renewed in the early nineteen eighties. This in part was motivated by strategic defense initiative research, and concern regarding impact of micro-meteoroid and orbital debris particles on space voyagers. Techniques that are either available or are under development include (i) electromagnetic techniques such as the rail-gun [5, 9], which can launch a few grams to 7.5 km/s, (ii) explosive techniques [79], which can launch 1-mm-thick steel plates to 9 km/s, (iii) inhibited shaped-charge techniques [49], which can launch a shaped charge jet tip of 0.4 gm to velocities of 11 km/s, (iv) electrical discharge techniques such as the electric gun [87], which can launch 43 mg kapton flyer plates to 18 km/s, (v) plasma accelerators [58] which can launch micron size (10^{-6}) gm glass beads to velocities of 18 km/s, (vi) Van de Graaff accelerators [59], which can launch sub-micron size (10^{-13} gm) particles to over 100 km/s, (vii) implosion techniques [104], which will accelerate spherical shells to over 22 km/s, (viii) an impact technique [29–31] in which a structured ~100 GPa pressure pulse is generated (upon impact) to launch 0.5–1.0-mm thick flyer plates to velocities up to 16 km/s, and (ix) structured ~300–400 GPa pressure pulses generated by magnetic fields to accelerate flyer plates to 27 km/s [67, 72]. It is the purpose of this chapter to describe the techniques that are currently used to launch flyer plates to velocities in excess of 8 km/s, which is the limit of the conventional two-stage light-gas-gun technology.

There are two main requirements to launch flyer plates to hypervelocities. First, very high pressures are needed to launch the flyer plate and, second, this loading must be nearly shockless and uniform over the entire surface. The kinetic energy of a flyer plate moving at very high velocity is quite large. For example, the kinetic energy of a plate moving at 30 km/s is 450 kJ/gm. For most metals, the melt or vaporization energy is of the order ~10 kJ/gm. Therefore, the energy deposition must be controlled so that it does not transform to internal energy. The shockless loading criterion prevents the plate from melting or vaporizing. The uniform loading criterion prevents the plate from fracture.

In the case of the modified two-stage light-gas gun (also referred to as the three-stage light-gas gun), a graded-density material referred to as a "pillow" [11] or a "multi-ply" [24, 26] is used to impact the flyer plate in order to achieve both these criteria. When this graded-density material is accelerated to a high velocity by a two-stage light-gas gun and impacted upon flyer plate, nearly shockless, 100 GPa pressure pulses are introduced into the flyer plate [25, 27]. Since the loading on the flyer plate is nearly shockless, excessive heating is minimized and melting of the flyer plate is prevented. The method has been used to launch a 2-mm thick titanium alloy Ti-6Al-4V plate to a velocity of 8.1 km/s, and 1-mm thick aluminum, magnesium, and titanium alloy plates to velocities in excess of 10 km/s, and 0.5-mm thick titanium and aluminum alloy plates to 12.2 km/s [29, 31]. With further improvements to this technique launch velocities approaching 16 km/s has been achieved [30].

A new technique that has been developed uses fast-pulsed electrical power at the Sandia National Laboratories Z accelerator [80] to launch macroscopic, planar flyer plates to ultra-high velocities. The technique uses smoothly increasing ~300–400 GPa magnetic pressure pulses to launch plates ~200–300 μm in thickness to velocities approaching 30 km/s. The absence of shocks in the magnetic loading drive is the key in achieving the enormously high flyer plate velocities while keeping the plate cool. Successful launches of aluminum to ~27 km/s and titanium to ~22 km/s have been demonstrated.

4.2 Technical Issues

The principle of gun operation is primarily based on using the energy release from highly compressed gases to propel projectiles to high velocities. This technique is schematically depicted in Fig. 4.1. If $P(t)$ is the time dependent driving pressure used to propel a projectile or a plate of mass M in a gun barrel having a cross sectional area A, then it can be shown that the velocity of the projectile at time t, $V(t)$, is given by

$$V(t) = \frac{A}{M} \int_0^t P(t)\, dt \, . \tag{4.1}$$

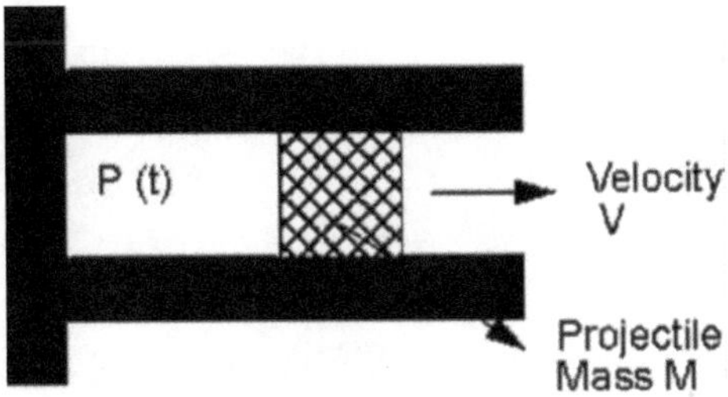

Fig. 4.1. Principle of using high pressures to accelerate projectile or plates

This principle of gun operation is applicable in the operations of smoothbore gun launchers developed for controlled impact studies. Specifically, in a 100-mm bore single-stage compressed-gas gun a 0.04 GPa gas pressure is used to drive a projectile weighing approximately 1 kg to a terminal velocity of ∼1.1 km/s. Gaseous combustion products from nitrocellulose propellants in a 89-mm bore diameter powder gun will yield a pressure of 0.3 GPa and propel a ∼1 kg projectile mass to a velocity of over 2.3 km/s. In a 29-mm bore two-stage light-gas gun the hydrogen gas is dynamically compressed to a peak pressure of ∼0.7–1.0 GPa to launch a projectile mass of 10–20 gm to a velocity approaching 8 km/s [6]. Note that, to achieve higher velocities, the mass of the projectile is decreased and the average driving pressure is increased. (The projectile mass is dependent on the bore diameter of the gun.) This relation is illustrated in Fig. 4.2 as a plot of projectile velocity as a function of projectile acceleration.

Notice that (4.1) implies a time-dependent pressure history. This is especially true for launching projectiles or plates to extremely high velocities. It would seem, therefore, that either gas loading pressures of ∼0.7 GPa be sustained for a longer duration, or yet higher driving-gas pressure and higher acceleration be used to accelerate projectiles to hypervelocities. The former has been attempted on the two-stage light-gas gun by using collapsible launch tubes (using explosive techniques) to achieve projectile velocities in excess of 11 km/s [15]. These attempts which, at the time were limited to launching of lexan sabots, were successful. These early attempts were abandoned either due to a lack of interest, or due to the expense associated with replacing launch tubes after every experiment. Other attempts to launch flyer plates to high velocities, using gas at the higher pressure of ∼100 GPa [13,14] and higher acceleration, have resulted in plate fragmentation [27,28]. Under these conditions the flyer plate is subjected to driving gas pressures that are orders of magnitude above its elastic limit, and severe deformation will occur. Because the flyer plate has limited tensile strength, the non-uniformity in gas pressure loading over its face can be sufficiently large to cause flyer fragmentation [27,28]. The non-uniformity in gas pressure loading over the entire face of the flyer plate should not exceed the fracture strength of the plate.

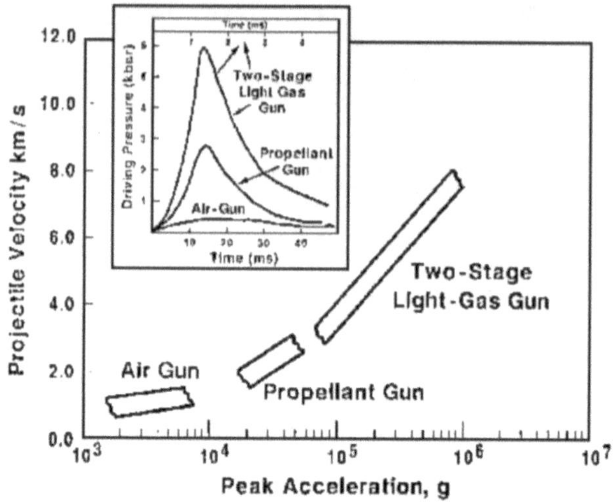

Fig. 4.2. Variation of projectile velocity with "g" loadings for current smoothbore gun operations. The inset shows the driving pressure loading history for single-stage compressed gas gun, propellant gun and the two-stage light-gas gun

Evidently, the loading pressure used to accelerate a flyer plate to high velocity must be extremely uniform and the compression of the plate must be isentropic; more so than is achieved through high dynamic gas pressure. Two successful schemes have been developed; a three-stage light-gas gun technique and a magnetic technique. Both of these techniques rely on isentropic compression of the flyer plate over significantly shorter timescales than those achieved in conventional gas gun techniques. These approaches will be briefly outlined below.

4.2.1 Isentropic Loading Techniques – Gun Systems

Dynamic, quasi-isentropic loading techniques using light-gas guns are not new. Techniques were first introduced by Russian investigators to synthesize very hard materials, such as diamond and boron nitride, for commercial grinding applications. They used graded-density layers on a target to introduce a time-dependent pressure pulse into the target. The use of graded-density layers on a target, however, limits the sample size that is maintained at uniaxial, one-dimensional loading conditions. Other isentropic-loading techniques include cylindrical compression schemes either using explosives or magnetic fields. The time-dependent collapse of a cylindrical liner induces the isentropic compressive waves into the material of interest. The former was used to make synthetic diamonds for commercial use, and the latter was used for a material property study exploring for metallic hydrogen [55]. However, no attempt was made to use these techniques for launching flyer plates to high velocities.

In the early 1980s interest was rekindled in development of isentropic loading techniques for material studies [11]. Concepts using particle sedimentation techniques were introduced to make "pillow" impactors. A pillow was fabricated using powder metallurgical techniques [11] such that a relatively smooth variation in its shock impedance was obtained through the thickness. The shock impedance of the impact surface of the graded-density pillow is similar to that of polyolefin, whereas the impedance of the back of the pillow is similar to that of copper or tantalum. Other concepts introduced included multi-ply layered materials of graded densities [26]. A multi-ply layered impactor is fabricated by bonding a series of thin plates in order of increasing shock impedance from the impact surface. By using the graded-density material as the impactor, samples having a larger lateral dimension could be investigated in material property studies under uniaxial strain conditions. Early studies were focused on determining isentropes for metals [11, 26] and the dynamic shear strength and the spall strength of materials under isentropic loading [26, 28, 29].

It was not until the early 1990s that the graded-density impactors were used to launch plates to velocities in excess of 10 km/s [29]. First, very high pressures are needed to launch the flyer plate to hypervelocity. Second, this loading must be nearly shockless, structured, and uniform over the entire plate surface. Shockless acceleration of the plate is crucial to prevent shock-induced heating and subsequent melting of the flyer plate during the loading process. A structured pressure pulse is required to prevent tensile failure of the flyer plate. Uniform loading across the entire face of the flyer plate is necessary to prevent plate fragmentation. A graded-density impactor provides these required conditions. When this impactor is used to impact a flyer plate at high velocities in a two-stage light-gas gun, nearly shockless, ~100 GPa pressure pulses are introduced into the flyer plate. This time-dependent loading subsequently propels the flyer plate to high velocities. The resulting acceleration of these plates is of order 10^{12} cm/s^2 (10^9 g). Incidentally, this concept of velocity multiplication using graded density materials was proposed in the late 1970s [46].

4.2.2 Isentropic Loading Techniques – Magnetic Systems

Recently a new capability for producing smoothly increasing pressure waves using fast, pulsed power as the energy source has been developed at the Sandia National Laboratories Z accelerator [50]. The Z accelerator [80] is a low inductance pulsed power generator capable of capacitively storing 11.6 MJ of electrical energy. The accelerator utilizes a combination of fast switches and transmission lines to deliver a ~20 MA, ~200 ns rise-time current pulse to a short circuit load at the center of the accelerator. This pulse generates a time-varying magnetic field between the anode and cathode of the short circuit. As a result of the Lorentz force, these high currents and magnetic fields produce nearly isentropic loading of the cathode and anode over the discharge time of the pulse [4]. Fields of the order several MG are produced, which result in magnetic pressure levels approaching 400 GPa.

With the aid of accurate magnetohydrodynamic (MHD) simulations, which will be described in the following sections, short circuit loads have been designed that are capable of generating planar stress waves with rise times of order 200 ns in centimeter sized material samples. The experimental configuration is shown schematically in Fig. 4.3. Typically, four anode panels are arranged about a central stainless steel cathode post, forming a symmetric anode-cathode (A–K) gap. The nominal panel size is 20–35 mm in width and 30–50 mm in height. A short circuit is created between the anode panels and cathode post through a shorting cap at the top of the coaxial load. The interaction of the current density and magnetic field produced in the insulating gap results in a time-dependent pressure, $P(t)$, that is applied to the inner surface of the electrodes. The magnitude of this loading is given by

$$P(t) = B^2/2\mu_0 = \mu_0 J^2(t)/2 = \mu_0 I^2(t)/2S^2 \,. \tag{4.2}$$

Here, $I(t)$ is time-dependent current at the sample location; S is the scale factor (typically ~2–10 cm) necessary to cvert $I(t)$ to the current density, $J(t)$ (amps/unit length); B is the magnetic field strength; and μ_0 is the magnetic permeability of fonree space.

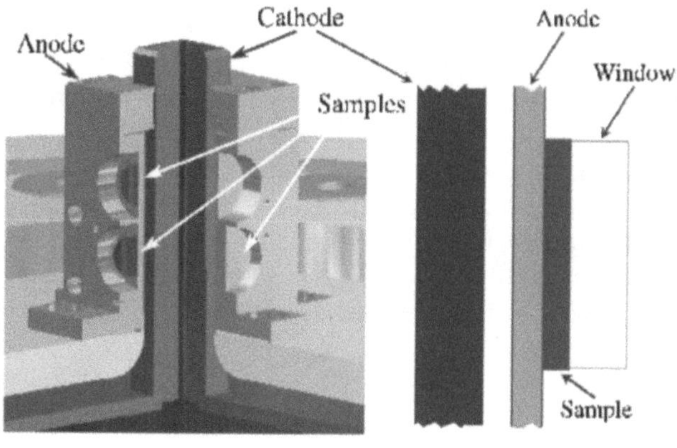

Fig. 4.3. Experimental configuration for isentropic loading on Z, in this case the square short circuit geometry. The square cathode and flat plate anode produce a planar magnetic field within the A–K gap

The magnitude of the magnetic field in the A–K gap, and thus the driving pressure pulse, can be adjusted by changing the geometry of the short circuit load, and therefore the current density. Two typical short circuit geometries are routinely used: a square coaxial geometry, in which four equal width anode panels surround a square cathode stalk, and a rectangular geometry, in which the width of two of the sides of the square geometry are reduced by roughly one-half. The square geometry is capable of producing up to a ~100 GPa driving pressure pulse. The rectangular or slab geometry, which increases the current density at the expense of two panels, is capable of producing up to a ~300–400 GPa driving pressure pulse. Besides changing the short circuit geometry, the current density for a given geometry can also be lowered by reducing the charge voltage of the accelerator. Since the peak current scales approximately linearly with the charge voltage, the driving pressure can be reduced by as much as 45%. These two parameters, the geometry and the charge voltage, allow the peak magnetic pressure to be varied continuously over the range of a few tenths of a GPa to several hundred GPa.

Since this initial development, a large number of applications and spin-offs of the technique have been developed. In particular, experiments have been performed to determine material isentropes [41, 50, 54, 89], to study solid–solid [39, 93] and liquid–solid [40] phase transformations, and to investigate the compressive yield strength of material at high pressure [107]. Furthermore, the ability to identically load multiple samples to high pressures makes the Z accelerator a particularly attractive test-bed for performing high-fidelity relative comparisons of dynamic material response [90, 91].

It has also been recognized that the magnetic loading achievable on the Z accelerator is very well suited for acceleration of flyer plates to ultra-high velocity [51, 72]. In particular, the timescales and peak accelerations achievable on Z are such that velocities approaching 30 km/s are attainable. Furthermore, the exceptionally smooth and uniform pressure pulse enables intact, high integrity plates to be launched in spite of the extreme accelerations (of order 10^{10}g). In fact, flyer plate velocities as high as 27 km/s have been measured in which the condition of the plate at impact is essentially ambient temperature and density [73].

We note that the technique is similar to conventional electromagnetic launcher (rail-gun) technology [5, 9], however the Z technique uses much higher currents, the loading is applied in approximately 200 ns instead of milliseconds, and the flyer attains maximum velocity in millimeters instead of meters. Furthermore, with the Z technique as many as to four plates can be launched simultaneously in a single firing of the accelerator.

4.3 Experimental Techniques

4.3.1 Gun Systems

This section briefly describes the experimental techniques employed to augment the launch capabilities of two-stage light-gas guns. The experimental impact configuration is indicated in Figs. 4.4a and 4.4b. The two-stage light-gas gun used in these studies had a bore diameter of 29 mm. As indicated in the figure, a two-stage light-gas gun projectile, which has a graded-density impactor facing is made to impact a thin flyer plate located at the gun muzzle.

Implementations of these techniques require that the barrel of the two-stage light-gas gun be extended [14], and that the flyer plate be laterally confined to minimize two-dimensional effects [29, 31]. Tungsten makes a good confinement fixture, but calculations have indicated that a cheaper material such as steel could provide a satisfactory confinement for the geometry indicated in Fig. 4.4a. Impact velocities over the range 6–7.5 km/s have been used to obtain flyer plate velocities from 10–12 km/s. Typically, the flyer plate velocity resulting from impact is ~1.6 times the impact velocity, when the experimental set up shown in Fig. 4.4a is employed [31]. The flyer plate velocity is further increased if the experimental configuration shown in Fig. 4.4b is used. In this case, the high pressure in the tungsten barrel is sustained for a longer time and increases the integrated loading pressure history that is driving the flyer plate. A flyer plate velocity that is over twice the velocity produced by the initial graded-density impactor design is obtained [30].

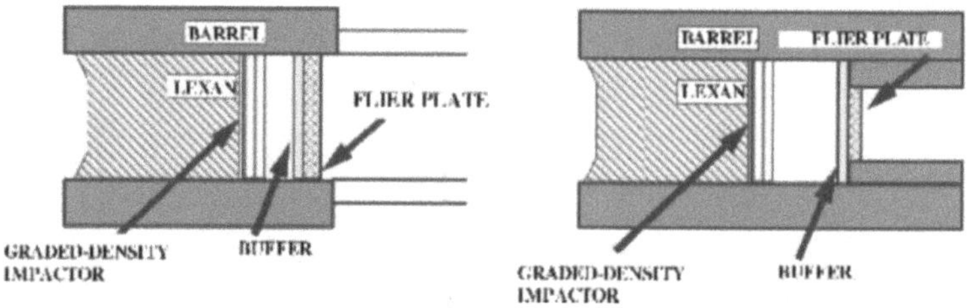

(a) (b)

Fig. 4.4. Experimental configuration that yields flier plates velocities up to 12 km/s is shown in (**a**), whereas the stepped down configuration shown in (**b**) propels flier plates up to 16 km/s

Flyer Plate Designs

The flyer plate used in these experiments consists of a center plate made to fit exactly into a guard ring. The outside diameter of the guard ring used in these studies was 29 mm, whereas the inner diameter of the guard ring and the diameter of the center plate can be varied. Typical center plate diameters are approximately 19 mm. Two-dimensional effects due to radial release waves (generated upon impact) emanating from the edges of the plate would cause a velocity gradient across the radius of the plate. Large velocity gradients across the radius of the plate would cause the flyer plate to bend and, perhaps, even fragment. As depicted in Fig. 4.5c, the guard ring geometry shown in Fig. 4.5a allows a controlled separation of the center plate from its edges without causing the entire flyer plate to fragment. Confinement in a target fixture is desirable for maximizing the diameter of intact flyer [29]. However, this may not be essential for launching the central region of the flyer.

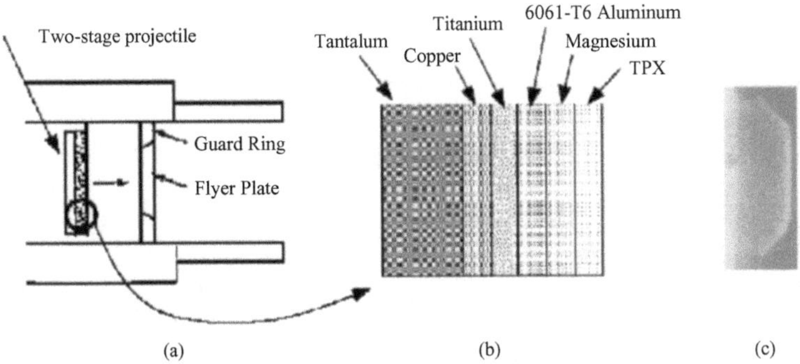

Fig. 4.5. Configuration of the three-stage, launcher; (**a**) shows a two-stage projectile with a graded density impactor and three-stage flyer plate with a guard ring. (**b**) Details of the materials used in the graded density impactor. The TPX facing impacts the stationary flier plate (TPX is a trademark for poly 4-methyl-1-pentene.) (**c**) X-radiograph of a typical three-stage flier plate traveling at 11.1 km/s

The flyer plate materials that have been used to date consist of titanium alloy Ti-6Al-4V, aluminum alloy 6061-T6, and magnesium alloy AZ31. As indicated in Fig. 4.4, a TPX plastic buffer with a density of 0.81 g/cm^3 is used because it further cushions the input pressure pulse and has a tendency to minimize the tensile stress induced in the flyer plate [29, 31].

Projectile Designs

A lexan projectile that has a facing of a graded-density material such as a pillow or a multi-ply impactor backed by tantalum is used in these studies. A pillow is fabricated using powder metallurgical techniques [11] such that a smooth variation in its shock impedance occurs through its thickness. The shock impedance of the impact surface of the graded-density material is that of polyolefin, and the shock impedance of the back surface of the pillow resembles that of copper or tantalum. A multi-ply impactor is fabricated by bonding a series of thin plates in order of increasing shock impedance from the impact surface. The series of layered materials used is shown in Fig. 4.5b. The impactor consists of layers of plastic TPX, magnesium, aluminum, titanium, copper, and tantalum. The thickness of each layer is precisely controlled to tailor the time dependent stress pulse required to launch the flyer plate intact [24, 26]. When these graded-density materials are used to impact a titanium alloy flyer plate at a velocity of ~6.4 km/s, an initial shock of approximately 50 GPa is introduced into the flyer plate, followed by a ramp wave to over 100 GPa. At higher impact velocities the input pressure profile would result in a higher peak pressure pulse, and thus the launching of flyer plates to yet higher velocities.

Velocity Measurements

Following impact, multiple flash x-rays are taken of the flyer plate while it is in motion. They are used to estimate the velocity of the flyer plate and also to assess its integrity following impact and subsequent acceleration by the shockless pressure pulse. These x-ray sources have a 300–450 keV energy and a 25 ns pulse duration. Due to the hypervelocities achieved in this study, the 25 ns pulse duration can cause up to 250 μm blurring of the flyer plate while in flight. Radiographic pictures of the flyer plate taken in flight over these large distances allow an accurate measurement of its velocity. The flyer plate velocity is determined to better than ±1%.

The velocity of the flyer plate can also be determined using a velocity interferometer, commonly referred to as VISAR [12]. Figure 4.6 shows the velocity history of a flyer plate launched to 11.1 km/s. However, depending on the configuration and nature of the experiment being conducted, it may not always be convenient to perform such a velocity measurement using VISAR. It has been shown [21] that, if the graded-density impact configuration is known exactly, then hydrodynamic codes such as CTH [81] can be used to predict the flyer plate velocity. The input parameters for the CTH code are the dimensions of the layers of the graded-density impactor, the TPX buffer, and the flyer plate, all of which can be determined to within ∼0.1%. The velocity of the graded-density impactor, which is the two-stage gun projectile velocity, can also be determined to ∼0.2% [88]. As indicated in the figure, the measured flyer plate velocity agrees with the CTH code predictions to within ∼1% [21]. The velocity profile shown in Fig. 4.6 also yields an average acceleration greater than 10^{12} cm/s^2 (10^9 g).

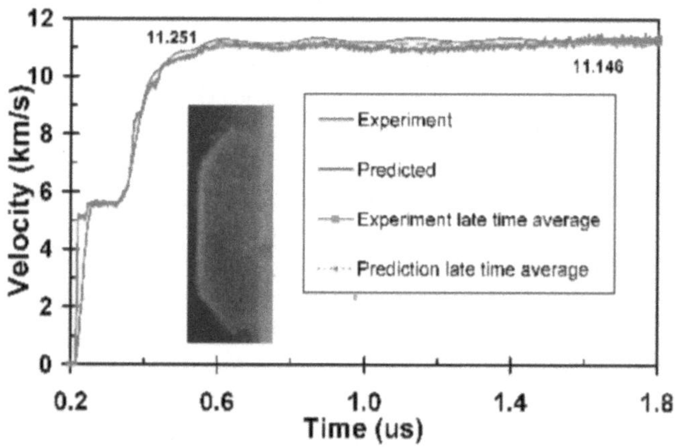

Fig. 4.6. Velocity history of a flyer plate launched at 11.1 km/s. Also shown are the CTH simulations of the experiment yielding a velocity of 11.2 km/s

Results

In this section, the results of research investigations where the projectile is allowed to impact the buffer-flyer plate combination (see Fig. 4.4a) are summarized. Plastic buffers are used because they further cushion the launch. The plastic buffer will also reduce the magnitude of the tensile stress generated within the flyer plate material as a result of wave interactions [29, 31]. Titanium and an aluminum alloy flyer plate are generally used because of their high fracture resistance [27, 28]. At an impact velocity of ~6.3 km/s, this technique has launched intact, 1-mm thick titanium alloy Ti-6Al-4V plates to 9.5 km/s, and 1-mm thick aluminum alloy 6061-T6 plates to 10.4 km/s. As the impact velocity is increased to ~7.4 km/s, velocities of 0.5 -mm thick aluminum and titanium flyer plates have been increased to over 12 km/s. An example of a radiograph of a flyer plate launched at 11.1 km/s is shown in Fig. 4.5c and in Fig. 4.6 as an inset.

The mass of the plate is given by $M = \rho x A$, where ρ is the density of the plate, x is its thickness, and A is its cross sectional area. Insertion of this relation into (4.1) results in

$$V(t) = \frac{1}{\rho x} \int_0^t P(t) dt \,. \tag{4.3}$$

Equation (4.3) suggests that V is inversely proportional to ρx, the mass density, and should be independent of the cross-sectional area of the gun. It is therefore useful to express the velocity of the flyer plate as a function of its mass density. This dependence is plotted in Fig. 4.7. As indicated in the figure, the terminal velocity of the flyer plate is dependent on the impact velocity of the graded-density driver. Also, as shown in Fig. 4.7, the presence of plastic buffers further increases the terminal velocity of the flyer plates. To increase the velocity of the flyer plate further, one needs to decrease the thickness of the plate or increase the time-dependent pressure loading pulse. This relation also suggests that scaling should apply, i.e., heavier masses can be launched on a larger bore-diameter gun.

Figure 4.8 indicates that the maximum mass that can be launched on a 29 -mm bore gun, in the absence of any two-dimensional effects, as a function of the mass of the center plate (currently a 19-mm diameter plate). At higher loading pressures, however, the two-dimensional loading times are reduced by a factor of ~two since the sample thickness is approximately one-half. This reduction in loading time would make it permissible to have a larger center plate diameter at higher velocities. As indicated in Fig. 4.8, a mass of approximately 0.5 gm can be launched to velocities of 12.2 km/s. If techniques were introduced to mitigate two-dimensional effects, then the potential of launching heavier flyer plates exists for the same bore diameter. This type of scaling has been demonstrated by launching a heavier mass (a titanium alloy Ti-6Al-4V plate that was 2 mm thick, 34 mm in diameter and had a mass of 7.6 gm) to 9.3 km/s on a 64 -mm bore diameter gun [30].

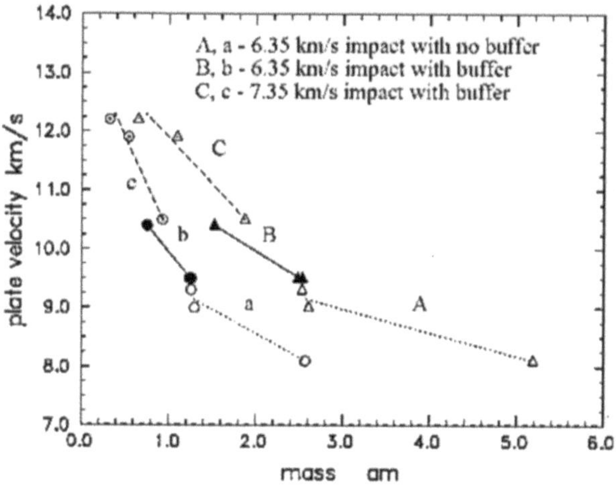

Fig. 4.7. Variation of mass density (areal density) versus velocity of the flier plate for the experimental configuration shown in Fig. 4.4a

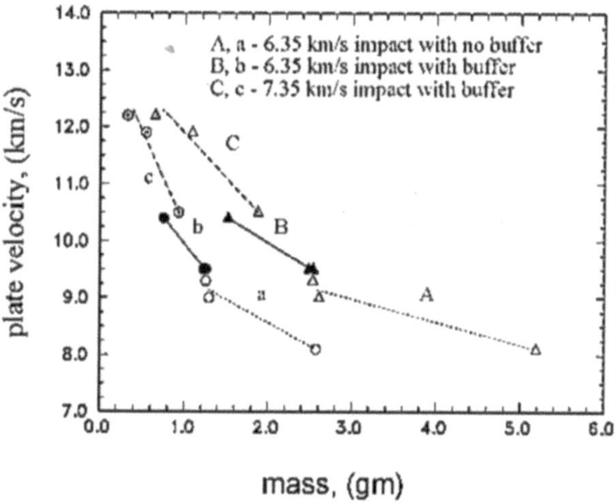

Fig. 4.8. Variation of mass versus velocity of the flier plate launched on a 29-mm bore gun. The mass launched will depend on the bore diameter of the two-stage gun

There has been a systematic study to characterize the flyer plate integrity, in particular for applications to equation of state (EOS) studies [88]. For flight distances of approximately seventy millimeters, the plate remains flat. In particular, the flatness, the non-simultaneity (tilt) of the impact, and the temperature rise of the flyer plate has been determined experimentally [88]. The results of this study indicate that the tilt and bow at impact is comparable to the two-stage light-gas gun experiments, thereby making it a valuable tool for EOS measurements at impact velocities not attainable with conventional two-stage light-gas gun technologies [88, 110]. The temperature of the flyer plate, based on the quasi-isentropic loading pulse that results from graded-density impact, suggests a temperature increase of ~300 K. For accurate EOS measurements it is necessary that the temperature increase of the impacting flyer plate be taken into account.

Higher flyer plate velocities can be obtained using the graded-density impactor technique by impacting the stationary flyer plate that is confined in a smaller diameter barrel compared to the diameter of graded-density impactor [30]. As shown in Fig. 4.4b, an undersized flyer plate is inserted in the tungsten barrel. A projectile that has a graded-density impactor embedded in the sabot impacts the TPX plastic buffer, which is attached to the tungsten barrel. Upon impact, the high pressures in the tungsten barrel confine the flyer plate (more efficiently than the geometry in Fig. 4.4a) and augment the loading pressure history that is accelerating the flyer plate. The resultant flyer plate velocity depends on the tungsten barrel diameter. An increased flyer plate velocity over twice the initial (graded-density) impact velocity can be obtained. Plastic buffers not only reduce the magnitude of tensile stress states induced in the material, but further increase the flyer plate velocities. Figure 4.9 shows the results of a graded-density impact on a 1-mm thick by 6-mm diameter titanium plate at 6.75 km/s. The flyer plate is travelling at a velocity of 14.4 km/s. As indicated in the figure, the flyer plate velocity is determined over a distance of 1400 mm. Unlike the experiments described previously, no guard ring is used in this experiment. Due to high pressure confinement, the edges of the flyer plate are not expected to be traveling slower than the center portion, as they do in the geometry illustrated in Fig. 4.4a [32]. Notice that, for long distance throws, the flyer plate does not remain flat. No x-ray measurements close to muzzle are available to evaluate the flyer plate flatness conditions because high-density tungsten barrels were used. This technique has also been adopted to form plates into chunky projectiles and launch them at velocities in excess of 10 km/s [34, 102].

Specifically, the flyer plate diameter, d_{fp}, is small and the diameter of the graded-density impactor, D_{gdi}, is significantly larger. In Fig. 4.4a, the flyer plate diameter is the same as the graded-density impactor diameter, i.e.,$d_{\mathrm{fp}}/D_{\mathrm{gdi}} = 1$. In Fig. 4.10a, the variation of flyer plate velocity is shown as a function of the ratio of the flyer plate diameter to the graded-density impactor diameter. The same results are shown in Fig. 4.10b, in which the flyer velocity is normalized with respect to the impact velocity. As shown in these figures, the flyer plate velocity increases with decreasing $d_{\mathrm{fp}}/D_{\mathrm{gdi}}$. The values shown at $d_{\mathrm{fp}}/D_{\mathrm{gdi}} = 1$ are calculated values based on previous studies [32]. Thus, to enhance flyer plate velocities, a necessary requirement is to impact a step-down barrel as indicated in Fig. 4.4b.

In this configuration, the barrel itself is loaded to high pressures upon impact. The stress states at the barrel/flyer boundary are now maintained

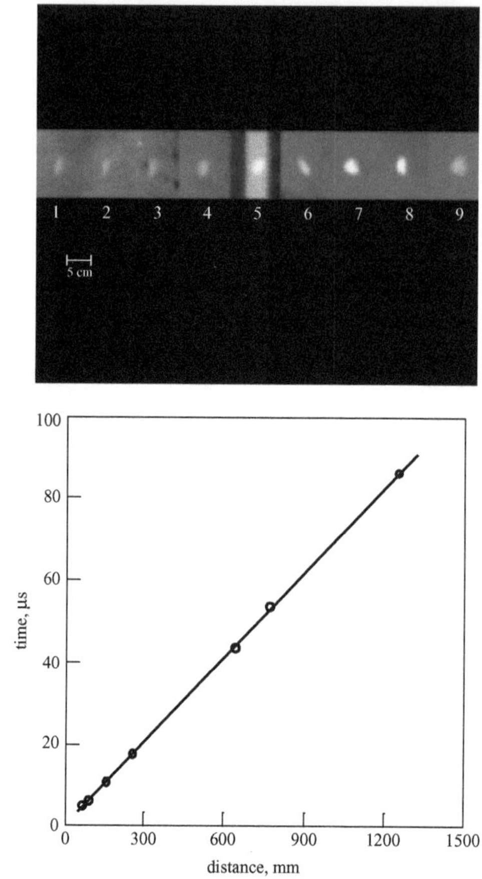

Fig. 4.9. Radiographs of a projectile/plate launched at 15.8 km/s using the experimental configuration shown in Fig. 4b

at high pressures (when compared to the configuration shown in Fig. 4.4a). This prevents the release of the driving pressure behind the flyer plate and sustains higher loading stresses over a longer duration. As the graded-density impactor material enters the step-down barrel, it accelerates further, maintaining an efficient push behind the flyer plate. All of these factors contribute toward accelerating the flyer plate to yet higher velocities. The experimental geometry indicated in Fig. 4.4b to launch flyer plates, therefore, operates as an (impact-generated) accelerated reservoir.

4.3.2 Magnetic Systems

In the magnetically accelerated flyer plate technique, the flyer plates comprise the anode of a short circuit load at the center of the Z accelerator, as shown

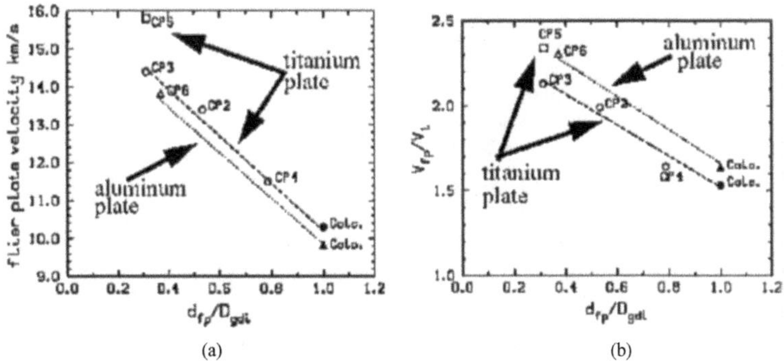

Fig. 4.10. (**a**) Variation of flyer plate velocity versus the ratio of the stepped-down barrel diameter to the graded-density impactor diameter. (**b**) Flyer-plate velocity normalized to impact velocity versus the ratio of the stepped-down barrel diameter to the graded density impactor diameter

schematically in Fig. 4.11a. The interaction of the current density and magnetic field produced in the insulating gap results in an impulsive ramp load, providing momentum to the anode, launching it as an effective flyer plate to high velocity. Each anode panel becomes a flyer plate; this is achieved by machining the entire current carrying portion of the aluminum anode panel to a prescribed material thickness of approximately 800–900 μm. To retain rigidity, and to allow the panels to be assembled together, the flyer frame is attached to a panel back as shown in Fig. 4.11b. The panel back also allows for mounting of the experimental target at a prescribed distance from the flyer plate, which is typically ∼3–4 mm, and the mounting of optical probes used to diagnose the target. The current carrying surface of each panel and the impact surface of the flyer are flat to ∼200 nm and parallel to ∼2 μm with ∼20 nm surface finishes. Each of the panel backs can hold two separate targets, allowing up to eight simultaneous shock wave experiments during a single firing of the accelerator.

To a crude approximation the final velocity of the flyer can be estimated analytically by applying conservation of momentum, and treating the magnetic pressure pulse as an impulsive load, integrating over time to determine the total momentum transferred to the plate. However, due to significant joule heating and ablation near the current carrying surface as well as magnetic field penetration into the metal associated with this technique, the details of the flyer plate launch are complicated, and cannot be approached through simple analytical means. Computer simulations, which include all the experimental parameters and necessary physics, are of great value in designing the experimental configuration that minimizes density gradients and accurately predicts the final flyer plate velocity and optimal flight distance [75,76]. These simulations will be described in the next sections.

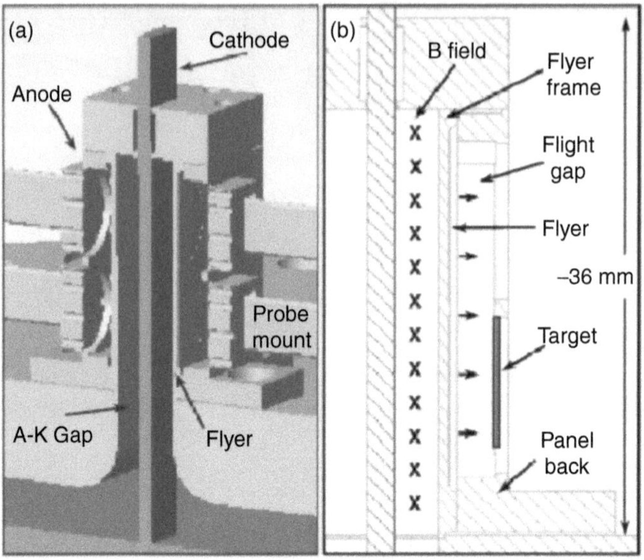

Fig. 4.11. Schematic of the load used to launch flyer plates on the Z accelerator, in this case the rectangular short circuit geometry. (**a**) Section view of the coaxial load. Also shown are the probe mounts used to hold the fiber optics used to diagnose the sample. (**b**) Section view showing more detail

Short Circuit Design

In order to optimize the short circuit load used to launch flyer plates on Z, both one- and two-dimensional Eulerian simulation schemes have been developed that use the finite element, arbitrary Lagrangian–Eulerian, magnetohydro-dynamic (MHD) code ALEGRA [99] to solve the MHD equations for a compressible material with material strength. EOS models valid for a wide range of pressures ($P < 500\,\text{GPa}$), densities ($\rho < 50.0\,\text{g/cm}^3$), and temperatures ($T < 10\,\text{eV}$) are used for the anode and cathode materials, in addition to models for the thermal and electrical conductivities [42, 43]. Density and internal energy are used in the EOS to obtain stress and temperature. The density and temperature are used in the conductivity model to obtain electrical and thermal conductivities. The form of Ohm's law used in ALEGRA is $\mathbf{J} = \sigma(\mathbf{E} + \mathrm{v} \times \mathbf{B})$, where $\mathbf{J}$ is the current density, σ is the electrical conductivity, $\mathbf{E}$ is the resistive electric field, and $\mathbf{B}$ is the applied magnetic field. More details concerning the MHD modeling of the flyer plate launch can be found in previous references [75, 76].

Results of one-dimensional MHD simulations driven by a prescribed load current are in very good agreement with the experiment, and have provided significant insight into the details of the flyer plate launch [75]. As an illustration, Fig. 4.12 shows a comparison between the predicted velocity profile of a $\sim$22 km/s aluminum flyer plate and the actual measured velocity profile

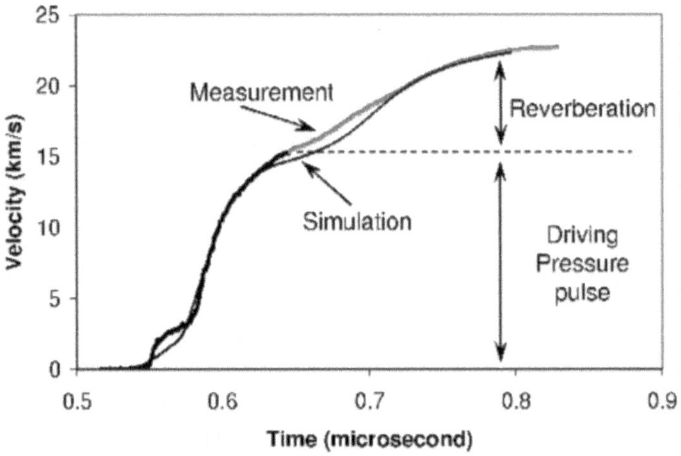

Fig. 4.12. VISAR measurement of an aluminum flyer plate launched to over 22 km/s

obtained using a conventional VISAR [12]. The agreement between the experiment and simulation is of the order of a few percent over the entire trajectory of the flyer plate launch, and the final velocity, corresponding to the impact velocity in a Hugoniot experiment, is in excellent agreement with the measurements.

The agreement between simulation and experiment over the entire trajectory indicates that several key details concerning the launch of the flyer plate are accurately reproduced in the simulations. In particular, the two portions of the flyer trajectory highlighted in Fig. 4.12 provide a good indication as to the peak current at the load, the magnitude of the current slightly beyond the peak, and the extent of magnetic diffusion and associated Joule heating. The initial velocity increase, over the first ~75–100 ns of the launch, is the result of the driving pressure pulse reaching the free surface (impact surface) of the flyer plate. This wave interaction results in acceleration of the free surface and the propagation of a rarefaction wave back into the flyer plate. Thus, the peak velocity at the end of this stage of the launch provides a measure of the magnitude of the driving pressure wave, and therefore the magnitude of the peak current at the load. Somewhat later in time the rarefaction reaches the portion of the flyer plate in which the majority of the current is flowing. This region is a few hundred μm into the flyer plate due to magnetic diffusion. The interaction of the rarefaction with the magnetic pressure results in a recompression wave that propagates back toward the free surface of the flyer. The magnitude of the magnetic pressure at the time of this interaction, which relates to the magnitude of the current at that time, dictates the magnitude of this recompression. When the recompression reaches the free surface, the free surface is again accelerated, and the flyer plate receives a second impulse driving it to higher velocity. Thus, the magnitude of the second increase in velocity provides a measure of the current at the load just beyond peak current. Finally, the extent of magnetic diffusion and Joule heating can be inferred from the sharpness of the reverberation. If there were very little diffusion and Joule heating, there would be a relatively sharp density gradient at the power-flow side of the flyer plate, and the reverberation would be quite pronounced. In contrast, if there were significant diffusion and Joule heating, the density gradient would be quite broad, and the resulting reverberation would be smeared out and difficult to identify. Given the level of agreement in the form of the reverberation between simulation and experiment, it appears that the extent of diffusion and Joule heating is well reproduced in the simulations.

In order to construct a fully predictive MHD capability that accurately predicts the performance of arbitrary anode/cathode configurations in the launch of high velocity flyer plates, two-dimensional MHD simulations must be performed [76]. Due to the significant deformation of the load that occurs during the rise time of the current pulse (the anode is accelerated across a ~3–4-mm gap during this time), the inductance of the load increases dramatically. Simulations must account for the coupling between the accelerator and the load in a self-consistent manner to accurately reproduce the load current, and therefore to accurately reproduce the time-dependent profile of the pressure wave that accelerates the flyer plate. This is accomplished by coupling an accurate circuit model of the accelerator to the two-dimensional simulation of the load [76].

The specific circuit, shown in Fig. 4.13, includes equivalent resistances and inductances for the accelerator, as well as a time-dependent model of current loss and a time-dependent resistance that emulates a short circuit, both of which occur upstream of the MHD load. Values for the inductances L_{m}, $L_{\mathrm{convolute}}$, and L_{im} are determined from geometrical considerations. The inductance of the MHD load is calculated self-consistently in the simulation using an effective transverse length representative of the actual experimental load. The circuit can be driven by a time-dependent accelerator voltage obtained from experimental measurements of accelerator performance where similar short circuit loads were used.

Results of these two-dimensional simulations are also in good agreement with experiment. In particular, both the simulated magnetically insulated transmission line current (MITL, the feed lines inside the vacuum section of the accelerator), denoted as I_{u} (upstream current), and the load current, denoted as I_{d} (downstream current), are accurately reproduced, as shown in Fig. 4.14. This level of agreement indicates that the circuit model accurately reproduces the details of the electrodynamics involved in the discharge of the accelerator. This is significant in that without explicitly taking the increase in inductance into account, the predicted peak current would be too large, and thus the driving pressure would be overestimated. Furthermore, as discussed

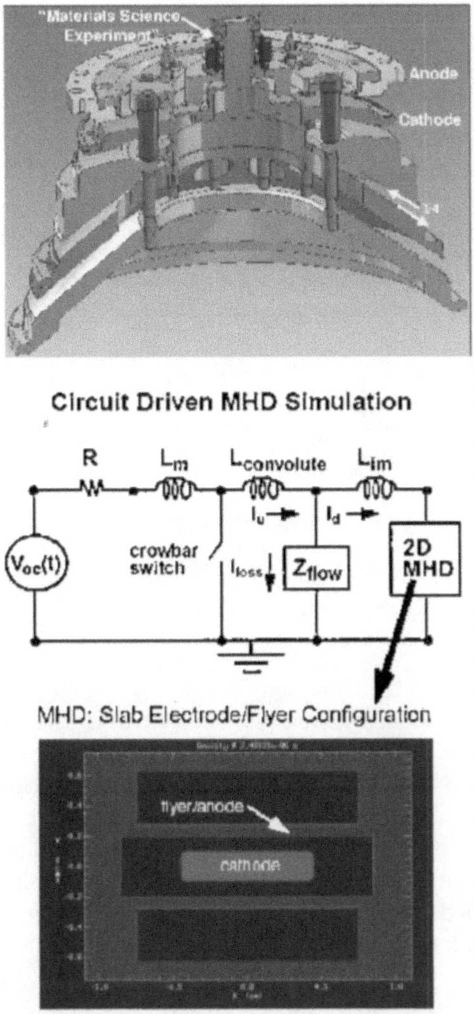

Fig. 4.13. Circuit model of the Z accelerator with a two-dimensional simulation as one of the circuit elements. In this case, the two-dimensional simulation is of a rectangular geometry short circuit load

above, the magnitude of the reverberation is dependent on the magnetic pressure just after peak current, when the release interacts with the drive surface. These key features are necessary in being able to accurately predict the performance of an arbitrary load configuration.

In developing the self-consistent, two-dimensional modeling capability for these high velocity flyer plate experiments, it was determined necessary to invoke a time dependent loss and crowbar model that emulates a short circuit at the convolute of the accelerator in order to obtain this level of agreement

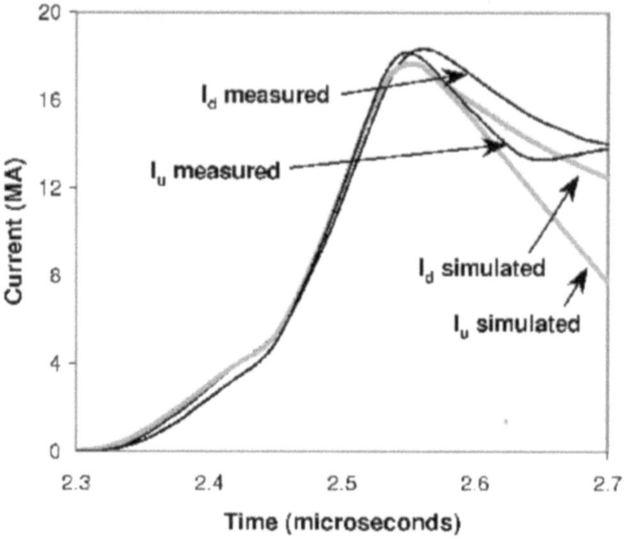

Fig. 4.14. Comparison of the simulated (*gray*) and measured (*black*) load current, I_{d}, and MITL current, I_{u}

between simulation and measurement. Evidently, due to the presence of magnetic nulls in the complex geometry of the convolute section (see Fig. 4.13), there is significant electron loss to the anode. Electron bombardment of the anode creates a plasma which eventually results in shorting of the A–K gap at a time just after peak current. The effect of this current loss and short circuit is twofold. First, the peak load current, I_{d}, is reduced somewhat due to the current loss upstream of the load. Second, after the short circuit occurs, flux is trapped downstream of the short circuit load, and the current subsequently decays with a time constant indicative of the inductance downstream of the short. This tends to result in a slower decay of the load current relative to the MITL current. Indeed this phenomenon is predicted in the two-dimensional simulations, as illustrated in Fig. 4.14. This slower decay of the load current is necessary to reproduce the magnitude of the reverberation observed experimentally. Thus, both experiment and simulation indicate the presence of a machine dependent loss impedance and short circuiting.

Using this simulation capability we have explored various load configurations in order to identify the optimal load geometry to launch high velocity flyer plates that are suitable for use in EOS experiments. The final design that is routinely used to launch flyer plates to the highest velocities (~22–27 km/s) is shown in Fig. 4.15. The nominal cathode outer dimensions are 9 mm wide by X mm thick (where $X = 4$, 2, or 1), and the nominal anode inner dimensions are 17 mm by X + 2 mm. Thus, the A–K gaps are 4 mm on the unused sides of the coaxial load and 1 mm on the flyer plate sides of the load. The thickness of the cathode is determined by the desired velocity of the

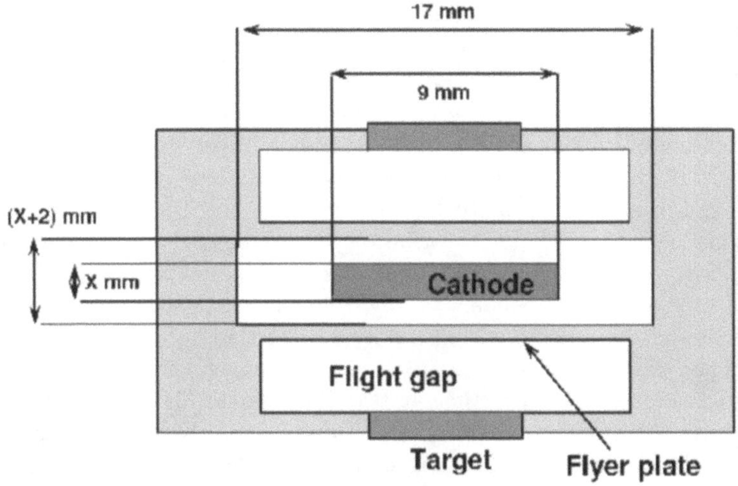

Fig. 4.15. Top view of the optimal load geometry used to launch flyer plates to over 22 km/s on the Z accelerator

flyer; 4-, 2-, and 1-mm cathodes result in approximately 22, 25, and 27 km/s aluminum flyer plate velocities, respectively. A few comments regarding this optimal geometry are warranted.

The optimal geometry is one that maximizes the current density on the power flow side of the flyer plate. The rectangular geometry with unequal A–K gaps satisfies this requirement. The load can be thought of as a circuit with four parallel inductance elements, i.e., the four sides of the rectangular geometry. The unequal A–K gaps result in unequal inductances for the sides of the load. By increasing the relative size of the A–K gaps on the unused sides of the load, the load inductance is significantly increased, resulting in a preference for the current to flow on the flyer plate sides of the coaxial load. To minimize the inductance increase, it is desirable to choose a high impedance material for the cathode stalk. The high impedance minimizes the deformation of the cathode, which results in the smallest possible increase in the load inductance during the launch of the flyer plate. Typically tungsten is used as the cathode material; however, stainless steel has also been used.

The velocity dependence on cathode thickness, with all other dimensions the same, can be understood by the fact that a thinner cathode corresponds to a lower overall current carrying surface on the anode, and a lower overall inductance. Both of these characteristics result in a larger driving pressure, and therefore a higher flyer plate velocity. The optimization of the cathode width is a compromise between the current density and two-dimensional effects resulting in flyer plate deformation. Clearly, a larger current density is achievable with a smaller cathode width. However, flyer plate deformation is also magnified with a smaller cathode width. The two-dimensional nature of the short circuit load results in magnetic pressure gradients across the horizontal direction of the flyer plate (perpendicular to the axis of the load). The uniformity of the magnetic pressure for the initial load geometry can be determined through use of the electromagnetic code, QUICKSILVER [95]. Examination of typical initial geometries indicate that the magnetic pressure is uniform to better than 1% over the central ~2/3 of the flyer plate. However, the pressure drops quite rapidly towards the corners of the cathode. These gradients cause velocity variations across the flyer surface, leading to non-uniform increase in the A–K gap that increases as the flyer is launched toward the target. Independent of any edge effects, this process gives rise to further velocity gradients within the flyer plate that result in further curvature of the plate. Consequently, the area over which the magnetic pressure is uniform is reduced. Nevertheless, two-dimensional MHD simulations that account for the time-dependent deformation of the A–K gap [76] suggest that the central ~3–4 mm width remains planar at impact for a configuration with a 9-mm-wide cathode.

In contrast, the magnetic pressure gradient hould be independent of height along the flyer portion of the coaxial short circuit load. There are variations as a function of height near the transition of the cathode from a radial feed to a coaxial feed (see Fig. 4.13), however, the current density is quickly redistributed, and becomes substantially uniform over the coaxial region of interest for these experiments. This suggests that there should be little to no curvature of the plate along the vertical direction; i.e., the principal curvature effects are limited to the horizontal direction.

Flyer Plate Design and Pulse Shaping

As mentioned above, a significant portion of the power-flow side of the flyer plate is vaporized due to magnetic diffusion and associated Joule heating. Thus, on the one hand, it is desirable to start with a relatively thick flyer plate to accommodate diffusion. However, the driving pulse also tends to steepen with propagation distance. Thus the flyer cannot be too thick or a shock will form in the driving pulse prior to the pulse reaching the free surface (impact surface) of the plate. These two considerations, diffusion and steepening of the driving pressure pulse, are conflicting requirements, and a suitable compromise must be found.

The process of optimizing the flyer plate thickness has been accomplished through one-dimensional MHD simulations. Representative snapshots of a typical simulation are presented in Fig. 4.16. Shown in the figure are the evolution of the hydrodynamic wave, the diffusion of the field and current into the flyer, and the density of the plate as a function of position. The snapshots in Fig. 4.16a, 4.16b, and 4.16c correspond to 0.69 μs, 0.74 μs, and 0.82 μs, respectively, in the 22 km/s launch shown in Fig. 4.12. At 0.69 μs the release wave has reduced the drive pressure to a local minimum, which occurs

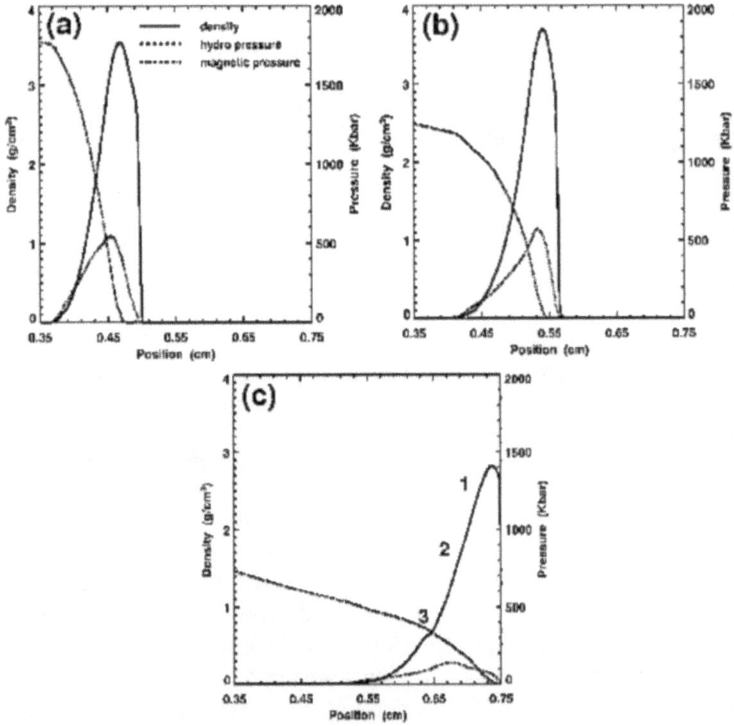

Fig. 4.16. Snapshots of the magnetic pressure, hydrodynamic pressure, and density profiles at three different times during the launch of a ∼22 km/s aluminum flyer plate. (**a**) $t = 0.69\,\mu$s, the time at which the release has reduced the hydrodynamic pressure to a local minimum. (**b**) $t = 0.74\,\mu$s, the time at which the reverberation is fully developed. (**c**) $t = 0.82\,\mu$s, the time just prior to impact. Also indicated in (c) are the solid–liquid (1), liquid–vapor (2), and vapor–plasma (3) transitions in the density profile

at 0.45 cm. The reverberation wave originates at this location, and is determined by the value of the magnetic field at that time. The fully developed reverberation wave is shown at 0.7 µs in Fig. 4.16b. Figure 4.16c corresponds to a time just prior to impact with a target. Note that the condition of the flyer plate at impact spans the states of plasma, vapor, liquid, and solid. The result of several simulations indicate that a rise time of ∼200 ns is optimal for the acceleration of the flyer plate; this rise time mitigates the magnitude of current diffusion, while maintaining the ability to propagate a ramp through the flyer plate without shock formation. Based upon these simulations we have settled on an initial flyer plate thickness of ∼850 µm. Of this, approximately 350–400 µm remains unaffected by diffusion and represents the usable portion of the plate for impact experiments.

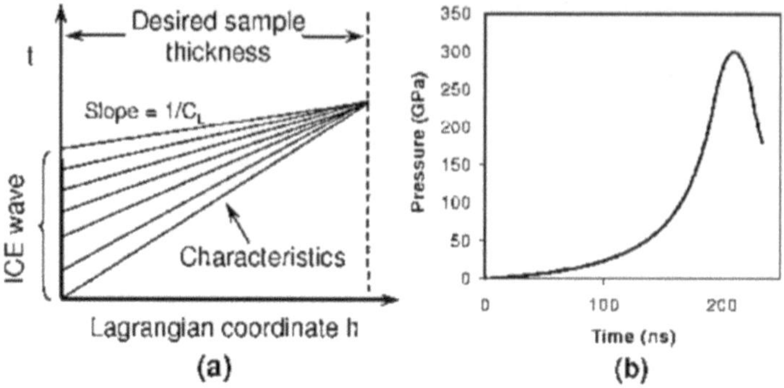

Fig. 4.17. (**a**) An x–t diagram showing characteristics emanating from a 200 ns shaped ramp. (**b**) Ideal pressure profile to launch an aluminum flyer plate to ultra-high velocity

The optimal pulse shape for accelerating the plate is shown in Fig. 4.17. This pulse is considered to be the ideal ramp wave profile in that the propagation of the ramp will result in an abrupt shock formation at a single plane in the material. Such an ideal profile maximizes the propagation distance prior to shock formation for a pressure pulse having a given rise time. The ideal profile can be determined by considering a position–time diagram, as illustrated in Fig. 4.17a. It is desired that all of the characteristics converge at a single time and position in the sample. Projecting these characteristics, which have a slope given by the inverse of the Lagrangian wave speed at the particular pressure, backwards in both space and time to a given location in the material results in the desired optimal pressure–time profile. As an example, the optimal profile determined in this way for a ~300 GPa ramp wave in aluminum is shown in Fig. 4.17b. In this case the rise time is 200 ns, and the shockless propagation distance is ~1.28 mm. Since the flyer plate thickness is 850 μm, the driving pressure pulse remains shockless for the entire propagation through the plate, and ensures that the plate experiences only quasi-isentropic loading and release. Hence, the condition of the plate at impact is essentially ambient temperature and density, as illustrated in Fig. 4.16c.

We note that we have also performed experiments with composite flyer plates of aluminum and titanium [72]. In this case the aluminum layer, which is typically 500-μm thick, is sacrificial, acting as the power-flow electrode; nearly the entire aluminum portion of the flyer plate is consumed by field diffusion and Joule heating. The titanium layer, typically 300-μm thick, remains unaffected by the field and is at essentially ambient temperature and density at impact. We note that this type of composite flyer should enable sound speed measurements to be performed, since the wave interaction at the aluminum titanium interface has been shown to provide a prominent rarefaction wave that propagates back toward the shocked sample.

These simulations also allow us to determine the optimal flight distance. In particular, the accurate reproduction of the flyer trajectory enables the acceleration profile to be predicted with confidence. The flight distance can then be chosen such that the acceleration of the plate is minimized. At that point the residual pressure in the plate is minimized, and the plate is nearly at ambient conditions. For flyer plate velocities of ~22–27 km/s the optimal flight distance is ~3.75–4.5 mm.

Results

Figure 4.18a shows the results achieved for aluminum flyer plates on the Z accelerator using the geometries discussed in the previous sections. Shown are representative velocity profiles for aluminum flyer plates for experiments from ~9–27 km/s. The details regarding the load configurations (anode and cathode dimensions) and charge voltages are listed in Table 4.1. As indicated in the table, the flyer velocity was varied by altering the geometry of the short circuit load and/or the charge voltage of the accelerator. Similar results for the aluminum/titanium flyer plates are shown in Fig. 4.18b. The experimental details for these composite flyers are also listed in Table 4.1.

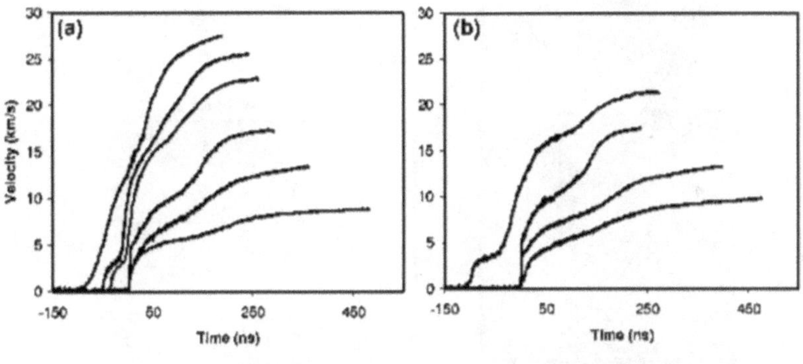

Fig. 4.18. VISAR measurements of aluminum (**a**) and titanium (**b**) flyer plates. Note that the velocity profiles are time shifted for clarity

The planarity of the flyer plate at impact has been verified experimentally. Figure 4.19 shows line-imaging VISAR [101] measurements of stepped aluminum target plates impacted by an aluminum flyer at ~20 km/s. The line-imaging VISAR diagnostic images the sample of interest through an interferometer and onto the slit of a streak camera. The interferometer is slightly de-tuned to produce a line interference pattern across the image, with the interference lines perpendicular to the slit of the streak camera. The streak camera then provides a time-resolved measure of fringe displacement, which is proportional to velocity, along a line of ~7 mm across the sample of interest. In these experiments the sample of interest was a stepped aluminum

Table 4.1. Flyer plate configurations. Cathode and anode geometry are in mm, charge voltage is in kV, and flyer velocity is in km/s

Short number	plate configuration	Cathode geometry	Anode geometry	Charge voltage	Flyer velocity
Z668	Al	11 × 11	15 × 15	60	9
Z698	Al	11 × 11	15 × 15	74	13
Z792	Al	11 × 11	15 × 15	90	16
Z1201	Al	9 × 4	17 × 6	90	22
Z1136	Al	9 × 2	17 × 4	90	25
Z1238	Al	9 × 1	17 × 3	90	27
Z1014	Al/Ti	11 × 11	15 × 15	65	9
Z1012	Al/Ti	11 × 11	15 × 15	90	13
Z1109	Al/Ti	9 × 4	17 × 6	90	17
Z1110	Al/Ti	9 × 2	17 × 4	90	21

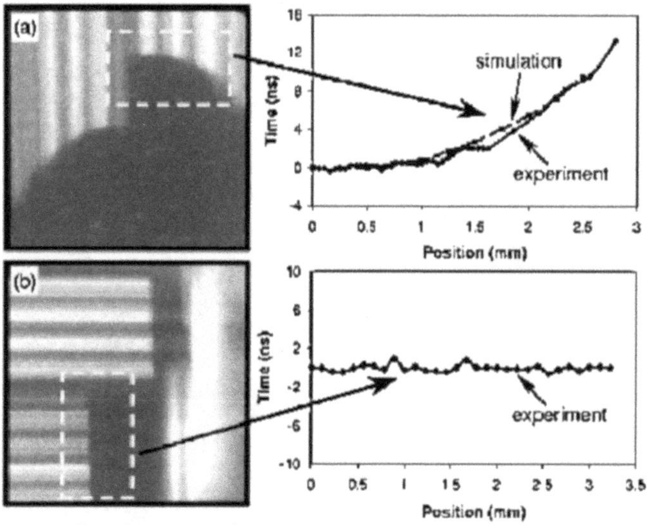

Fig. 4.19. Streaked line-imaging VISAR records of shock break out in stepped aluminum targets impacted by an aluminum flyer plate at ~20 km/s. Orientations of the streak camera slit relative to the target image was (**a**) perpendicular to the coaxial load, and (**b**) parallel to the coaxial load. In each case the spatial extent of the image was ~7 mm

target with dimensions of 300 and 900 μm on the thin and thick portion of the step, respectively.

In both images shown in Fig. 4.19, the parallel nature of the fringes prior to loss of reflectivity indicates that there was no motion of the rear of the stepped targets before arrival of the main shock (in this case ~500 GPa). This result suggests that the impact of the flyer with the target produced a sharp shock input to the sample. As expected, curvature was observed in the shock arrival at the rear of the stepped target when the streak camera slit was oriented perpendicular to the coaxial load (Fig. 4.19a). Analysis of the extent of curvature after the shock traversed the 300 μm step of the aluminum target indicates that the flyer plate was essentially planar over roughly 3–4 mm at impact, in very good agreement with the two-dimensional MHD simulations of the experiment [76]. Also, as expected, no curvature was observed when the slit was oriented parallel to the coaxial load (Fig. 4.19b). Based upon these findings, diagnostics in EOS experiments are positioned along a vertical strip in the center of the target to ensure that measurements are being made within the rectangular area encompassing the planar region of the flyer plate (~3–4 mm width by ~25 mm height).

The best indication of the constancy of stress produced at impact by these magnetically accelerated flyers is from data obtained in experiments performed on cryogenic liquid deuterium [69, 70, 73]. These experiments will be described in more detail in the following section. Both VISAR measurements made directly off the deuterium shock front and measurements of self-emission intensity as a function of time indicate that the drive pressure remains constant to better than ~1–2% over the timescale of the experiment, approximately 30–40 ns.

As indicated by the data, the use of pulsed power techniques to launch flyer plates to high velocity is a rapidly maturing technology. Experimental data indicate that the quality of the impact is excellent; impact occurs after the plate acceleration has essentially ended, the plates are intact and planar, and the resulting pressure is essentially constant for times of order several 10s of ns. The plates are relatively large, with a planar region of dimensions ~3–4 mm by essentially 25 mm. Furthermore, two to four plates are launched for each firing of the accelerator The multiple flyer plates and the relatively large size enables multiple targets and numerous diagnostics to be fielded, resulting in unprecedented accuracy in EOS measurements at several hundred GPa.

4.4 Applications

4.4.1 Gun Systems

The development of the high-speed launcher, also referred to as the three-stage light-gas gun, with capabilities approaching 16 km/s, can be used for well-controlled EOS studies of materials shocked to high pressure and temperature regimes never before accessible in the laboratory. Since titanium alloy Ti-6Al-4V flyer plates are used routinely, the three-stage launcher has been used to extend its shock EOS to 250 GPa [88]. New experimental techniques and diagnostic capabilities have also been developed to allow time-resolved measurements of shock loading and release wave profiles resulting from impact at over 10 km/s [35,47]. Melting was inferred from these studies.

These capabilities were also applied for use with applications such as shock-induced vaporization studies [19,36] and to assess impact damage to substructures [34, 36]. In this section, the two applications that will be highlighted are the techniques that have been established for EOS [88] and shock-induced vaporization measurements [19, 36].

EOS Measurements

The experimental premise for the EOS experiments is determined by the Rankine–Hugoniot relations [92], which are based on the conservation of momentum, mass, and energy across the shock front as it travels through matter. These relations are

$$P - P_0 = \rho_0 U_s u_p \tag{4.4}$$

$$V = V_0 \left[1 - (u_p/U_s)\right] \tag{4.5}$$

$$E - E_0 = \tfrac{1}{2}(P + P_0)(V_0 - V) \,, \tag{4.6}$$

where P_0, ρ_0, V_0, and E_0 are initial pressure, density, specific volume, and specific internal energy of the material. The quantities P, V, and E are final shock pressure, specific volume, and specific internal energy, respectively, and U_s is the velocity of the shock wave and u_p is the mass velocity behind the shock front. Thus an accurate determination of the shock velocity U_s and the mass velocity u_p are essential to determine the EOS, namely the pressure–volume compression curve of the material.

Shock wave velocities are determined by measuring shock transit times over a known distance between two parallel planes in the target. In these experiments, the planes are about 2 mm apart and shock transit times are of the order of 150–200 ns measured with two types of time-of-arrival detectors (anodized pins and needles). To achieve these objectives a "two-stepped-target" (Fig. 4.20) is used. As shown in the figure, and unlike previous studies [83, 84], two different sized holes were drilled into the target on two separate diameters. The outer diameter, which is referred to as the A plane, has twelve locations for six anodized self-shorting pins and six 100-μm diameter optical fiber "needles". The fibers are inserted into a 500 μm hypodermic tube (hence the reference to fibers as "needles"). The center of the target accepts either a pin or a needle, which bottoms out on the A plane. Inside the A plane diameter are twelve more pin and needle locations, distributed over the B plane and located at the same angular positions as the A-plane pins and needles.

The locations of the pins and needles are such that the edge release disturbance starting from the A plane pin or needle does not interfere with those detectors on the B plane. Electrical pins work extremely well in high-pressure regimes attained from conventional two-stage and three-stage light-gas gun impact velocities. In addition to the anodized pins, fiber-optic needles were

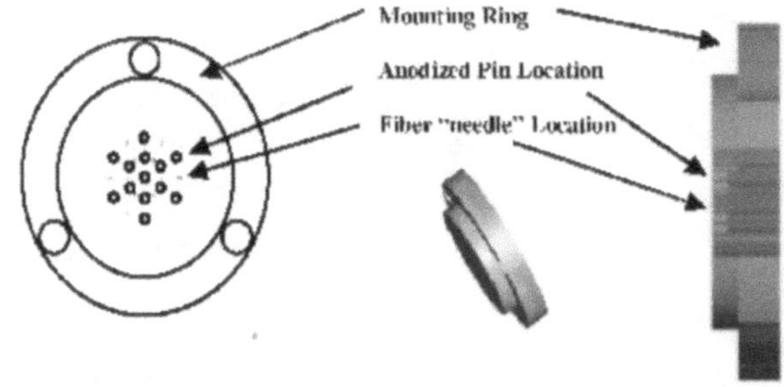

Fig. 4.20. Top-hat configuration used for EOS measurements on the three-stage light-gas gun. Two different types of pins are used – fiber optic and anodized pins

utilized as passive time-of-arrival detectors. Since the diameter of the fiberoptic cables is smaller than that of the electrical anodized pins, they become ideal detectors, and have the potential of being employed in smaller diameter samples to measure transit times smaller than those used in this study – and to the same degree of precision.

The fibers consist of Polymicro 100-μm diameter, pure fused silica core inserted into a 500-μm tube and its end coated with a thin metal layer, normally aluminum, thick enough to shield any extraneous light. As the shock front arrives at the tip of the fiber, the fiber will luminescence [105, 106] and is diagnosed using low-noise photomultiplier units electronically modified to obtain 1 ns rise times. Embedding the fiber/needle assembly in the target ensures that all of the fibers have the same external configuration so that the mechanism of luminescence when the stress front arrives is identical in all instances. Anodized pins are placed in a tight tolerance drilled hole to shield them from extraneous sources such as ionized gases or ejecta from the material.

The outputs of the fiber-optic signals are shown in Fig. 4.21 indicating the arrival of the shock front arrival on both the A and B planes. Notice that the time of closure resulting from tilt over an impact diameter of 13 mm is less than 12 ns. Likewise, the B plane fiber closure time indicates less than 6 ns closure time over a diameter of 6.5 mm.

The experimental records suggest that the flyer is axially symmetric; the two circular arrays of needles produce the same tilt angle of the impactor (with respect to the target), and the impact distortion is parabolic. The data from the fibers indicate a flyer plate bow of less than 5 ns and a corresponding tilt of about 8.5 mrad. The precision of our diagnostic system is better than a nanosecond, which translates to a shock velocity measurement of approximately 11 ± 0.05 km/s for this experiment. Comparing the fiber-optic-needle data with the anodized pin data provided redundancy on the experiment while validating the new technology of fiber-optic needles to obtain shock velocity to within 0.1%.

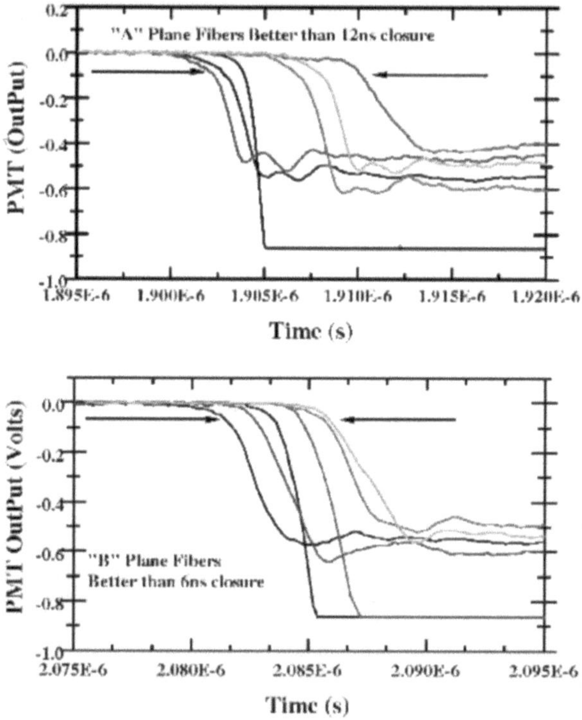

Fig. 4.21. Output histories of the fiber-optic luminescence signals that are generated upon shock arrival. The two measurements at A and B pin locations yield the same tilt and bow information

As mentioned above, a time-dependent pressure pulse is used to launch titanium alloy Ti-6Al-4V flyer plates. As a result of using quasi-isentropic compression pressures approaching a ~100 GPa, the flyer plate will experience a temperature increase. This temperature increase was measured for the flyer plate that attained a peak terminal velocity of 9.8 km/s and 10.8 km/s, respectively, using a four channel fiber-coupled infrared pyrometer. Conservatively, the temperature did not exceed 850 K when the flyer reached a terminal velocity of 10.8 km/s, or 650 K when it reached a velocity of 9.8 km/s. To use impedance matching techniques, a Hugoniot for the warm flyer plate material must be used to determine the particle velocity in the target plate. Based on the measured thermal expansion coefficients, the new density of the warm flyer plate can be determined. With this value, the predicted densities are 4.368 and 4.341 g/cm^3 at approximately 650 and 850 K, respectively. A detailed analysis [88,110] has indicated that it is accurate to use the new density, and a new coefficient c, the coefficient in the $U_s - u_p$ relation to represent the Hugoniot for the warm material. The slope of the $U_s - u_p$ curve is very nearly the same in each case.

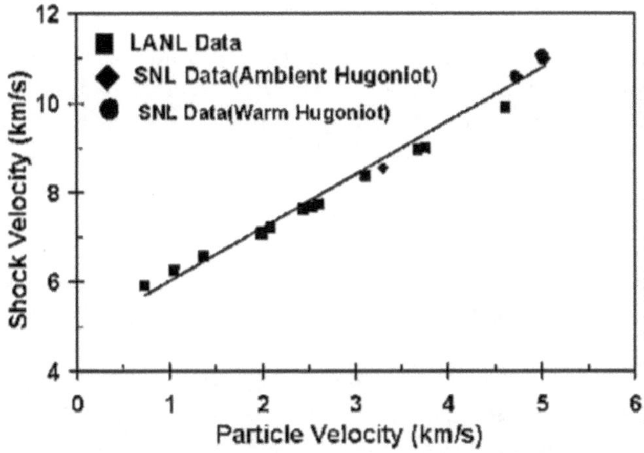

Fig. 4.22. Shock velocity vs. particle velocity measurements obtained on the three-stage light-gas gun for a titanium alloy Ti-6Al-4V plate for impact velocities over 11 km/s

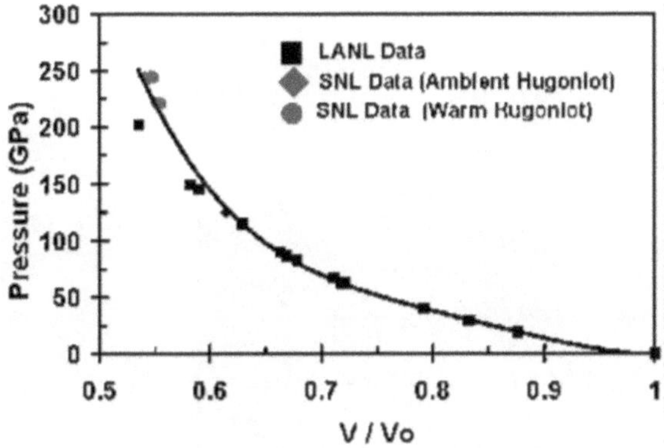

Fig. 4.23. Stress strain behavior of titanium alloy Ti-6Al-4V as obtained from three-stage light-gas-gun measurements

To summarize, the three-stage launcher has been used to estimate the EOS of materials at impact velocities over 11 km/s (see Figs. 4.22 and 4.23.) Note the higher pressure response deviates from previous measurements up to 200 GPa. This has been attributed to melting in the shocked state [63], and is also consistent with the recent results obtained on the Z machine using magnetic flyer plate techniques [70].

Shock-Induced Vaporization Measurements

Historically, experimental efforts to understand high-pressure melting and vaporization have been hindered by unavailability of experimental launchers that are capable of speeds needed to induce vaporized states [20]. This problem has been circumvented to some extent by studying materials such as lead, cadmium, and zinc, which have relatively low melting and boiling points [7, 8, 85]. For materials of greater interest (such as aluminum), an alternative is to shock porous samples for which irreversible pore collapse enhances heating of the matrix material [65,112]. Using the three-stage light-gas gun, a tantalum flyer plate was launched to speeds from 8 km/s to 10.1 km/s to impact a thin target plate of zinc, producing shock pressures of about 300 to over 550 GPa, and temperatures as high as 39000 K (~3.4 eV). Zinc was chosen, because the zinc liquid–vapor co-existence diagram [19] suggests the feasibility of approaching the critical region using the three-stage light-gas gun technology. The release isentropes computed from these states pass near the thermodynamic critical point of zinc, and it is therefore believed that significant, perhaps complete, vaporization of the zinc target occurred (see Fig. 4.24).

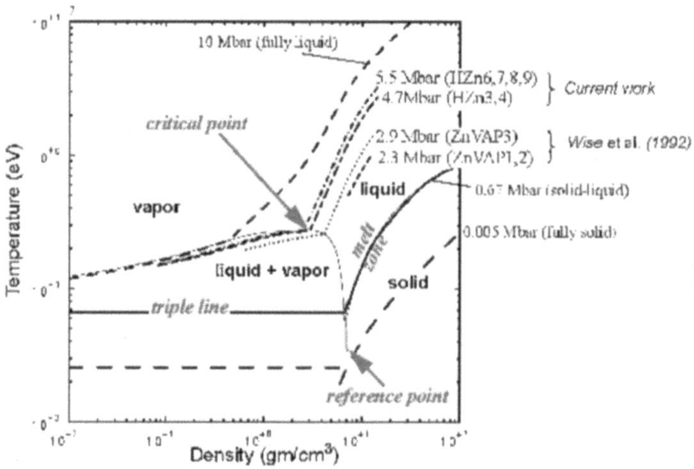

Fig. 4.24. Phase diagram for Zinc suggesting that full vaporization is achieved at a stress level of 5.5 Mbar

The vapor states are characterized by measuring the stagnation history of the expansion products against a witness plate using a velocity interferometer. The experimental configuration is shown in Fig. 4.25. The target and witness plate were approximately 35 mm in diameter. The LiF window was approximately 22 mm in diameter and 25 mm thick in the direction of motion. To maintain one-dimensional impact conditions, the tantalum flyer traversed a gap of 30 mm prior to impacting the zinc target plate. Impact of the flyer plate against the zinc plate produced a debris cloud of rarefied liquid–vapor zinc, which was permitted to traverse a gap of known dimensions before stagnating against an aluminum witness plate. The subsequent

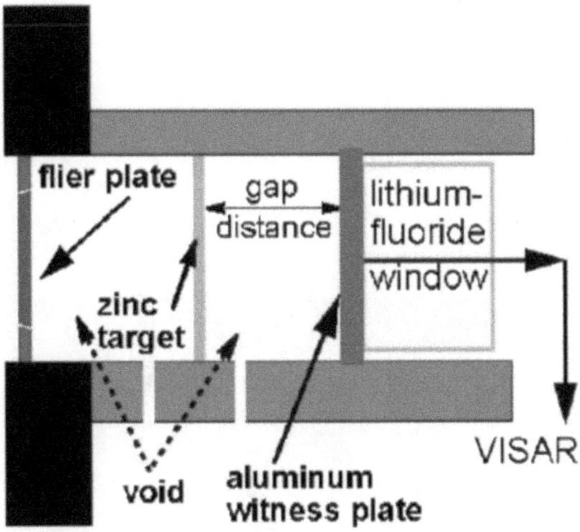

Fig. 4.25. Experimental configuration used for shock-induced vaporization studies on the three-stage light-gas gun

particle velocity history at the witness plate/window interface [111] was measured using a velocity interferometer [12]. Measurements of stagnation velocity history of experiments conducted at impact velocities of 10 km/s (shocked to 550 GPa) are shown in Fig. 4.26. Note that the results show that the peak velocities decrease whereas the rise times increase with increasing gap. This is because the expansion products consist of mixed aerosol/liquid exhibiting both a velocity and density distribution. The leading edge of the debris cloud consists of fast-moving low-density vapor products followed by slower-moving higher-density products, leading to a time-dependent loading on the aluminum witness plate and consequent finite rise time measurements. The phase separation as a function of propagation distance results in increased rise times. The amount by which peak witness plate velocity decreases for increasing distance between the zinc target and witness plate is an indicator of the degree of vaporization.

The aluminum witness-plate/LiF window can be regarded as a target with which the liquid/vapor debris cloud interacts. The peak interface velocity measurement, u_{wp}, is an indicator of the maximum stress resulting from this interaction. Although not shown, the measured peak velocities are higher for the lower-speed 9.1 km/s shots than for the corresponding 10.1 km/s shots [19], which suggests that greater vaporization occurred in the 10.1 km/s shots. Figure 4.27 shows that the peak witness-plate velocity, u_{wp}, and therefore the target/debris interaction stress, decrease monotonically with increasing propagation distance (gap size). All curves in Fig. 4.27 must – in one-dimensional theory – asymptote to some constant value as the gap size is increased. When such a curve asymptotes very nearly to zero, the sample must have vaporized completely. When a curve asymptotes to some

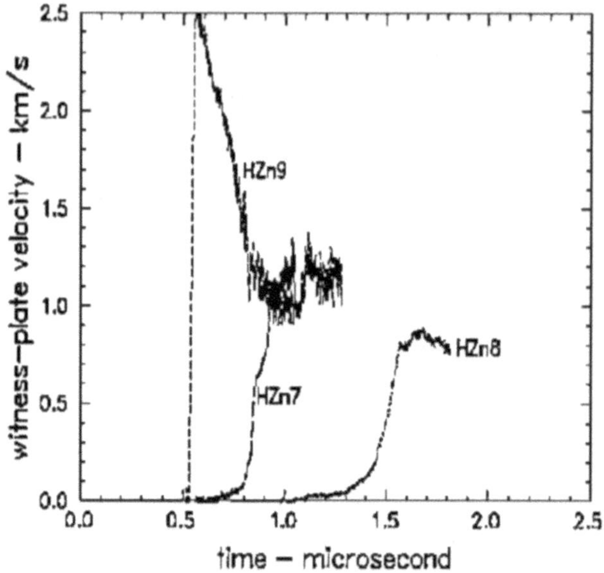

Fig. 4.26. Stagnation particle velocity history measurements for zinc for various gap dimensions at the aluminum/window interface. The experimental configuration is shown in Fig. 4.25

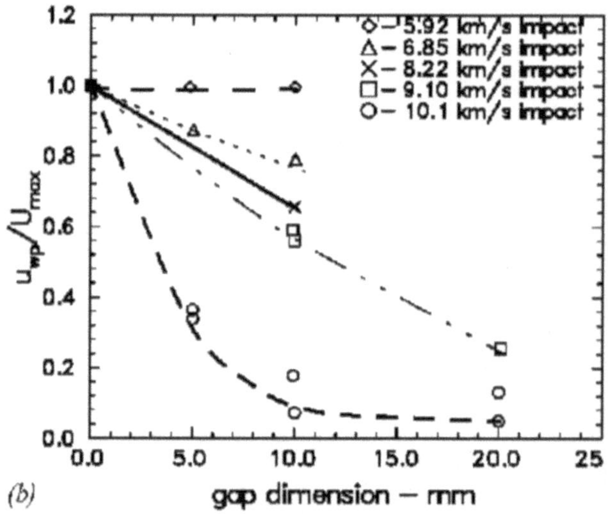

Fig. 4.27. The ratio of the measured particle velocity history to the anticipated peak particle velocity for zero gap vs. gap dimension. No variation implies lack of vaporization kinetics. Any departure from unity implies the concentration of vapor falls in the mixed phase. The slope of the line determines the rate of the vaporization process

non-zero value of u_{wp}/U_{max}, the sample must have vaporized only partially. The velocity U_{max} is the anticipated peak particle velocity measurement for zero gap distance. When a curve is constant, no vaporization occurred and the maximum interaction stress will be independent of gap size. In Fig. 4.27, the lowest speed experiment (5.92 km/s) [112], exhibits negligible expansion. That is, the zinc target remains essentially intact as it crosses the gap. In contrast, the highest speed (10.1 km/s) experiment shows considerable expansion of the zinc, which corresponds to a much lower stress on the buffer than the lower-speed lower-vaporization experiments. Even though loading stress can be reduced substantially by vaporization, the survivability of any target nevertheless depends on many other parameters, including the duration of the pressure pulse, the thickness of the target, and the yield and fracture strength of the target. For zinc, the rapid approach to an asymptotic limit suggests that boiling occurs more rapidly from super-critical states. A detailed discussion of both the experiments and the limitations of the current vaporization model, the reader should refer to the article by [19].

For symmetric impact velocities of 10 km/s, aluminum is shocked to 160 GPa and is molten in the shocked state [35,47]. When impacted by titanium at a velocity of 11.1 km/s, aluminum is shocked to over 230 GPa [36]. In this study, the velocity history produced by stagnation of the aluminum expansion products against an aluminum witness plate is measured using velocity interferometry, as shown in Fig. 4.25. In addition, x-ray measurements of the debris cloud are recorded prior to stagnation against the witness plate. Shocks followed by rapid releases are observed at an aluminum/lithium fluoride window interface [36], suggesting that thin "plate-like" debris is interacting with the witness plate even though the debris was allowed to propagate through a 50-mm gap. The stagnation particle velocity history is shown in Fig. 4.28. The magnitude of the loading suggests peak stresses approaching 230 GPa. This is confirmed by the radiographic measurements of the impact-generated products, which confirm that the expansion debris is plate-like (see Fig. 4.29). This relatively high density, plate-like debris generates stresses of several hundred GPa when allowed to interact with the witness plate. Even though the debris plate expands over 40 mm, it does not disintegrate, suggesting that the vaporization process is time dependent. However, when aluminum is shocked to lower pressures, results have indicated little or no vaporization products when expanded over similar distances. Both radiographs and witness-plate velocity measurements suggest that the vaporization process is both time dependent and heterogeneous when the material is released from shocked states at or below 230 GPa. Experiments strongly suggest that the vaporization kinetics in aluminum should become dominant when expanded from shocked states over 230 GPa.

For zinc, the latent heat of vaporization is approximately 2 kJ/g. The latent heat of vaporization for aluminum, however, is 11 kJ/g. Unlike zinc, symmetric impact velocities approaching 30 km/s would be necessary to obtain

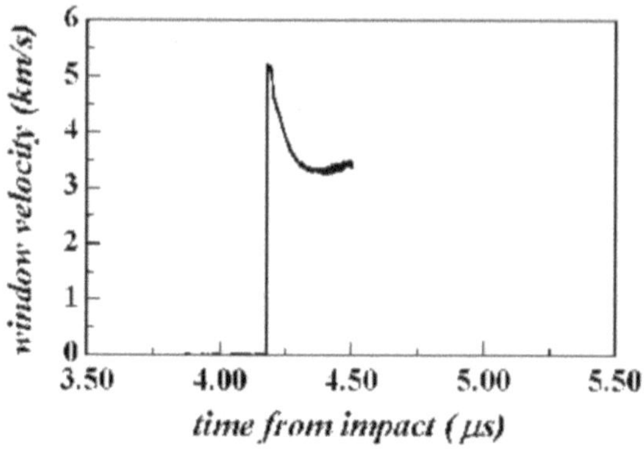

Fig. 4.28. Measured stagnation particle velocity history measurements for aluminum are shown. The shock loading and release waveform is characteristic of a thin flier plate impact

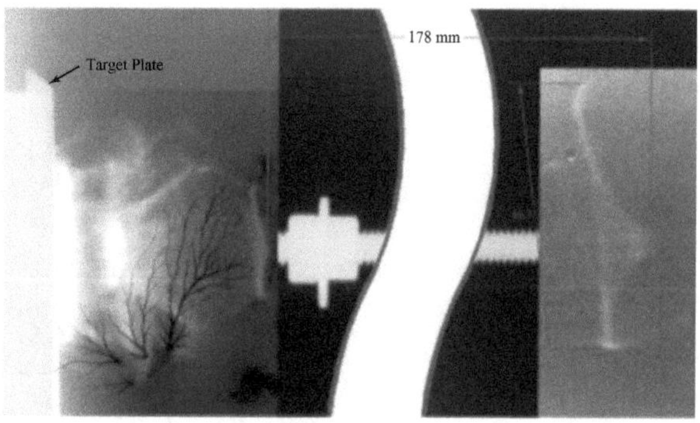

Fig. 4.29. X-radiographs of a "vaporized" aluminum plate after traversing a gap dimension of 45 mm and over 210 mm. Impact-generated flier-plate-like debris suggests time dependent vaporization kinetics

full vaporization of aluminum. This is achievable using the magnetic flyer plate techniques described above. These are the first quantitative measurements of shock-induced vaporization processes made using the three-stage light-gas gun. For zinc, the time-dependent kinetics leading to full vaporization is observed whereas, for aluminum, the threshold pressure of 230 GPa appears to be the onset for time-dependent vaporization kinetics. For detailed discussions of both the experiments and the limitations of the current vaporization model, the reader is referred to the original articles both for zinc [19] and aluminum [36]. At 16 km/s impact, high pressures of 2 TPa (20 Mbar) and temperatures of 10 eV are attainable in (high-impedance) engineering materials such as tungsten or steel. Isentropic expansion of these engineering materials shocked to such a high pressure molten state would result in partial vaporization. Taking advantage of the nearly 30 km/s impact velocity available using magnetic flyer plate techniques would allow us to probe the vapor dome for most engineering materials.

4.4.2 Magnetic Systems

To date, the predominant application of the high velocity magnetically accelerated flyer plate technique has been for accurate EOS experiments in the several hundred GPa regime. The ability to launch intact, well characterized, and relatively large flyer plates to velocities approaching 30 km/s enables experiments to be performed with accuracies near those achieved using gas guns at pressures of the order of five times that achievable with conventional gun techniques, and over two times that achievable with the three-stage light-gas gun technique [30, 32].

The initial experimental campaign was focused on symmetric impact experiments on aluminum [71]. Due to the extensive study of the response of aluminum at high stress [1–3, 48, 74, 83, 95, 108], the results obtained in the present study could be independently checked against techniques used previously. Agreement with existing high-stress data would help to establish this technique as a suitable shock wave driver. In conjunction with the aluminum experiments, Hugoniot experiments were also performed on liquid deuterium [69, 70, 73]. Results of these experiments have helped to resolve a discrepancy in the dynamic response of deuterium in the pressure range of ~50–150 GPa. These applications will be described in the following sections.

We have also begun to explore the use of this high-velocity flyer capability to study relatively hot matter at densities near ambient – so called warm dense matter. The relatively large size of the magnetically accelerated flyer plates allow shock and release experiments to be performed with multiple diagnostics to probe the released state. Due to the significant temperature increase associated with high-pressure shock compression, the released states achieved are near the liquid–vapor critical point. In particular, release states from Hugoniot pressures of ~1 TPa (10 Mbar) in aluminum (corresponding to ~30 km/s symmetric impact) are predicted to intersect the vapor dome very near the critical point. This region of the EOS surface is not well characterized for most materials. Preliminary shock and release experiments in aluminum aimed at probing this region of the EOS surface will be discussed briefly in the next section.

Aluminum Symmetric Impact

The experimental configuration used to obtain symmetric impact data on aluminum using the magnetically driven impact technique [71] is shown schematically in Fig. 4.30. Shock waves were generated by planar impact of an aluminum alloy 6061-T6 flyer plate onto a target of the same material mounted in the anode panel. The rectangular flyer plate was accelerated across the

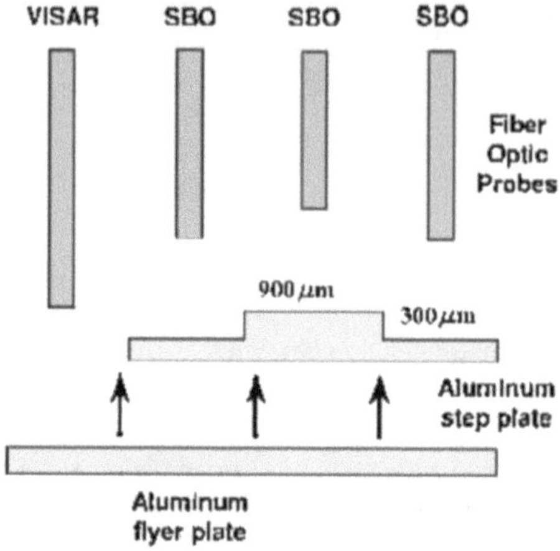

Fig. 4.30. Experimental configuration for symmetric impact experiments on aluminum

nominal 3–4 mm flyer gap by the magnetic field. A conventional, single-point VISAR [12] was used to directly measure the velocity history of the flyer plate, from launch to impact, with an accuracy of ~0.5–1%.

Two different target designs were used in these experiments; a top-hat design or a single-step design. The top-hat design, illustrated in Fig. 4.30, consisted of a ~10-mm diameter plate of 300 μm nominal thickness, with a raised central area ~3 mm in diameter and 900 μm nominal overall thickness. The single-step design consisted of a ~10 mm diameter disc, half of which had a nominal 300 μm thickness and half a nominal 900 μm thickness. In all cases, the targets were prepared by diamond turning the surfaces to the desired dimensions. The surfaces were flat to ~200 nm and parallel to ~2 μm with ~20 nm surface finishes. Step heights were measured to an accuracy better than 1 μm using interferometry techniques. Nominal initial densities of the targets were 2.699 g/cc.

The shock response of the aluminum samples was diagnosed with a fiber-optic-coupled-shock-break-out (FOSBO) diagnostic. The FOSBO diagnostic is sensitive to the sudden change in reflectivity that occurs upon emergence of a strong shock. Light is reflected off the surface of interest and collected into an optical fiber that delivers the reflected light to a streak camera. The streak camera provides high-time-resolution recording of the signal. Typically, the streak rate of the camera was set to provide ~100–200 ns of recording time, resulting in a temporal resolution of ~0.5–1 ns. In some cases, a line-imaging VISAR diagnostic [101] was used for shock timing. Here, the target image was directly coupled to the streak camera. For the line-imaging VISAR experiments the streak camera was set to provide ~50–100 ns of total recording time, resulting in a temporal resolution of ~0.25–0.5 ns.

Several pairs of 100-μm diameter fibers were used, allowing FOSBO measurement of shock arrival at several locations on the aluminum sample. Multiple FOSBO measurements were necessary to assess and account for impact tilt (typically less than ~4 mrad), mainly due to slight misalignment of the target relative to the flyer plate. For the top-hat-design target, FOSBO measurements were made on either side of the raised central area in addition to the top of the central area. For the single-step design two FOSBO measurements were made on each side of the step. In both cases, differences in the time of shock break out could be compared over distances of several mm, enabling the magnitude of the tilt to be accurately determined and accounted for in the determination of the shock velocity in the target. In the case of the line-imaging VISAR measurements, impact tilt was clearly observed and accounted for by noting the break out as a function of time across the target. These measurements enabled the shock velocity to be ascertained to an accuracy ranging from 1–3%.

The $U_s - u_p$ Hugoniot data for aluminum are plotted in Fig. 4.31a. Also plotted in the figure are data from the compendium published by Los Alamos National Laboratory [78] (chemical-explosive driver), [84] (gas-gun), [1–3,48, 72] (explosively driven striker plate), [97, 108](nuclear explosion). The solid line is the predicted $U_s - u_p$ response from SESAME 3700 [61, 62]. The σ–ρ Hugoniot data are plotted along with predictions from several SESAME EOS models [56] in Fig. 4.31b.

Several points are apparent from these comparisons. First, Fig. 4.31 illustrates the fact that the various SESAME models exhibit rather significant differences in the high-stress response, particularly above the limit of the gas-gun data(~170 GPa) [84]. Thus, the data generated in these experiments are useful in distinguishing among the various models. Second, the results obtained using this technique are in excellent agreement with the previous data over the pressure range of 100–500 GPa. In particular the lowest stress results from this study (~120–130 GPa) agree very well with the results reported by [84], obtained in very accurate two-stage light-gas-gun experiments. In addition, data from the present just above the gas-gun regime (about 160 GPa) show a decrease in the slope of the $U_s - u_p$ response, in good agreement with previous results [1–3, 48, 74, 97, 108]. This decrease in slope is likely due to the melting of aluminum on the Hugoniot, as expected from the EOS calculations. This conclusion is further reinforced from a previous study utilizing release measurements on a three-stage gun, which suggests that shock-induced melting occurs in the stress range of 125–150 GPa [35, 47]. The Hugoniot below this stress range is indicative of the response of solid aluminum, whereas the Hugoniot above this stress range is indicative of liquid aluminum, with a correspondingly lower slope in $U_s - u_p$. Finally, as shown in Fig. 4.31, the present data exhibit higher accuracy and less scatter than previous data obtained in the stress range of 200–500 GPa, lending higher confidence to the understanding of the high-stress response of aluminum. We also note that

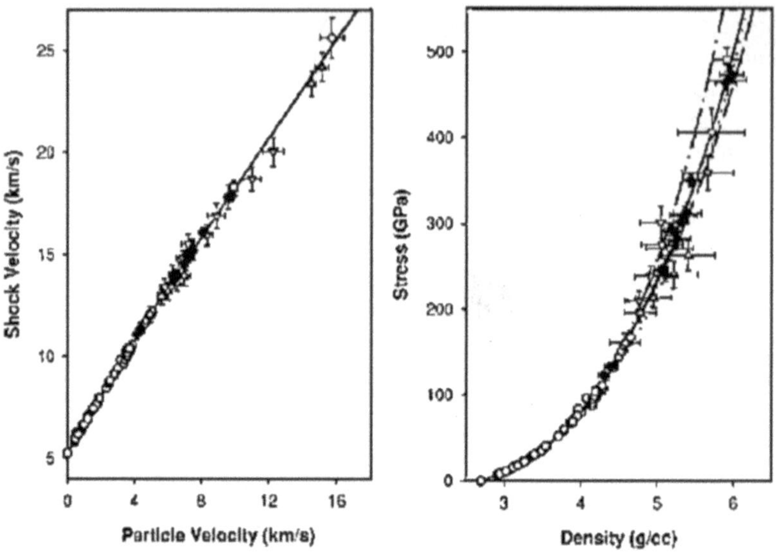

Fig. 4.31. $U_s - u_p$ and $P - \rho$ Hugoniots for aluminum: This work (*black diamonds*); LASL compendium (*open circles*); Mitchell and Nellis (*gray circles*); Al'tshuler and Kormer (*gray squares*); Kormer (*open square*); Al'tshuler and Chekin (*open triangles*); Al'tshuler and Kalitkin (*gray inverted triangle*); Glushak (*open inverted triangles*); Volkov (*open diamond*); and Simonenko (*gray triangles*). Theory: SESAME 3700 (*solid line*); 3711 (*dot-dashed line*); 3715 (*dashed line*); and 3719 (*dotted line*)

recent experiments have extended the absolute Hugoniot data for aluminum to ∼830 GPa [67]. These results further substantiate the conclusions drawn above.

Liquid Deuterium Hugoniot

The experimental configuration used to obtain Hugoniot data for liquid deuterium (LD_2) with the magnetically driven impact technique is shown in Fig. 4.32. The necessary cryogenics are provided by an expendable cryocell connected to a survivable cyrostat [52, 53]. The cavity of the cryocell is defined by a stepped aluminum alloy 6061-T6 pusher plate and a Zcut sapphire window, with cavity dimensions of approximately 5 mm in diameter and 300 and 600 μm in thickness. The LD_2 samples are condensed in the cryocell by filling the cavity with high purity deuterium gas at 0.124 MPa (18 psi), cooling the cryocell to its equilibrium temperature of 16–18 K, and then warming the cell to 22.0 ± 0.1 K [52,53]. This produces a quiescent LD_2 sample below the boiling point of about 25 K, with nominal initial density of 0.167 g/cm^3.

Shock waves were generated by planar impact of either an aluminum alloy 6061-T6 or a titanium alloy Ti-Al6V4 flyer plate onto the aluminum pusher

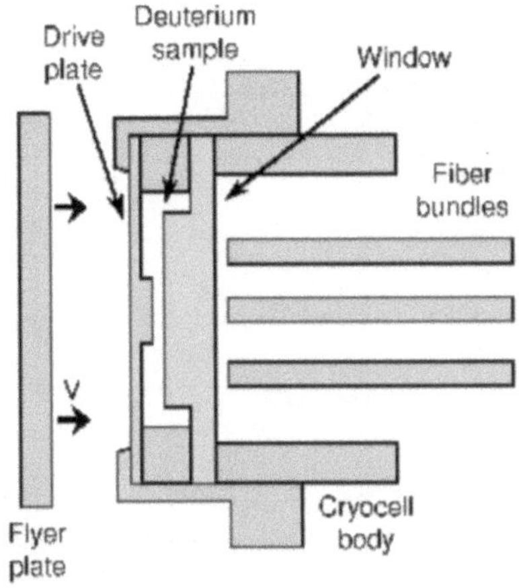

Fig. 4.32. Experimental configuration for Hugoniot experiments on liquid deuterium

plate at the front of the cryocell. The rectangular flyer plate was accelerated across a nominal 3–4 mm vacuum gap by the magnetic field. Titanium flyer velocities as high as ~22 km/s have been achieved, which are capable of generating shock states to ~700 GPa in the aluminum drive plate and transmitting up to ~100 GPa shock waves into LD_2. Conventional velocity interferometry [12] (VISAR) is used to directly measure the velocity history of the flyer plate from launch to impact with an accuracy of ~0.5%.

The shock response of LD_2 is diagnosed with a number of fiber-optic-coupled diagnostics. Typically, several optical fiber bundles of 100 and 200 µm diameter fibers are used, allowing multiple, redundant diagnostics, including (i) conventional VISAR, (ii) FOSBO, and (iii) temporally-resolved spectroscopy. Figure 4.33 shows sample data obtained from a typical LD_2 experiment. In all, up to 16 channels of data are obtained for each experiment, allowing up to 16 independent measurements of U_s in LD_2 and up to four independent measurements of U_s in the aluminum drive plate. The uncertainty in U_s is ~2–3% from the measured transit time through the cell and the initial cell dimensions. Since the uncertainties were due to random errors, statistical techniques can be used to decrease the uncertainty in U_s to approximately 1% and 2% for the LD_2 sample and the aluminum drive plate, respectively.

The VISAR records for the higher-pressure experiments confirm the constancy of the pressure drive obtained from the flyer-plate impact, as shown in Fig. 4.33. In this case the VISAR velocity is indicative of U_s in the LD_2 because LD_2 becomes reflective at shock pressures above ~30 GPa [37]. From these records it is determined that the shock pressure is constant to better than 1% as the shock traverses the cryocell. For the lower-pressure experiments, the shock front is not sufficiently reflective to obtain VISAR

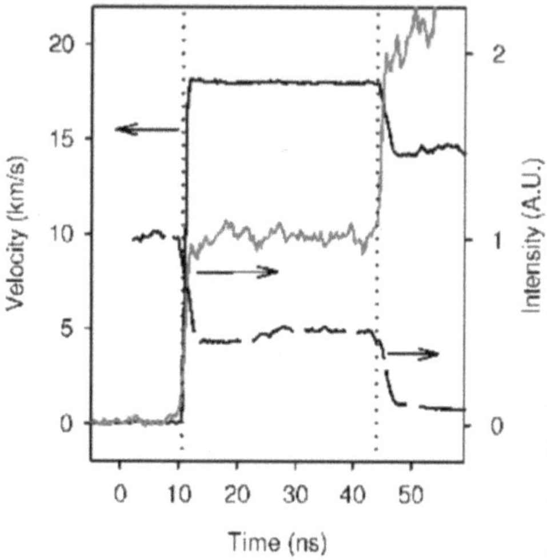

Fig. 4.33. Typical data obtained in a LD_2 experiment. (i) VISAR record of the shock front (solid black line), (ii) FOSBO record (dashed black line), and (iii) self-emission record (gray line). Vertical dotted lines indicate break out of the shock from the aluminum/LD_2 interface and the arrival of the shock at the LD_2/window interface

measurements. However, U_s is obtained for all experiments using the FOSBO and self-emission data. As seen in Fig. 4.33, both of these measurements provide a clear signature of shock arrival at the aluminum/LD_2 and the LD_2/sapphire interfaces. Also, in all experiments, high quality spectra are obtained over the continuous wavelength region between 250 and 700 nm. The spectral dependence of the self-emission can be analyzed to provide a measure of the temperature of the shocked LD_2. We emphasize that the constancy of the emission signal during the traversal of the shock through the cryocell further verifies the constancy of the pressure states achieved with the flyer-plate impact, as the intensity of emission is proportional to the pressure of the LD_2 to the ~1.75 power [69].

An impedance matching method, utilizing the Hugoniot jump conditions [45], was used to obtain Hugoniot points for the shocked LD_2. As shown in Fig. 4.34a, the initial shocked state of the aluminum drive plate is described in the pressure–particle velocity ($P - u_p$) plane by the point labeled A, and the shocked state of LD_2 is constrained to lie on a straight line, with the slope of the line given by $\rho_0 U_s$, where ρ_0 is the initial density of the LD_2 sample. An EOS model for aluminum [61, 62] was used to calculate the release isentrope from the shocked state in the aluminum drive plate. The intersection of the calculated release isentrope and the line defined by the LD_2 shock velocity determines u_p of the shocked deuterium sample. The uncertainty in u_p for LD_2, typically 2–3%, was determined from the uncertainty in the shocked state of the aluminum drive plate, and thus from the uncertainty in U_s for

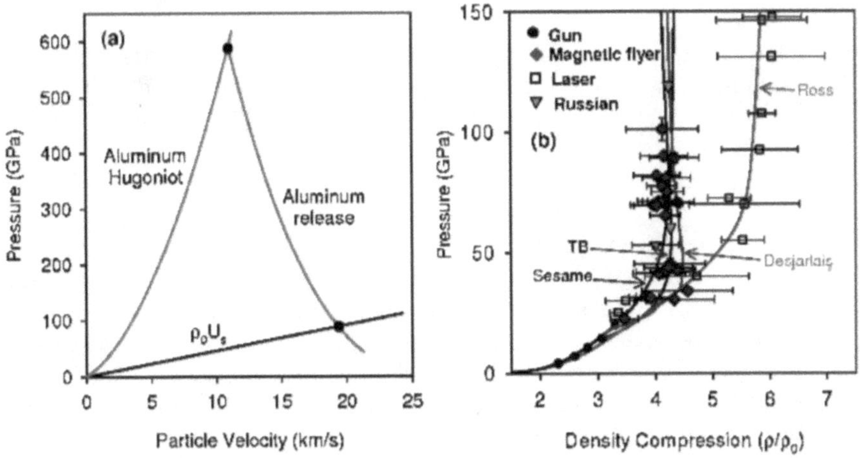

Fig. 4.34. (**a**) Impedance matching technique used to determine the Hugoniot of deuterium. (**b**) Experimental Hugoniot results for liquid deuterium

the aluminum drive plate. The density compression was determined using the jump conditions given by (4.5–4.6).

The pressure–density compression states determined in this way are displayed in Fig. 4.34b. The lowest pressure experiment was found to be in good agreement with the results reported from the earlier gas-gun experiments and the lower pressure laser experiments. However, at higher pressures, particularly the data above ~70 GPa, there is a distinct deviation of the present results from those reported from the laser-driven experiments. Further, the data obtained from our study are in quite good agreement with the predictions from the ab-initio models throughout the entire range of pressures investigated.

Also shown in Fig. 4.34b are the results of recent LD_2 Hugoniot measurements obtained using a convergent geometry technique [16–18]. The results, shown as a weighted average point in the figure, appear to confirm the stiffer response observed in the present experiments. Given the fact that these experiments utilize completely independent experimental configurations, the agreement of the inferred density compression makes a strong case for a ~fourfold limiting compression for the equilibrium response of LD_2 along the principal Hugoniot.

Aluminum Release Measurements

As a first step towards performing measurements of aluminum near its liquid–vapor critical point, release experiments were performed using a low density (200 mg/cm^3) silica aerogel. We note that these release measurements into aerogel are also relevant to the liquid deuterium experiments described above;

this particular density aerogel has a shock impedance very close to that of liquid deuterium, and thus the release into the aerogel simulates the unloading of aluminum in the deuterium impedance matching experiments. It is also noted that the technique described here is similar to that used by Holmes to measure the release response of aluminum from ~80 GPa [57].

To determine the Hugoniot response of the aerogel, several direct impact experiments were performed using high velocity aluminum flyer plates. The shock speed in the aerogel was measured using a passive shock break out technique. The aerogel is highly distended silica ($\rho_{\mathrm{silica}}/\rho_{\mathrm{aerogel}} \approx 13$) and thus, due to pore collapse upon shock compression to several tens of GPa, the temperature of the aerogel approaches ~7 eV. This provides copious self-emission that can be used as a fiducial for the onset of shock compression. The aerogel was backed by LiF, both to provide physical support for the low density material, and to provide a clear fiducial of the shock reaching the rear of the sample; since the LiF has a considerably higher shock impedance, the several tens of GPa shock in the aerogel results in a several hundred GPa shock in the LiF, which causes the LiF to become opaque. These two clear fiducials provided an accurate measure of the shock transit time in the aerogel. Given the shock velocity in the aerogel, the particle velocity could then be determined through impedance matching using the measured flyer plate velocity and the known compression response of aluminum.

The release measurements were performed using a similar configuration; the only difference being the inclusion of an aluminum drive plate on the impact side of the aerogel. Thus, the flyer plate impacts aluminum and the shock releases from aluminum into the aerogel. The method described above was again used to determine the shock velocity in the aerogel. In this case, however, the particle velocity in the aerogel was determined from the measured aerogel shock velocity and the Hugoniot obtained from the direct impact experiments. The measured shock velocity and inferred particle velocity then determine a point in $P-u_{\mathrm{p}}$ space that the aluminum release isentrope must pass through. Results from a series of experiments in which release points were measured from initial shock states in the range of ~250–500 GPa are shown in Fig. 4.35. Also shown in the figure are the predicted release isentropes obtained from the Sesame 3700 model for aluminum [61, 62]. The agreement between experiment and calculation is within experimental uncertainty over this entire range. Thus, these release measurements validate the Sesame 3700 release behavior over this pressure range, and indicate that no significant errors in the inferred deuterium density are the result of the aluminum EOS. In particular, these release results, along with the near-absolute Hugoniot measurements, make a strong case for the use of the Sesame 3700 EOS model for aluminum in performing impedance matching experiments.

We have also performed several full-release experiments for aluminum. In these experiments a LiF window was placed a known distance from the rear of an aluminum target, and the shock response of aluminum was diagnosed

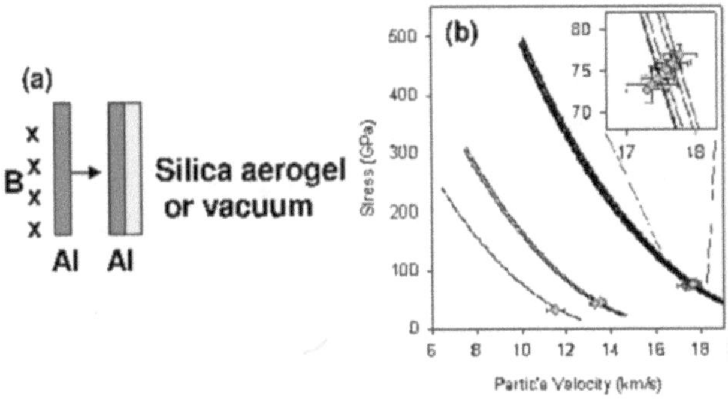

Fig. 4.35. Aluminum unloading experiments. (**a**) Experimental configuration. (**b**) Release data obtained using low density aerogel

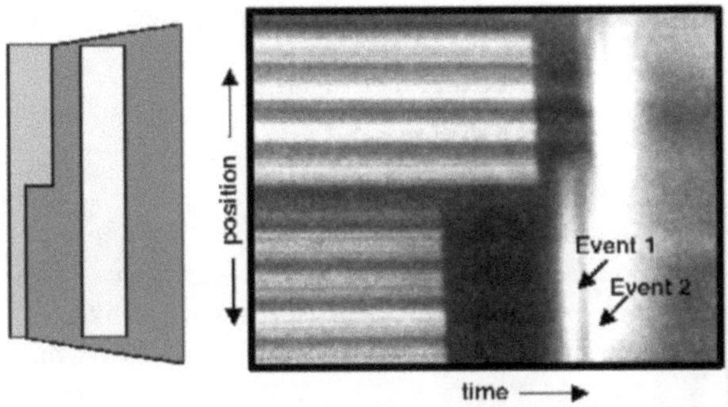

(a) (b)

Fig. 4.36. Full-release experiments in aluminum. (**a**) experimental configuration. (**b**) Data obtained from a ~500 GPa experiment. Note that two distinct events are recorded at the LiF witness plate

through the LiF window, as shown in Fig. 4.36. When the shock reached the rear surface of the aluminum target, the pressure fully released to zero, and the aluminum was accelerated toward the LiF window. A seen in the figure, two distinct impacts at the LiF window were observed. These two impact times were analyzed to determine two nominal velocities for the released aluminum. The results of three experiments released from ~250, ~300, and ~500 GPa are shown in Fig. 4.37, plotted as the measured velocity less twice the particle velocity of the shocked state (this velocity is referred to as the expansion velocity in that it represents the increment of velocity above that one would obtain by approximating the release path as the reflected Hugoniot) as a function of the particle velocity of the shocked state. This apparent two-phase flow is being interpreted as a fast vapor component followed by a slower, possibly liquid component, and is preliminary evidence that the release adiabats are intersecting the vapor dome. Also plotted in the figure

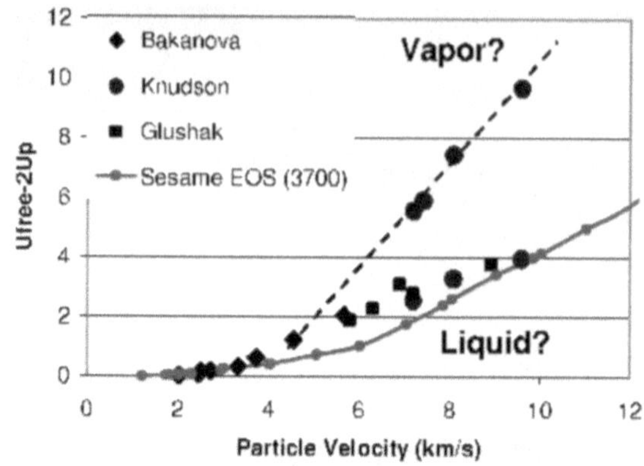

Fig. 4.37. Full-release experiments in aluminum. Plotted are the measured expansion velocities as a function of particle velocity in the shocked state

are results obtained from Russian investigators [10, 48] in experiments in which aluminum was released into an atmosphere of air (as opposed to vacuum). Note the slower component observed in our experiments correlates well with the Russian results, whereas the faster component observed in our experiments apparently was not observed in the Russian experiments. This absence of the faster component in the Russian work is likely due to the fact the air tamped the release, preventing the release path from entering the vapor dome.

4.5 Discussions

Historically, it was not until the early nineteen fifties that smooth-bore guns were developed to launch projectiles to high velocities. Prior to the early fifties, the propellant gun was the fastest gun available, and allowed projectiles to be launched at about 2 km/s. In particular, the development of two-stage light-gas gun in the early fifties was a significant achievement in which projectile velocities around 5 km/s were first obtained. Since then the two-stage light-gas gun capability has been enhanced to achieve velocities approaching 8 km/s, even though many attempts were made to get yet higher velocities. High pressures approaching 1 GPa, combined with Mega "g" acceleration are used to launch lexan projectiles or projectile packages to obtain the high two-stage light-gas gun velocities. Much higher megabar pressures and Giga "g" accelerations are necessary to obtain velocities faster than 10 km/s. The high pressures required exceed the structural strength of both the gun barrel and projectile materials and is the main limiting factor in achieving high velocities on a traditional two-stage light-gas gun. The energy required to launch a flyer plate to 30 km/s is approximately 50 times the energy required to melt and vaporize the projectile or the plate. The energy must, therefore, be deposited in a well-controlled quasi-isentropic manner to prevent melt or vaporization. The high-pressure loading must also be uniform to prevent fracture. On the modified two-stage light-gas gun (the three-stage light-gas gun), this is achieved by using a graded-density assembly to impact a stationary flyer plate. Upon impact, time-dependent, structured, ~100 GPa pressure pulses are generated and used to propel the plates to hypervelocities approaching 16 km/s without melt or fracture. On the Sandia Z accelerator, time-dependent and smoothly increasing 400 GPa magnetically-generated pressure pulses are used to launch plates to ultra-high velocities in excess of 27 km/s without melt or fracture. For both systems, titanium alloy Ti-6Al-4V and aluminum alloy 6061-T6 are used for the flyer plate material. These are the highest metallic projectile-plate velocities ever achieved for masses in the range of 0.1–1.0 g. Both the technologies have matured to allow dynamic material property measurements to accuracies comparable to traditional light-gas gun techniques. Most significantly, dynamic material property measurements can now be obtained at pressures and temperatures that were not previously accessible in the laboratory.

Very high pressures are needed to launch flyer plates to hypervelocities. The loading must be very nearly shockless, time-dependent, and uniform. Figure 4.38 displays an example of a stress loading history driving the flyer plate to high velocities on a three-stage gun. The stress loading history of a flyer plate driven by magnetic fields is shown in Fig. 4.17b. There are two main differences when compared to the pressure histories shown for driving projectiles on conventional smooth bore guns. First, the peak driving pressure levels are 100 to 400 times higher and, second, the time durations are three orders of magnitude less than the conventional two-stage light-gas gun. This results in the very high plate accelerations 10^9–10^{10} g as shown in Fig. 4.39. Not only must the loading history be time dependent, but it must also be very rapid. This also necessitates that the flyer plates be thin – otherwise the time dependent stress pulse would shock up and melt or even vaporize the flyer plate. At these high stresses, the plate is compressed to approximately half its thickness and then released to zero stress. For uniform deformation and release to occur over the entire plate, its lateral dimension has to be large compared to its thickness so that one-dimensional loading conditions persist.

There is what appears to be a sudden transition between the peak accelerations observed on the traditional two-stage gun versus those indicated for the three-stage and the magnetic flyer plate technique. This is because, for peak accelerations at or below those achieved in the two-stage light-gas gun, the peak loading stresses are lower than the structural strength of the barrel or projectile material. At high peak accelerations, 10^9 g or greater, the peak loading stresses are significantly above the dynamic yield and fracture strength of the material and factors such as pulse uniformity, shape, and width become more significant in preventing melt and fracture. It should also be noted that both the impact technique used on the three-stage gun and the

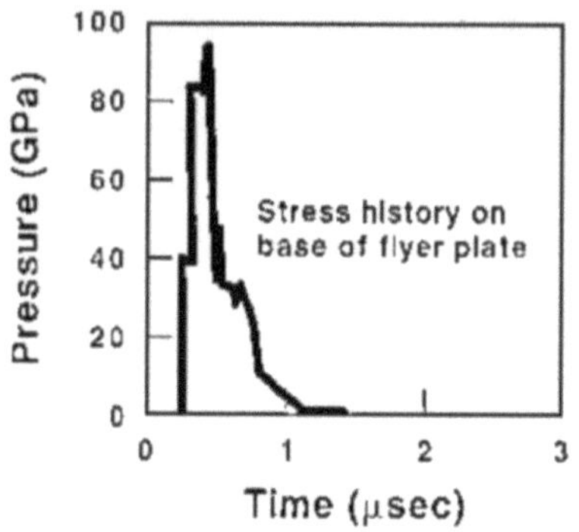

Fig. 4.38. Typical driving pressure history at the base of the flyer plate for a flyer-plate launched on the three-stage light gas gun at around 10 km/s

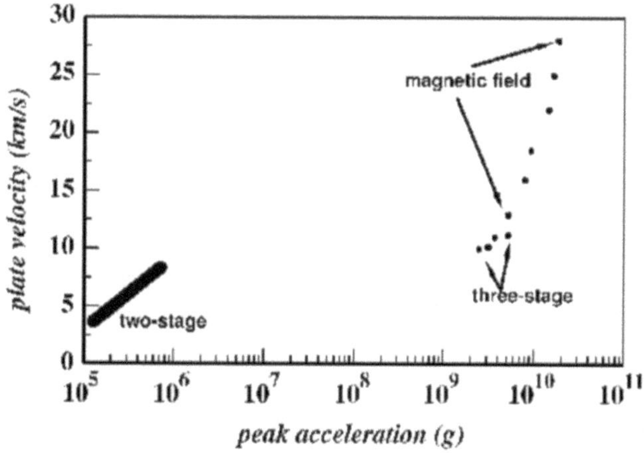

Fig. 4.39. Flyer plate velocity versus peak accelerations for flier plates launched to ultra high velocities using the three-stage light-gas gun and using magnetic field on the Z accelerator

magnetically accelerated flyer plate technique can be used to propel plates to velocities lower than those reported herein, even though the emphasis of this article is on achieving higher velocities than those so far obtained on conventional two-stage guns.

The mass of the plate that can be launched is directly proportional to the volume of the material that is launched. This reduces to the areal density or the mass density, ρx, where ρ and x are the density and the thickness of the flyer plate. Figure 4.40 shows the dependence of the velocity of the flyer plate on its mass density and, as expected, it is inversely proportional to its mass density and is independent of its lateral cross section because one-dimensional conditions are maintained during plate launch. This also implies

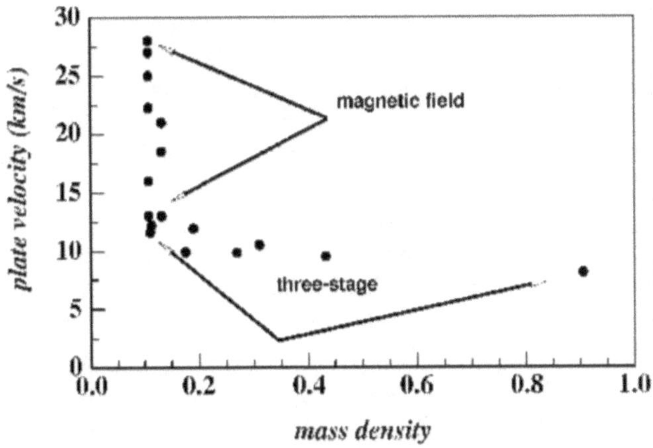

Fig. 4.40. Variation of mass density versus velocity for flier plates launched on the three-stage light-gas gun and the on the Z accelerator using magnetic fields

that scaling is easily applicable as long as one-dimensional loading conditions are maintained.

As indicated in this chapter, the principle and the techniques developed to achieve velocities in excess of traditional two-stage light-gas gun are summarized. The flyer plates have been well characterized to determine their "initial" state prior to using them as impactors in subsequent applications. It is crucial, especially for EOS determination, that the initial conditions be very well known and are repeatable. The use of this technology to obtain the EOS of metals [68,88], foam, and deuterium [69,70,73] are also described. Other examples include characterization of shock and release states [35,47,68], shock-induced vaporization processes [19,36], and debris propagation [34,36]. These are just a few examples that suggest the maturity of the technology for material property studies to stresses in excess of 500 GPa in the laboratory.

Further improvements and enhancement to the technology are being pursued. Impact techniques used on three-stage light-gas guns indicate that the use of frozen hydrogen buffers, replacing TPX, has the potential of achieving velocities around 20 km/s. Furthermore, significant progress has been made in both the understanding of the magnetically driven flyer plate technique and in the optimization of the experimental configuration. As a result, velocities in excess of 27 km/s have been realized. We anticipate that flyer velocities of ~30 km/s can be achieved on the Z accelerator with further load optimization.

In the next few years the infrastructure of the Z accelerator will be modernized and upgraded. The resulting accelerator, referred to as ZR, is expected to have twice the energy storage capacity, and should produce nearly 28 MA in a short circuit load configuration. This upgrade should substantially increase the capability of the magnetically driven flyer technique. In particular, load designs have been identified that should result in flyer velocities in excess of 40 km/s, while still maintaining intact flyer plates at essentially ambient temperature and density. This increase in capability will further expand the pressures and temperatures achievable in EOS experiments. Furthermore, through shock and release studies, it should be possible to access material states near the liquid–vapor critical point. These types of experiments may allow determination of the location of the vapor dome, as well as measurements of transport properties for materials in these hot, slightly expanded states for most engineering materials, not merely low-boiling-point materials such as zinc, lead, and cadmium.

4.6 Acknowledgements

The success of both the three-stage light-gas gun and the magnetically accelerated flyer plate technologies is a result on many significant and valuable contributions from experimentalists, theorists, code analysts, and technicians. We are especially thankful to James Asay, Lynn Barker, Bill Reinhart, Clint Hall, Tim Trucano, and Luba Kmetyk for their cheerful and active participation during many phases of the three-stage light-gas gun technology development and applications. We would also like to acknowledge several individuals who played a crucial role in the development of the magnetically accelerated flyer plate technology. They include Clint Hall, who engineered the original configuration used for isentropic loading on the Z accelerator; Ray Lemke and Mike Desjarlais, who were instrumental in developing and defining the magnetic loading conditions necessary to shocklessly accelerate flyer plates to high velocity; Randy Hickman, who was instrumental in developing the large array of diagnostic techniques at the Z accelerator that are necessary to perform the experiments described; and the superb operations and diagnostic team at the Z facility whose dedication and constant support were critical to the success of this technology.

Sandia is a multiprogram laboratory operated by Sandia Corporation, a Lockheed Martin Company, for the United States Department of Energy's National Nuclear Security Administration under Contract DE-AC04-94AL85000.

References

1. Al'tshuler, L.V., and B.S. Chekin (1974), "Metrology of high pulsed pressures," in: *Proc 1 All-Union Pulsed Pressures Simposium.* VNIIFTRI, Moscow **vol. 1**, pp. 5–22 (in Russian).
2. Al'tshuler L.V., S.B. Kormer, A.A. Bakanova, and R.F. Trunin (1960), "The isentropic compressibility of aluminum, copper, lead, and iron at high pressures," *Zh. Eksp. Teor. Fiz.* **38**, p. 790. (in Russian) *Sov. Phys. JETP* **11,** p. 573.

3. Al'tshuler, L.V., N.N. Kalitkin, L.V. Kuz'mina, and B.S. Chekin (1977), "Shock adiabats for ultrahigh pressures," *Zh. Eksp. Teor. Fiz.* **72**, p. 317. (in Russian) *Sov. Phys. JETP* **45**, p. 167.
4. Asay, J.R. (2000), "Isentropic compression experiments on the Z accelerator," in: *Shock compression of condensed matter – 1999*, (eds. M.D. Furnish, L.C. Chhabildas, and R.S. Hixson). AIP Press, NY, p. 261.
5. Asay, J.R., and T.G. Trucano (1990), "Studies of density distributions in one-dimensional shock-induced debris clouds," *Int. J. Impact Engng.* **10**, pp. 35–50.
6. Asay, J.R., L.C. Chhabildas, and L.M. Barker (1985), *Projectile and impactor designs for plate impact experiments*, Sandia National Laboratories Report, SAND85-2009.
7. Asay, J.R., T.G. Trucano, and L.C. Chhabildas (1988), "Time-resolved measurements of shock-induced vapor-pressure profiles," in: *Shock waves in condensed matter – 1987*, (eds. S.C. Schmidt, and N.C. Holmes). Elsevier Science Publishers, BV, pp. 159–162.
8. Asay, J.R., T.G. Trucano, and R.S. Hawke (1990), "The use of hypervelocity launchers to explore previously inaccessible states of matter," *Int. J. Impact Engng.* **10**, p. 51.
9. Askew, R.F., B.A. Chin, B.J. Tatarchuk, J.L. Brown, and D.B. Jensen (1986), "Rail and insulator erosion in railguns," *IEEE Trans. Magn.* **22**, p. 1380.
10. Bakanova, A.A., I.P. Dudaladov, M.V. Zhernokletov, V.N. Zuberev, and G.V. Simakov (1983), "On evaporation of shock-compressed metals under expansion," *Zh. Prikl. Mekh. Tekh. Fiz.* **2**, pp. 76–81. (in Russian) *J Appl. Mech. Tech. Phys.* **24**, p. 204.
11. Barker, L.M. (1984), "High-pressure quasi-isentropic impact experiments," in: *Shock Waves in Condensed Matter – 1983* (eds. J.R. Asay, R.A Graham, and G.K. Straub). Elsevier Science Publishers, p. 217.
12. Barker, L.M., and R.E. Hollenbach (1972), "Laser interferometer for measuring high velocities of any reflecting surface," *J. Appl. Phys.* **43**, pp. 4669–4675.
13. Barker, L.M., L.C. Chhabildas, and T.G. Trucano (1990a), "High gas pressure acceleration of flyer plates – experimental techniques," *Int. J. Impact Engng.* **10**, p. 67.
14. Barker, L.M., L.C. Chhabildas, T.G. Trucano, and J.R. Asay (1990b), "Gas-accelerated plate stability study," in: *Shock Waves in Condensed Matter – 1989* (eds. S.C. Schmidt, J.N. Johnson, L.W. Davison). Elsevier Science Publisher, p. 989.
15. Baum, D.W. (1973), "Explosively driven hypervelocity launcher – second stage augmentation techniques," Physics International Report No. PIFR-245-1.
16. Belov, S.I., G.V. Boriskov, A.I. Bykov, R.I. Il'kaev, N.B. Luk'yanov, A.Ya. Matveev, O.L. Mikhailova, V.D. Selemir, G.V. Simakov, R.F. Trunin, I.P. Trusov, V.D. Urlin, V.E. Fortov, and A.N. Shuikin (2002), "Shock compression of solid deuterium," *Sov. Phys. JETP Lett.* **76**, p. 433.
17. Boriskov, G.V., A.I. Bykov, R.I. Il'kaev, V.D. Selemir, G.V. Simakov, R.F. Trunin, V.D. Urlin, V.E. Fortov, and A.N. Shuikin (2003), "Shock-wave compression of solid deuterium at a pressure of 120 GPa," *Dokl. Phys.* **48**, p. 553.
18. Boriskov GV, et al. (2004), manuscript submitted for publication. Private communication.

19. Brannon, R.M., and L.C. Chhabildas (1995), "Experimental and numerical investigation of shock-induced vaporization in zinc," *Intl. J. Impact Engng.* **17**, p. 109.
20. Bushman, A.V., I.K. Krasyuk, P.P. Pashinin, A.M. Prokhorov, V.Ya. Ternovoi, and V.E. Fortov (1984), "Dynamic compressibility and thermodynamics of a dense aluminum plasma at megabar pressures," *Sov. Phys. JETP Lett.* **39**, p. 411.
21. Carroll, D.E., L.C. Chhabildas, W.D. Reinhart, N.A. Winfree, and G.I. Kerely (2002), "Computational characterization of three-stage gun flyer plate launch," in: *Shock Compression of Condensed Matter – 2001* (eds. M.D. Furnish, N.N. Thadhani, Y. Horie) AIP Press, NY, p. 620.
22. Charters, A.C. (1986), "Development of the high-velocity gas-dynamics gun," *Int. J Impact Engng.* **5**, p. 181.
23. Chhabildas, L.C. (1992), "Hypervelocity Launch capabilities to over 10 km/s," in: *Proceedings of the XIII AIRAPT Conference on High Pressure Science and Technology.*
24. Chhabildas, L.C., and L.M. Barker (1988), "Dynamic Quasi-isentropic compression of tungsten," in: *Shock Waves in Condensed Matter – 1987* (eds. S.C. Schmidt and N.C. Holmes) Elsevier Science Publishers, BV, p. 111.
25. Chhabildas, L.C., and J.R. Asay (1992), "Dynamic yield strength and spall strength measurements under quasi-isentropic loading," in: *Shock Wave and High-Strain-Rate Phenomena in Materials* (eds. M.A. Meyers, L.E. Murr, and K.P. Staudhammer) Marcel Dekker Inc., NY.
26. Chhabildas, L.C, J.R. Asay, and L.M. Barker (1988), *Shear strength of tungsten under shock- and quasi-isentropic loading to 250 GPa*, Sandia National Laboratories Report No. SAND88-0306.
27. Chhabildas, L.C., L.M. Barker, J.R. Asay, and T.G. Trucano (1990a), "Relationship of fragment size to normalized spall strength for materials," *Int. J. Impact Engng.* **10**, p. 107.
28. Chhabildas, L.C., L.M. Barker, T.G. Trucano, and J.R. Asay (1990b), "Spall strength measurements on shock-loaded refractory metals," in: *Shock Waves in Condensed Matter – 1989* (eds. S.C. Schmidt, J.N. Johnson, and L.W. Davison) Elsevier Science Publisher, p. 429.
29. Chhabildas, L.C., L.M. Barker, J.R. Asay, T.G. Trucano, and G.I. Kerley (1991), *Sandia's New Hypervelocity Launcher, HVL*, Sandia National Laboratories Report SAND91-0657.
30. Chhabildas, L.C., L.N. Kmetyk, W.D. Reinhart, and C.A. Hall (1995), "Enhanced hypervelocity launcher - capabilities to over 16 km/s. *Int. J. Impact Engng.* **17**, pp. 183–194.
31. Chhabildas, L.C., L.M. Barker, J.R. Asay, T.G. Trucano, G.I. Kerley, and J.E. Dunn (1992), "Launch capabilities to over 10 km/s," in: *Shock Waves in Condensed Matter – 1991* (eds. S.C. Schmidt, R.D. Dick, J.W. Forbes, D.G. Tasker) Elsevier Science Publishers, BV, p. 1025.
32. Chhabildas, L.C., J.E. Dunn, W.D. Reinhart, and J.M. Miller (1993), "An impact technique to accelerate flyer plate velocities to over 12 km/s," *Int. J. Impact Engng.* **14**, pp. 121–132.
33. Chhabildas, L.C., T.G. Trucano, W.D. Reinhart, and C.A. Hall (1994), *Chunk projectile launch using the Sandia hypervelocity launcher facility*, Sandia National Laboratories Report No. SAND94-1273.

34. Chhabildas, L.C., M.B. Boslough, W.D. Reinhart, C.A. Hall (1994), "Debris cloud characterization at impact velocities of 5 to 11 km/s," in: *High-Pressure Science and Technology – 1993* (eds. S.C. Schmidt, J.W. Shaner, G.A. Samara, and M. Ross) AIP Press, NY, p. 1841.
35. Chhabildas, L.C., M.D. Furnish, and W.D. Reinhart (2000), "Shock induced melting in aluminum: wave profile measurements," in: *Shock Compression of Condensed Catter – 1999* (eds. M.D. Furnish, L.C. Chhabildas, and R.S. Hixson) AIP Press, NY, p. 97.
36. Chhabildas, L.C., W.D. Reinhart, T.F. Thornhill, G.C. Bessette, W.V. Saul, R.J. Lawrence, and M.E. Kipp (2003), "Debris generation and propagation from hypervelocity impacts on aluminum from 6 to 11 km/s. *Intl. J. Impact. Engng.* p. 29.
37. Collins, G.W., L.B. DaSilva, P. Celliers, D.M. Gold, M.E. Foord, R. J. Wallace, A. Ng, S.V. Weber, K.S. Budil, and R. Cauble (1998), "Measurements of the equation of state of deuterium at the fluid insulator-metal transition," *Science* **281**, p. 1178.
38. Da Silva, L.B., P. Celliers, G.W. Collins, K.S. Budil, N.C. Holmes, T.W. Barbee Jr, B.A. Hammel, J.D. Kilkenny, R.J. Wallace, M. Ross, R. Cauble, A. Ng, and G. Chiu (1997), "Absolute equation of state measurements on shocked liquid deuterium up to 200 GPa (2 Mbar)," *Phys. Rev. Lett.* **78**, p. 483.
39. Davis, J.P. (2004), "Quasi-isentropic compression of iron," Private communication.
40. Davis, J.P., and D.B. Hayes (2003), "Isentropic compression experiments on dynamic solidification in tin," in: *Proceedings of the 13*th *APS topical group conference on shock compression science of condensed matter-2003,* July 20–25, 2003, Portland, Oregon.
41. Davis, J.P. et al. (2004), "Quasi-isentropic response of aluminum," Private communication.
42. Desjarlais, M.P. (2001), "Practical improvements to the Lee–More conductivity near the metal-insulator transition," *Contrib. Plasma Phys.* **41**, p. 267.
43. Desjarlais, M.P., J.D. Kress, and L.A.Collins (2002), "Electrical conductivity for warm, dense aluminum plasmas and liquids," *Phys. Rev. E* **6**, p. 025401.
44. Desjarlais, M.P. (2003), "Density functional calculations of the liquid deuterium Hugoniot, reshock and reverberation timing," *Phys. Rev. B* **68**, p. 064204.
45. Duvall, G.E., and R.A. Graham (1977), "Phase transitions under shock wave loading," *Rev. Mod. Phys.* **49**, p. 523.
46. Fowles, G.R., C. Leung, R. Rabie, and J. Shaner, (1980) "Acceleration of Flat Plates by Multiple Staging", *Proceedings of the sixth AIRAPT conference, high pressure science and technology, V2, pp. 911*, Boulder Colorado.
47. Furnish, M.D., L.C. Chhabildas, and W.D. Reinhart (1999), "Time-resolved particle velocity measurements at impact velocities of 10 km/s. *Int J Impact Engng* **23**, p. 261.
48. Glushak, B.L., A.P. Zharkov, M.V. Zhernokletov, V.Ya. Ternovoi, A.S. Filimonov, and V.E. Fortov (1989), "Experimental investigation of the thermodynamics of dense plasmas formed from metals at high energy concentrations," *Zh. Eksp. Teor. Fiz.* **96**, p. 1301. (in Russian) *Sov. Phys. JETP* **69**, p. 739.

49. Grosch, D.,J.Walker, S.A. Mullin, R. Tullos (1991), "Development of an inhibited explosive hypervelocity launcher," Southwest Research Institute Report No. 06-3513.
50. Hall, C.A., J.R. Asay, M.D. Knudson, W.A. Stygar, R. Spielman, T.D. Pointon, D.B. Reisman, A. Toor, and R.C. Cauble (2001), "Experimental configuration for isentropic compression of solids using pulsed magnetic loading," *Rev. Sci. Instrum.* **72**, p. 3587.
51. Hall, C.A., M.D. Knudson, J.R. Asay, R. Lemke, and B. Oliver (2001), "High velocity flyer plate launch capability on the Sandia Z accelerator," *Int. J. Impact Engng.* **26**, p. 275.
52. Hanson, D.L., J.R. Asay, C.A. Hall, M.D. Knudson, J.E. Bailey, K.J. Flemming, R.R. Johnston, B.F. Clark, M.A. Bernard, W.W. Anderson, G. Hassall, S.D. Rothman (2000), "Progress on deuterium EOS measurements on Z," in: *Shock Compression of Condensed Matter – 1999* (eds. M.D. Furnish, L.C. Chhabildas, and R.S. Hixon) AIP Press, NY, p. 1175.
53. Hanson, D.L. R.R. Johnston, M.D. Knudson, J.R. Asay, C.A. Hall, J.E. Bailey, and R. J. Hickman (2002), "Advanced cryogenic system capabilities for precision shock physics measurements on Z," in: *Shock Compression of Condensed Matter* – 2001 (eds. M.D. Furnish, N.N. Thadani, and Y. Horie) AIP Press, NY, p. 114.
54. Hare, D.E., D.B. Reisman, F. Garcia, L.G. Green, J.W. Forbes, M.D. Furnish, C. Hall, and R.J. Hickman (2003), "The isentrope of unreacted LX-04 to 170 kbar," in: *Proceedings of the 13th APS topical group conference on shock compression science of condensed matter–2003*, July 20–25, 2003, Portland, Oregon.
55. Hawke, R., D.E. Duerre, J.G. Huebel, J.G. Klapper, D.J. Steinberg, and R.N. Keeler (1972), "Method of isentropically compressing materials to several megabars," *J. Appl. Phys.* **43**, p. 2734.
56. Holian, K.S. (1984), "T-4 Handbook of Materials Properties Data Bases," Los Alamos Scientific Report LA-10160-MS.
57. Holmes, N.C. (1994), "Shock compression of low-density foams," in: *High-pressure science and technology–1993* (eds. S.C. Schmidt, J.W. Shaner, G.A. Samara, and M. Ross) AIP Press N.Y. p. 153.
58. Igenbergs, E., S. Aigner, A. Hudepohl, H. Iglesder, H. Kuczerz, M. Rott, and U. Weishaupt (1987), "Launcher technology, in-flight measurement and impact diagnostics at the TUM/LRT. *Int. J. Impact.Engng.* **5**, p. 371.
59. Keaton, P.W., G.C. Idzork, L.J. Rowton, J.D. Seagrave, G.L. Stradling, S.D. Bergeson, M.T. Collopy, H.L. Curling, D.B. McColl, and J.D. Smith (1990), "A hypervelocity microparticle impacts laboratory with 100 km/s projectiles," *Int. J. Impact Engng.* **10**, p. 295.
60. Kerley, G.I. (1983), "In: Molecular based study of fluids," American Chemical Society, Washington, DC, p. 107.
61. Kerley GI (1987) Theoretical equation of state for aluminum. Int J Impact Engng 5:441.
62. Kerley, G.I. (1998), "Equations of state for aluminum and beryllium," Kerley Publishing Services Report No. KPS98-1.
63. Kerley, G.I. (2003a), "Equation of state for titanium and Ti6Al4V alloy," Sandia National Laboratories Report SAND2003-3785.
64. Kerley GI (2003b) Equations of state for hydrogen and deuterium. Sandia National Laboratories Report SAND2003–3613.

65. Kerley, G.I., and J.L. Wise (1988), "Shock-induced vaporization of porous aluminum," in: *Shock waves in Condensed Matter – 1987* (eds. S.C. Schmidt, and N.C. Holmes). Elsevier Science Publishers, BV, pp. 155–158.
66. Knudson, M.D. (2003), "Equation of state of titanium," Private communication
67. Knudson, M.D. (2004a), "High velocity flyer plate launch on Z," Private communication.
68. Knudson, M.D. (2004b), "Unloading response of aluminum to impact stresses of 500 Gpa," manuscript in preparation.
69. Knudson, M.D., D.L. Hanson, J.E. Bailey, C.A. Hall, J.R. Asay, W.W. Anderson (2001), "Equation of state measurements in liquid deuterium to 70 GPa," *Phys. Rev. Lett.* **87**, p. 225501–1.
70. Knudson, M.D., D.L. Hanson, J.E. Bailey, C.A. Hall, and J.R. Asay (2003a), "Use of a wave reverberation technique to infer the density compression of shocked liquid deuterium to 75 GPa," *Phys. Rev. Lett.* **90**, p. 035505–1.
71. Knudson, M.D., R.W. Lemke, D.B. Hayes, C.A. Hall, C. Deeney, and J.R. Asay (2003b), "Near-absolute Hugoniot measurements in aluminum to 500 GPa using a magnetically accelerated flyer plate technique," *J. Appl. Phys.* **94**, p. 4420.
72. Knudson, M.D., C.A. Hall, R. Lemke, C. Deeney, and J.R. Asay (2003c), "High velocity flyer plate launch capability on the Sandia Z accelerator," *Int. J. Impact Engng.* **29**, p. 377.
73. Knudson, M.D., D.L. Hanson, J.E. Bailey, C.A. Hall, J.R. Asay, and C. Deeney (2004), "Principal Hugoniot, reverberating wave, and mechanical reshock measurements of liquid deuterium to 400 GPa using plate impact techniques," *Phys. Rev. B* **69**, p. 144209.
74. Kormer, S.B., A.I. Funtikov, V.D. Urlin, and A.N. Kolesnikova (1962), "Dynamical compression of porous metals and the equation of state with variable specific heat at high temperatures," *Zh Eksp Teor Fiz* **42**, p. 626 (in Russian) Sov Phys JETP 15:477.
75. Lemke, R.W., M.D. Knudson, C.A. Hall, T.A. Haill, M.P. Desjarlais, J.R. Asay, and T.A. Mehlhorn (2003a), "Characterization of magnetically accelerated flyer plates," *Phys. Plasmas* **10**, p. 1092.
76. Lemke, R.W., M.D. Knudson, A.C. Robinson, T.A. Haill, K.W. Struve, J.R. Asay, and T.A. Mehlhorn (2003b), "Self-consistent, two-dimensional, magnetohydrodynamic simulations of magnetically driven flyer plates," *Phys. Plasmas* **10**, p. 1867.
77. Lenosky, T.J., J.D. Kress, and L.A. Collins (1997), "Molecular-dynamics modeling of the Hugoniot of shocked liquid deuterium," *Phys. Rev. B* **56**, p. 5164.
78. Marsh, S.P. (1980), *LASL Shock Hugoniot Data*, University of California Press, Berkeley.
79. Marsh, S.P., and T.H. Tan (1992), in: *Shock Waves in Condensed Matter – 1991* (eds. S.C. Schmidt, J. Forbes, and R. Dick). Elsevier Science Publishers, BV.
80. Matzen, M.K. (1997), "Z pinches as intense x-ray sources for high-energy density physics applications," *Phys. Plasmas* **4**, p. 1519
81. McGlaun, J.M., S.L. Thomson, and M.G. Elrick (1990), "CTH: A three-dimensional shock wave physics code," *Int. J. Impact Engng.* **10**, pp. 351–360.
82. Militzer, B., and D.M. Ceperley (2000), "Path integral Monte Carlo calculation of the deuterium Hugoniot," *Phys. Rev. Lett.* **85**, p. 1890.

83. Mitchell, A.C., and W.J. Nellis (1981a), "Diagnostic system of the Lawrence Livermore National Laboratory two-stage light-gas gun," *Rev. Sci. Instrum.* **52**, p. 3
84. Mitchell, A.C., and W.J. Nellis (1981b), "Shock compression of aluminum, copper, and tantalum," *J. Appl. Phys.* **52**, p. 3363.
85. Mullin, S.A., C.E. Anderson Jr, and J.S. Wilbeck (1990), *Dissimilar material scaling relationships for hypervelocity impact*, Report prepared by Southwest Research Institute for the Defense Nuclear Agency under contract DNA001-86-C-0037, Project 06-1039-001, DNA-TR-89-112.
86. Nellis, W.J., A.C. Mitchell, M. van Thiel, G.J. Devine, and R.J. Trainor (1983), "Equation-of-state data for molecular hydrogen and deuterium at shock pressures in the range 2–76 GPa (20-760 kbar)," *J. Chem. Phys.* **79**, p. 1480.
87. Osher, J.E., H.H. Chau, G.R. Gathers, R.S. Lee, and R.C. Weingart (1986), "Applications of a 100-kV electric gun for hypervelocity impact studies," *Int. J. Impact Engng.* **5**, p. 501.
88. Reinhart, W.D., L.C. Chhabildas, D.E. Carroll, T.K. Bergstresser, T.F. Thornhill, and N.A. Winfree (2001), "Equation of state measurements of materials using a three-stage gun to impact velocities of 11 km/s," *Intl. J. Impact Engng.* **26**, p. 625.
89. Reisman, D.B., A. Toor, R.C. Cauble, C.A. Hall, J.R. Asay, M.D. Knudson, and M.D. Furnish (2001), "Magnetically driven isentropic compression experiments on the Z accelerator," *J. Appl. Phys.* **89**, p. 1625.
90. Reisman, D.B., et al. (2003a), "Ramp loading of uranium alloys," Private communication.
91. Reisman, D.B., W.G. Wolfer, A. Elsholz, and M.D. Furnish (2003b), "Isentropic compression of irradiated stainless steel on the Z accelerator," *J. Appl. Phys.* **93**, p. 8952.
92. Rice, M.H., R.G. McQueen, and J.M. Walsh (1958), "Compression of solids by strong shock waves," in: *Solid State Physics* **Vol. 6**. (eds. F. Seitz, D. Turnbull) Academic Press, NY.
93. Rigg, P.A., W, Greeff, M.D. Knudson, D.B. Hayes, R.S. Hixson, and G.T. Gray III (2003), "Investigation shock-induced phase transitions in zirconium," in: *Proceedings of the 13th APS topical group conference on shock compression science of condensed matter–2003*, July 20–25, 2003, Portland Oregon.
94. Ross, M. (1998), "Linear-mixing model for shock-compressed liquid deuterium," *Phys. Rev. B* **58**, p. 669.
95. Seidel, D.B., M.L. Kiefer, R.S. Coats, T.D. Pointon, J.P. Quintenz, and W.A. Johnson (1991), "The 3D, electromagnetic, particle-in-cell code, QUICKSILVER," in: *Europhysics conference on computational physics* (ed. A Tenner) CP90. World Scientific, Amsterdam, pp. 475–482.
96. Seigel, A.E. (1979), in: *Interior Ballistics of Guns* (eds. H. Krier, M. Summerfield) AIAA Publishers, NY, p. 135.
97. Simonenko, V.A., N.P. Voloshin, A.S. Vladimirov, A.P. Nagibin, V.N. Nogin, V.A. Popov, V.A. Sal'nikov, and Yu. A. Shoidin (1985), "Absolute measurements of shock compressibility of aluminum at pressures • 1 Tpa," *Zh. Eksp. Teor. Fiz.* **88**, p. 1452 (in Russian) Sov. Phys. JETP **61**, p. 869.
98. Stilp, A.J. (1986), "Review of modern hypervelocity impact facilities," *Int..J. Impact Engng.* **5**, pp. 613–621.

99. Summers, R.M., J.S. Peery, M.W. Wong, E.S. Jr, Hertel, T.G.Trucano, and L.C. Chhabildas (1997), "Recent progress in Alegra development and applications to ballistic impacts," *Int. J. Impact Engng.* **20**, p. 779.
100. Taylor, J.W. (1984), "Thunder in the mountains," in: *Shock Waves in Condensed Matter* (eds. J.R. Asay, R.A. Graham, and G.K. Straub). Elsevier Science Publishers, BV, p. 1.
101. Trott, W.M., M.D. Knudson, L.C. Chhabildas, and J.R. Asay (2000), "Measurement of spatially resolved velocity variations in shock compressed heterogeneous materials using a line-imaging velocity interferometer," in: *Shock Compression of Condensed Matter – 1999* (eds. M.D. Furnish, L.C. Chhabildas, and R.S. Hixson) AIP Press, NY, p 993.
102. Trucano, T.G., and L.C. Chhabildas (1995), "Computational design of hypervelocity launchers," *Intl. J. Impact Engng.* p. 17.
103. Trunin, R.F., and W.J. Nellis (2003), *Shock Compression of Liquid Deuterium at 54 GPa*, Lawrence Livermore National Laboratory Report UCRL-JC-152886.
104. Trunin, R.F. (2004), "Hugoniot Curves of Metals," in *High-pressure shock compression of solids VII-Shock Waves and Extreme States of Matter*, (eds. V. E. Fortov, L.V. Al'tshuler, R. F. Trunin, and A. I. Funtikov) Springer-Verlag, N.Y., pp. 77–108.
105. Vesser, L.R., and J. Stokes (1993), "Fiber optic diagnostics for high explosives," Paper submitted to DNA conference on instrumentation for nuclear weapons effects testing, LA-CP-93–232.
106. Vesser, L.R., M.J. George, R. Menikoff, and B. Papatheofanis (1987), "Sensing of high pressure by shock heating of quartz optical fibers," Fiber optic and laser sensors V, SPIE, **vol. 838**.
107. Vogler, T.J. (2004), "Determining compressive strength of materials under ramp loading," Private communication.
108. Volkov, L.P., N.P. Voloshin, A.S. Vladimirov, V.N. Nogin, and V.A. Simonenko (1981), "Shock compressibility of aluminum at pressure 10 Mbar," *Pis'ma Zh. Eksp. Teor. Fiz.* **31**, p. 623 (in Russian) *Sov. Phys. JETP Lett.* **31**, p. 588.
109. Wenzell, A.B. (1987), "A review of explosive accelerators for hypervelocity impact," *Int. J. Impact Engng.* **5**, p. 681.
110. Winfree, N.A., L.C. Chhabildas, W.D. Reinhart, D.E. Carroll, and G.I. Kerley (2002), "EOS data of Ti-6Al-4V to impact velocities of 10.4 km/s on a three stage gun," in: *Shock Compression of Condensed Matter – 2001* (eds. M.D. Furnish, N.N. Thadhani, and Y. Horie) AIP Press, NY, p. 75.
111. Wise, J.L., and L.C. Chhabildas (1986), "Laser interferometer measurements of refractive index in shock compressed materials," in: *Shock Waves in Condensed Matter* (ed. Y.M. Gupta) Plenum Press, NY, pp. 441–454.
112. Wise, J.L., G.I. Kerley, and T.G. Trucano (1992), "Shock-vaporization studies on zinc and porous carbon," in: *Shock Compression of Condensed Matter – 1991* (eds. S.C. Schmidt, R.D. Dick, J.W. Forbes, and D.G. Tasker) Elsevier Science Publishers, BV, pp. 61–64.

5 Sabot Designs for Launching Penetrators and Projectiles

Alois J. Stilp

5.1 Introduction, Function and Basic Construction

Sabot is the French word for a wooden shoe. It is a successful word choice in the present case because some important features of a functioning projectile sabot are included in the word: The wooden shoe must be light and strong, must conform to the foot, and should be easily removable.

These and other requirements play an important role in the design of sabots for impact and free-flight tests. The design and fabrication of sabots is a small but important part of the complex operation of impact and free-flight ranges, because a satisfactory sabot is necessary if success of an experiment is to be guaranteed.

What is a sabot? It is a technical device conforming on its outer surface to the launch tube wall and on its inner surface to the projectile. It carries the projectile through the launch tube under the action of the driver gas. In a weapon system it is designed to deliver a specific projectile to a target with an acceptable dispersion and velocity. With a sabot it is possible to launch most regular and irregular shapes and subcaliber projectiles over a wide range of velocities.

The functions of sabots are diverse:

- The sabot positions and supports the projectile during launch.
- It seals, or obturates, the driver gases within the launch tube. Imperfect sealing permits blow-by of the driver gas, resulting in asymmetric loading of the sabot–projectile assembly and disturbance of the sabot as it is separated from the projectile.
- It should minimize undesired lateral or yawing motions produced by bore and/or projectile asymmetries, variations in bore friction, and asymmetries in pressure loading. Imperfect machining of the sabot results in a tight or loose fit in the launch tube, with the consequences of either excessive friction or blow-by of driver gases. Imperfect honing of the launch tube and cleaning after firing can also cause variation in bore friction. Undesired lateral effects can be caused by disturbances of the propellant burning rate in powder guns or by shock waves in light gas guns.

- The sabot should separate from the projectile without disturbing its flight. It should not induce yaw or pitch and should separate without creating hazards to personnel or equipment.

Three phases of sabot/projectile interactions can be distinguished:

- The interaction during the acceleration phase. The roundness, straightness, inner wall roughness, and cleanliness of the launch tube, on one hand, and the quality of sabot fabrication and the maintenance of tolerances, on the other hand determine this part of the interaction. Careful cleaning of the launch tube is often ignored with fatal results, especially at high velocity.
- The second stage of sabot–projectile interaction is the sabot separation phase. At low ambient pressure the separation proceeds smoothly, which means that the separation process takes place over a long flight path, and leads to a correspondingly long period of interaction of the sabot and projectile that is caused mainly by shock waves. At higher ambient pressure the separation occurs more rapidly.
- Depending upon the construction of the impact or free-flight range, a strong sabot–projectile interaction can occur as the sabot is caught or deflected. At most ranges it is necessary to use a catcher plate to stop or deflect the sabot parts after separation from the projectile. The simplest construction is a strong steel plate with a hole in the middle to allow passage of the projectile. As discussed later, some elaboration of this design is necessary to ensure that particles of shattered sabot parts cannot hit the passing projectile.

The sabot is a parasitic part of a sabot–projectile assembly, and serves no purpose after completion of the separation process. Because of this, the weight of the sabot should be as small as possible. However, the sabot must be able to withstand the large acceleration and centrifugal forces imposed on it during its passage through the launch tube. Therefore a compromise must be found between low weight and high dynamic strength. This compromise restricts the selection of sabot materials. For all high velocity operations a further restriction on the material selection is imposed by the fact that no metallic sabot part can slide along the metallic wall of the launch tube without damaging the tube. Plastic materials and fiber reinforced plastics are the best choice, and one of the best of these is polycarbonate, known as Lexan in the US and Makralon 4tlg® in Germany. This material is also attractive because it is inexpensive.

A sabot consists, with few exceptions, of three major parts: the base or obturator, the driver or pusher, and the finger petals.

The base or obturator is a single or a serrated piece of plastic. Its function it to form a seal to prevent blow-by of the driver gas. The driver or pusher plate is usually a high strength steel, titanium, or aluminum alloy disc or cup that is thick enough to prevent the projectile punching through it as it transfers the gas pressure to the projectile.

The sabot finger petals (two–four fingers) support and stabilize the projectile during its passage through the launch tube. The sabot fingers are usually made from plastic. Special sabot finger petals for kinetic energy (KE) penetrators and rods are fabricated from high strength aluminum or titanium alloys.

Diverse elements and features determine the design of sabots. The mass, shape, and velocity range of projectiles, the type of launch facility, the propellant (powder, gas, or light gas), the gun caliber, and the ratio of maximal to average base pressure in the launch cycle are important parameters [1]. Their role depends upon whether the projectile is being launched for impact or aerodynamic studies or is part of a weapon system.

The development of sabots for aerodynamic or impact studies is usually accomplished through empirical approaches and by intuition. Adoption of this procedure is possible because of the long history of successful experimental experience. The exchange of results, both good and bad, and experimental know-how in meetings of the Aeroballistic Range Association (ARA) has made an important contribution to the step-by-step optimization of sabot–projectile assemblies for many applications.

The experiments cover a wide range of velocities and employ projectiles of various shapes and materials and having various values of mass. Usually only a small number of tests of a given kind are conducted. For this reason, computer-aided design (CAD) methods have been little used. The situation is completely different when sabots for KE weapon systems are designed and fabricated. In this case it is reasonable to use CAD methods.

5.2 Sabots for Rods and Kinetic Energy Projectiles

Rods and KE penetrators are slender bodies having circular, rectangular, or triangular cross sections, and for which the axial dimension is larger than the greatest lateral dimension. If the ratio of axial dimension to greatest lateral dimension (called the length-to-diameter or L/D ratio) is less than ten we speak of short rods and if it exceeds ten we speak of long rods, although this classification is arbitrary. Because the mass per unit cross-sectional area of these projectiles is large, special sabot designs are necessary.

Short rods are usually launched with a sabot that pushes the projectile from the rear, whereas long rods are usually launched with a sabot that pulls it from the front. Kinetic energy penetrators used in weapon systems are always accelerated with a pulling sabot. The principle of each of these designs is illustrated in Fig. 5.1.

Because the sabot diameter is several times larger than the rod diameter, a specially designed pusher plate is placed at the back of the rod to transfer the accelerating force to its rear surface. This device is usually a truncated cone made of high strength steel or titanium alloy. Its base has a diameter nearly as large as the bore. The height of this cone depends on the peak

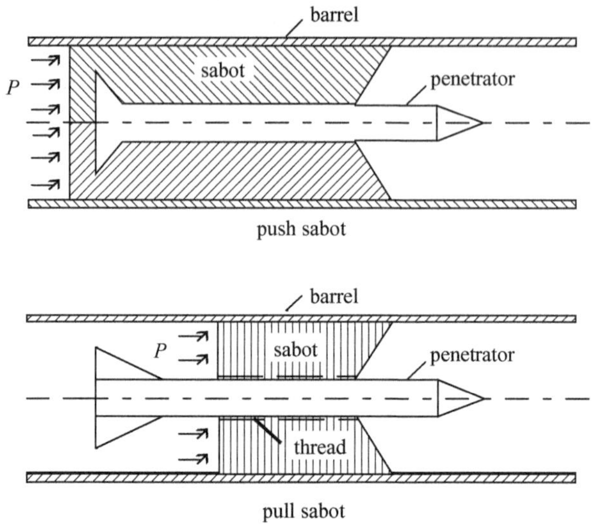

Fig. 5.1. Design principle of push and pull sabots

pressure applied to its base. If the pusher is too thin, it can fail in shear allowing it and the obturator to accelerate past the projectile so that the total sabot package is pushed past the rod. Fatal damage to the launch tube is often a consequence of this event.

Depending upon its density and strength, the rod can experience plastic flow at extremely high peak pressures. This leads to a large increase of the friction between launch tube wall and the sabot. The sabot–rod assembly may fail under high compressive load.

An example of a typical push sabot used for steel and sintered tungsten alloy rods, proved in many tests for different caliber launch tubes, is shown in Fig. 5.2. This is a sabot using four serrated polycarbonate fingers and having an inverted hollow cone at the front to improve the aerodynamic separation. The rod itself is pressed into a cylindrical cavity in the steel cone, which serves as a flight stabilizer. A thick steel or titanium alloy disc acts as the driver plate. The rear of the polycarbonate obturator has a spherical shape to improve the seal formed to prevent passage of the driver gases. If all sabot parts are well fabricated a symmetric aerodynamic separation from the rod is guaranteed, as demonstrated in Fig. 5.3. The shock waves form a symmetric pattern.

The acceleration of rods with length to diameter ratios exceeding about ten can only be performed with the pull technique. The acceleration force is transferred to the outer surface of the rod by a threaded or corrugated surface on the inner wall of the sabot.

The dimensions of this thread, its length, pitch, and the angle or shape (sharp, trapezoidal, or round) of the thread form depend mainly upon the

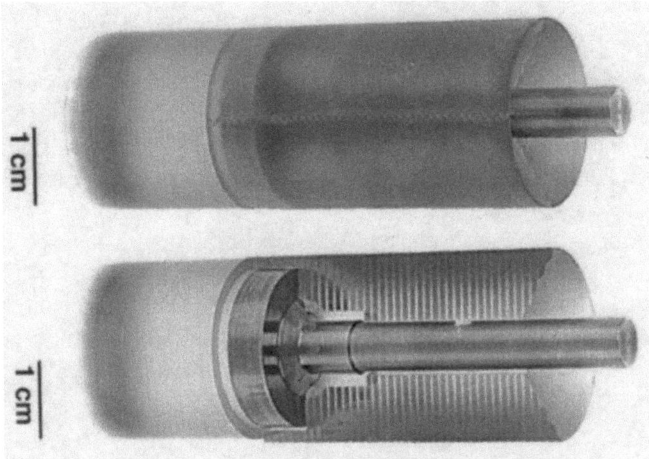

Fig. 5.2. Push sabot construction for steel and sintered tungsten alloy rods

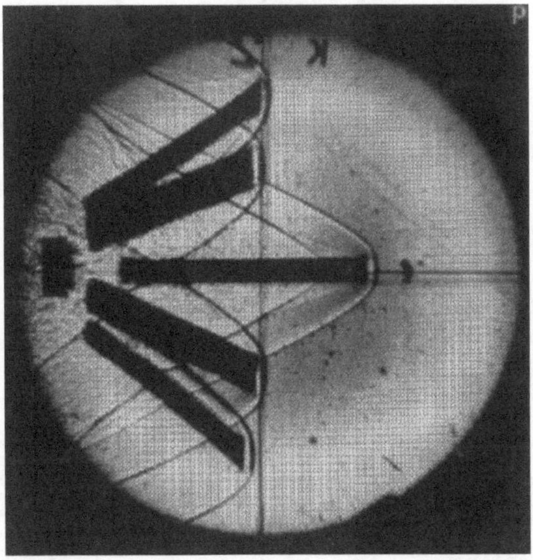

Fig. 5.3. Shadowgraph picture of a symmetric sabot separation

test velocity, the length of the rod, its diameter, and the strength and density of the sabot and rod. The stresses produced at the sabot–rod interface can be withstood only by high strength metals such as aluminum or titanium alloys. A plastic ring used as a bore rider at the front and a plastic cup used as an obturator at the rear of the metal sabot to prevent its coming into direct contact with the launch tube.

Three failure modes can be identified for pull sabots launching long rods: The part of the rod extending in front of the sabot is subjected to large compressive loading and can buckle or bend and the rear part of the rod can fail in tension. Finally the threaded or corrugated part of the sabot or rod can fail by shearing.

A typical pull sabot design, proven satisfactory in many tests for $L/D = 15$ rods, is shown in Fig. 5.4. This sabot is fabricated from a high quality aluminum alloy. It is a four piece aluminum body with a forward bore rider. A four piece plastic obturator fabricated from Polycarbonate with a serrated interface seals the assembly against the base pressure. In addition, the rear of the obturator is covered with a silicone rubber. These sabots were designed and constructed empirically, and have been improved over a long period of experience.

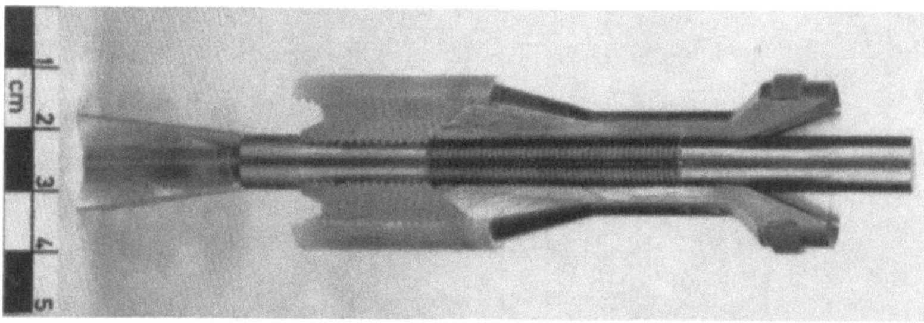

Fig. 5.4. Pull sabot for $L/D = 15$ rods

If long rods with $L/D = 25$ or greater are accelerated to velocities up to 2.5 km/s, many problems can arise due to the large acceleration loads. Small imperfections in the sabot design and tolerances in the launch tube can cause catastrophic damage to the expensive tube.

It is an important requirement for terminal ballistic tests that the rod hit the target with a yaw angle near zero. Details of a successful sabot design are reported in [2]. Figure 5.5 shows a basic sabot design for rods with different L/D ratios.

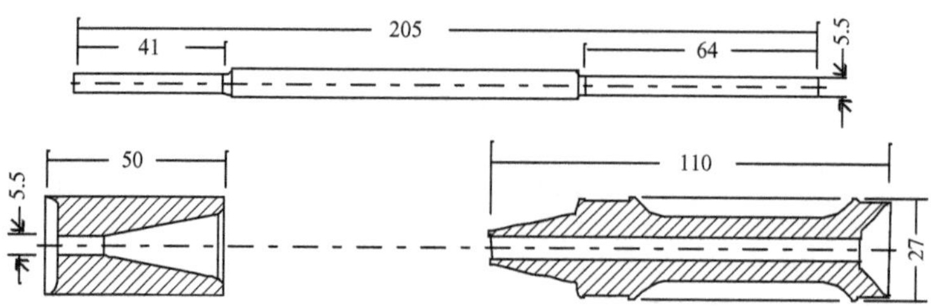

Fig. 5.5. Typical long-rod penetrator ($L/D = 35$) and the sabot designed to launch it

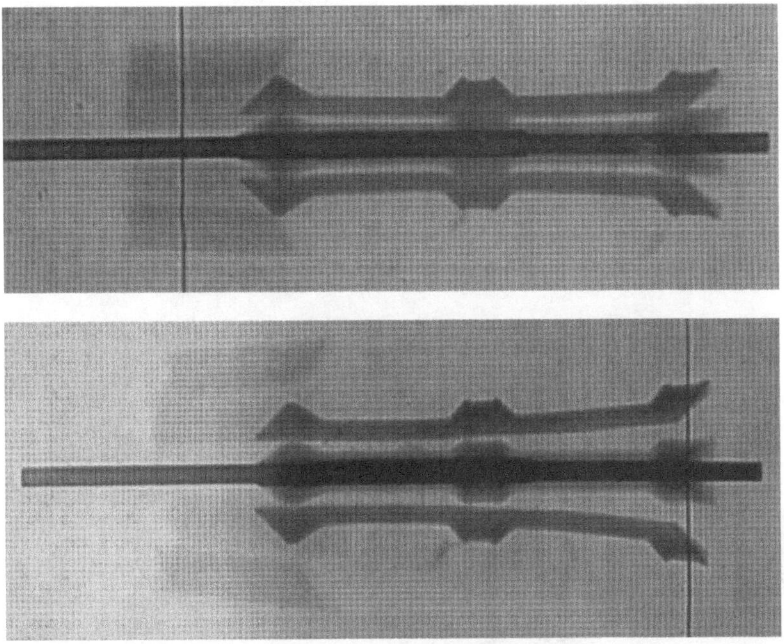

Fig. 5.6. Flash x-ray photographs of two phases of a perfect separation

Two phases of a perfect sabot release are shown in the flash x-ray photographs in Fig. 5.6.

Large yaw and pitch angles are generated and, in the worse case, rod bending can occur if the launch tube is not perfectly straight. The flash x-ray picture in Fig. 5.7 shows an example of rod bending due to use of an imperfectly straight launch tube.

A thick steel catcher plate with a hole and a short steel tube in the middle stops all parasitic parts when placed at the correct distance from the muzzle. The short tube on the catcher plate prevents the impacting sabot parts from influencing the passing rod. If metallic jets from the sabot impact area hit the rod and the stabilizing fin they can cause damage to the rod. Incorrectly positioned catcher plates induce yaw and pitch angles of the rod. The imprint of the impacted sabot fingers on the catcher plate show clearly the profile of the sabot thread. The sabot must be turned as each shot is loaded to prevent its parts impacting the catcher plate at the same position on successive shots. The catcher plate must be replaced after two or three rounds. The photograph in Fig. 5.8 shows an impacted catcher plate.

Kinetic energy penetrators with large L/D ratios are an efficient means of defeating heavy armor. The sabot for these penetrators was designed employing CAD methods with the goal of minimizing the sabot weight and maximizing the sabot strength. The basic form of these sabots is similar to

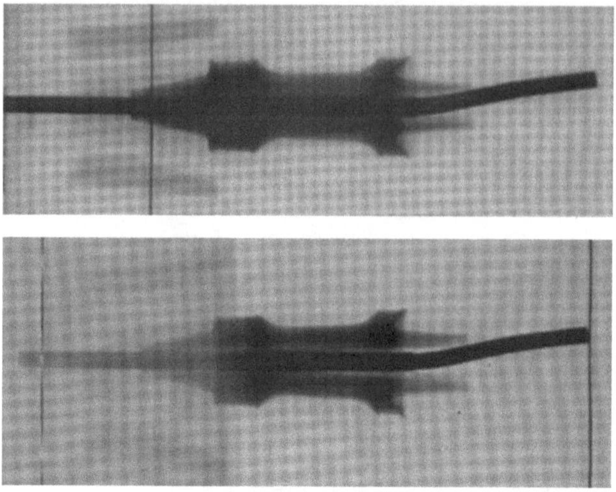

Fig. 5.7. Bending of a long rod launched from an imperfectly straight tube

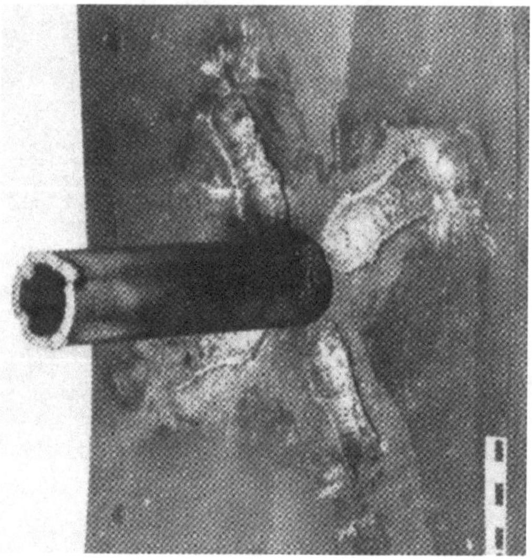

Fig. 5.8. Impacted catcher plate

the that of the sabot used for laboratory ballistic tests. However, one should keep in mind that sabots for KE weapons are optimized for one velocity, one caliber, and one L/D ratio. Sabots for laboratory experiments are used over a wide range of parameters, so their design must be much more flexible.

An example of an actual KE weapon consisting of a depleted uranium (DU) penetrator and an aluminum alloy sabot is shown in Fig. 5.9 [3].

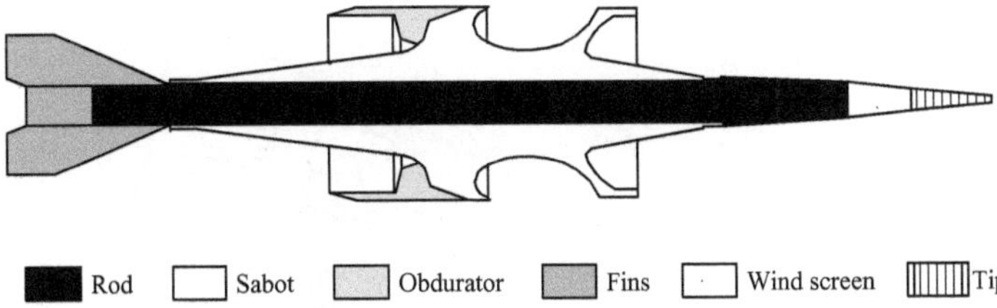

Fig. 5.9. Schematic illustration of an M829 sabot–penetrator assembly

5.3 Sabots for the Simulation of Warhead Fragment Impact

Missile warheads with preformed fragments are used for extended-range air defense to defeat theater ballistic missiles (TBM) and aircraft.

The defense against TBMs with conventional, chemical, or biological payloads, especially with chemical or biological submunitions, is a tremendous technical challenge if total destruction (hard kill) of the payload is demanded.

Of the several defense concepts that have been proposed, the one considered here is fragment impact. If the fragment shower is to be effective enough to destroy the TBM payload, each individual fragment must be able to penetrate deep into the inner layers of the TBM submunition.

Several anti-TBM warheads with preformed fragments such as high density cubes, spheres, discs, rods, and high performance shapes are being tested. Field tests often ignore the real encounter situation because neither the warhead nor the target is moving in the test arena.

The performance of light-gas guns and highly sophisticated sabot designs allow the acceleration of differently shaped warhead fragments such as cubes, cylinders, spheres, discs, and rods as single or multiple fragments over wide velocity and mass ranges. Sequential or simultaneous multiple impacts by fragments of specified orientation at the moment of impact can be produced under controlled test conditions.

For several years, the author was the head of the Ernst Mach Institute Impact Physics Division and was concerned with many aspects of the lethality endgame in different joint programs. The TBM defense endgame is hypervelocity impact research (Fig. 5.10). One part of this research program is the development of an efficient sabot for launching high density fragments.

To prevent the sabot from being perforated and pushed past a high density projectile (steel, sintered tungsten alloy, DU, etc.) by the propelling gas, the back end of the projectile must be supported in the sabot by an inner metal cup or plate made from high strength aluminum or titanium alloys as shown in Fig. 5.11.

The terminal ballistic effectiveness of cubes, discs, and rods depends strongly on their orientation at the moment of impact. The sabot developed

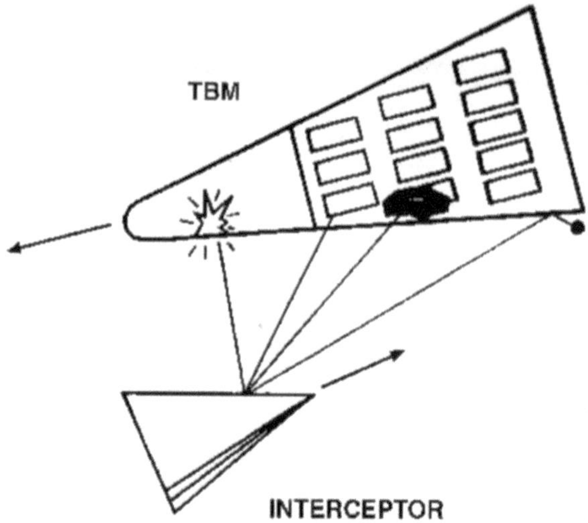

Fig. 5.10. Encounter situation of TBM and interceptor. The relative velocities of the TBM and the interceptor are in the range 2–4 km/s. The target TBM is usually a complex structure comprising multiple widely-spaced plates of differing materials. The fragments impact the target at angles ranging from 0 to 90 degrees

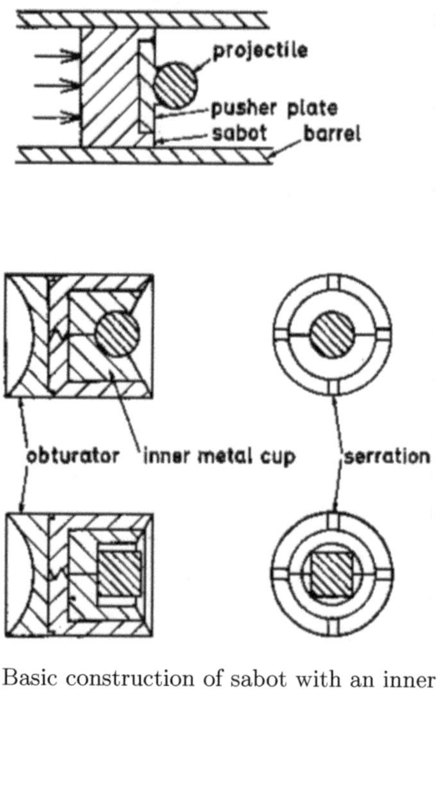

Fig. 5.11. Basic construction of sabot with an inner metal cup

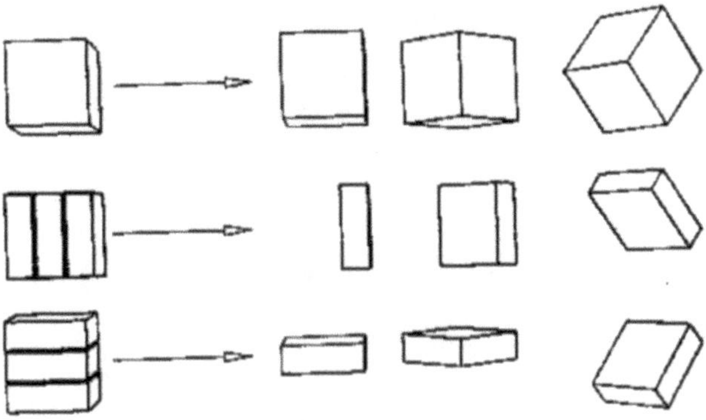

Fig. 5.12. Orientation variety of cubes and discs

makes it feasible to achieve the desired orientation for cubes and discs, as shown schematically in Fig. 5.12.

A four-piece serrated polycarbonate sabot with a thick metal driver plate designed to launch high-density discs for side and head-on impact is shown in Fig. 5.13. This sabot design has been tested successfully for 15–40-mm caliber launch tubes and fragment masses up to 100 g. Flash x-ray pictures in Fig. 5.14 show that the discs keep their orientation after separation of the sabot.

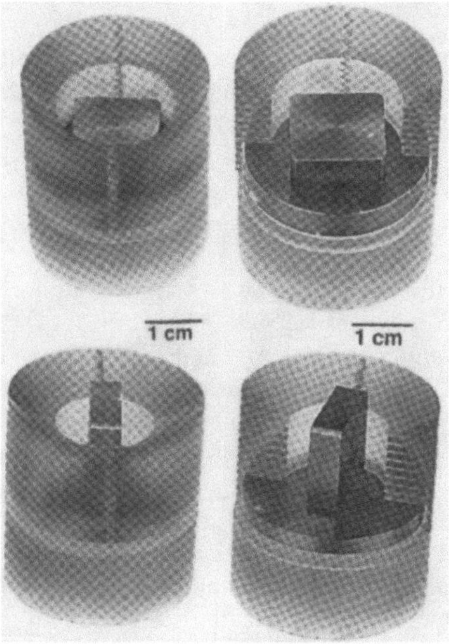

Fig. 5.13. Sabots to deliver fragments oriented for side and head-on impact

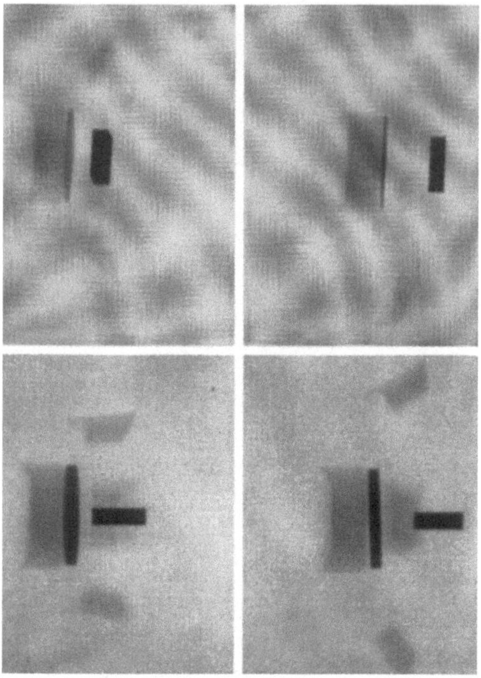

Fig. 5.14. Separation of the sabot parts from 20 g discs launched at 2.3 km/s with two orientations

The impact of a fragment by sabot parts, especially the metal parts, alters the effect of its impact on the target. To avoid this, double-plate sabot catchers are usually used. The plates are positioned close together with small holes in the middle that allow only the fragment to pass through. Sometimes the very stable flying driver plates are fabricated with asymmetries, such as holes or cuts, to produce instabilities so they can be captured by the sabot catcher.

Modified sabot designs with inner metal cups and driver plates are also applied for multiple-fragment impacts with discs, $L/D = 1$ cylinders, cubes, and rods. Some examples are shown in Fig. 5.15 for three-rod fragments, three discs with two orientations, and cylindrical fragments in two layers. Spacers or disc springs between the individual fragments or layers produce rapid dispersion of the fragments. These and other types of sabot–fragment assemblies have been used with great success in several lethality and target hardening projects. The impact of a shower of six cylindrical fragments on a double-plate steel target is shown in the flash x-ray picture in Fig. 5.16. Each fragment forms its own debris cloud which can interfere with the other debris clouds and increase the effect on the second target plate, producing perforation or spallation.

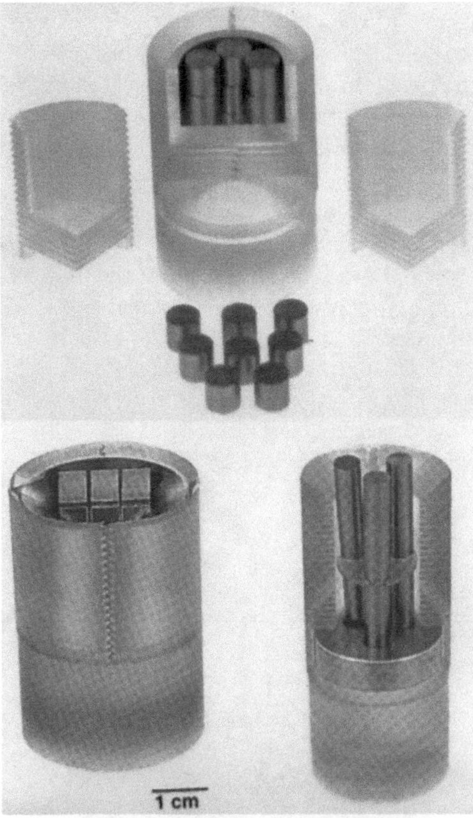

Fig. 5.15. Sabots for multiple-fragment impact studies

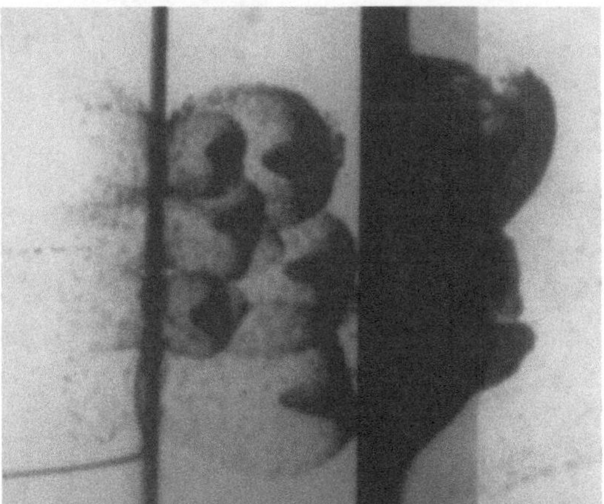

Fig. 5.16. Impact of six cylindrical fragments on a double-plate target (3.1 km/s)

Many missiles have warheads that produce explosively-formed projectile fragments (EFP) to defeat ground vehicles and aircraft. Softly-recovered real EFPs formed from Armco iron, copper, or tantalum as well as replica EFPs are developed and tested to study the influence of the actual impact velocity and other relevant parameters.

Figure 5.17 shows a sabot with a replica of an Armco iron EFP at the top and flash x-ray pictures of two phases of the separation process at the bottom. The greatest care was taken to catch all sabot parts with the double-plate sabot catcher technique. The alignment of the plates between the launch tube end and the aiming point must be performed very carefully using a laser beam.

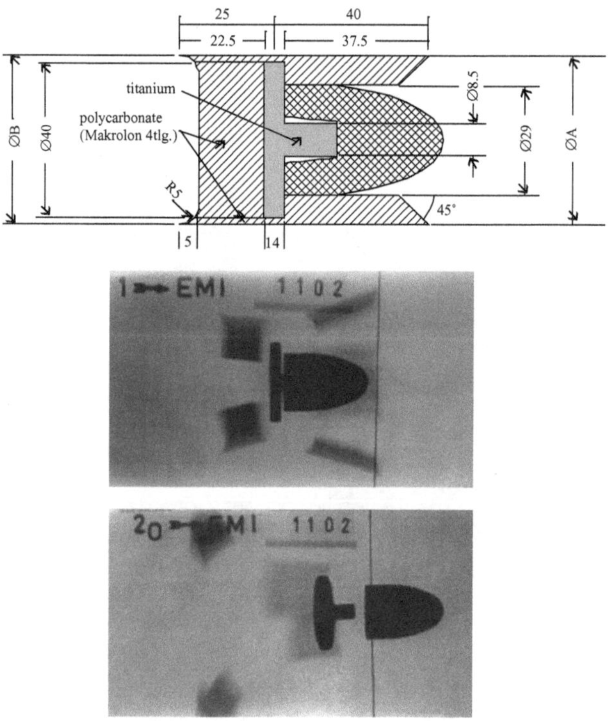

Fig. 5.17. Sabot with replica EFP and flash x-ray pictures of two phases of the sabot separation process. Makrolon 4tlg. is a polycarbonate material

Two-flash x-ray pictures in Fig. 5.18, one taken at the moment of impact on a composite target and the second taken behind the target show clearly that all parasitic sabot parts have separated and been stopped. The evaluation of EFP performance and the stopping power of the target is not influenced by secondary impact effects.

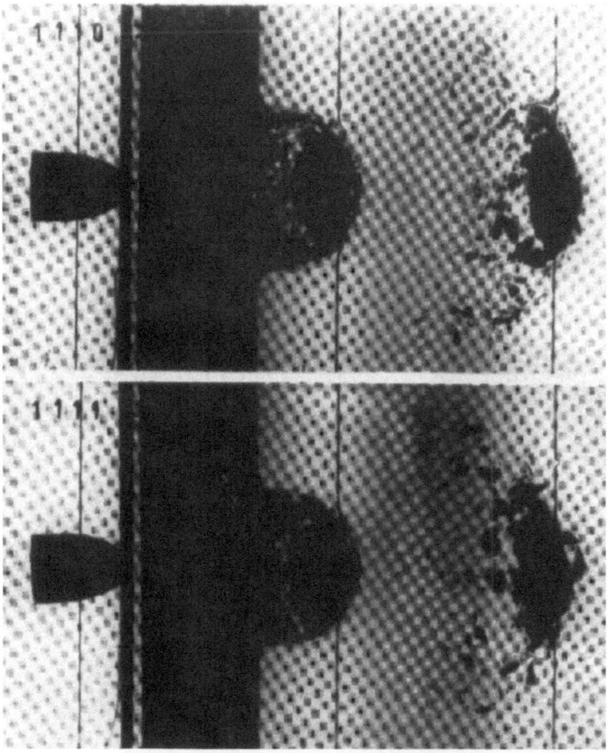

Fig. 5.18. Flash x-ray pictures of the impact of an EFP replica

5.4 Sabots for Reverse Ballistic Experiments

A number of impact experiments can be performed only in the so called *reverse ballistic mode*, which means that the projectile is at rest and the target is fired against it. This method is required to conduct impact tests in which the projectiles have extreme dimensions or are flat discs, needles, or of other unusual shapes.

Many of the segmented rod studies were performed using this method, as were most of the rain, ice, or dust erosion studies. With multiple flash x-ray systems or laser stroboscopes, the impact process, including the penetration process that occurs inside the target, can be observed. Highly sophisticated soft recovery methods are used to catch the target without producing any secondary damage.

Sabots for reverse ballistic tests are usually cup shaped. The cups carry the target (ceramics, metals, foams, layers of different materials, etc.) These cups are fabricated from plastic materials, often with an inner cup made from aluminum or titanium alloys.

5.5 Sabots for the Simulation of Micrometeoroid and Space Debris Impact

Earth-orbiting satellites and space stations encounter a fairly continual flux of natural meteoroids and man-made orbital debris. The meteoroid flux is not a serious hazard to most spacecraft in earth orbit. Only sensitive components such as thermal surfaces, optics, mirrors, windows, etc. can be damaged by the very small meteoroids. To simulate a meteoroid impact in the laboratory, advanced launchers such as plasma drag launchers, capacitor discharge launchers, electrostatic launchers, and other devices are used to accelerate microparticles (both metals and non-metals) to velocities as great as 100 km/s. None of these launchers requires the use of sabots. Larger projectile masses can be accelerated to lower velocities by shaped-charge and explosive techniques. No sabot can be used for these methods. The varying and uncertain shape and mass of objects launched by explosive or shaped-charge techniques complicates the data analyses.

The best choice for simulation of hypervelocity impact is the long proven light-gas-gun technique. These are the only accelerators capable of firing projectiles with well defined masses, shapes, and material consistency up to velocities as great as 11 km/s. The illustration in Fig. 5.19 gives an overview of the capabilities of hypervelocity launchers.

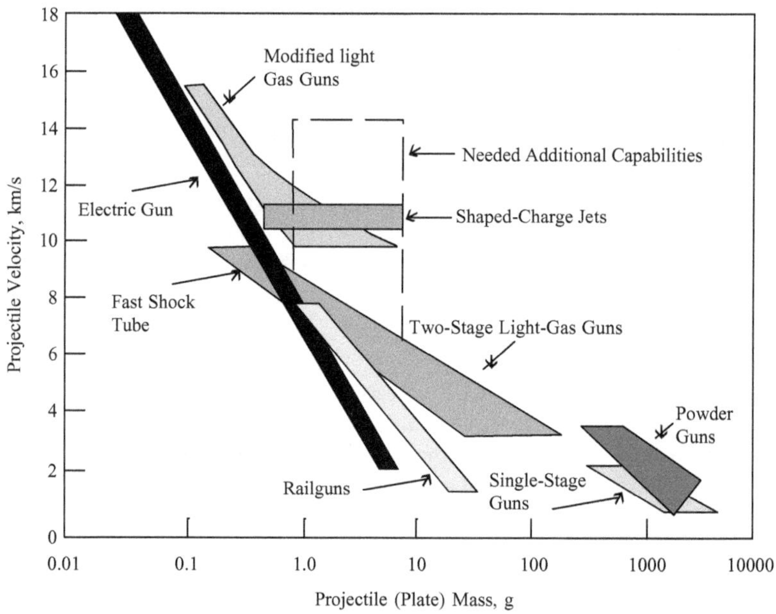

Fig. 5.19. Capabilities of hypervelocity launchers [4]

The total weight of the sabot–projectile assembly depends on the launch tube diameter. It is uneconomic to accelerate small projectiles with large-caliber guns, because the sabot must be so much larger than the projectile.

Actual debris objects have complex shapes but, for shield design, acceptance tests, and validation of spacecraft components, aluminum, glass, or other low density spheres or discs are usually used for the simulation.

The sabots for these low density projectiles are simple constructions because no special pusher plate is necessary to prevent punch through. Usually a two- or four -inger sabot, serrated or not, with an obturator is used. Another sabot construction consists of a polycarbonate cylinder with an opening for the sphere or disc. The cylinder is cut across its diameter (or across two perpendicular diameters) by a razor blade installed in a precision cutting device that includes a dial indicator to monitor the depth of the cut.

For launching high-density projectiles, a two-piece aluminum cup fit into the polycarbonate cylinder as shown in Fig. 5.20 is used. A hole for the projectile is made in the aluminum cup. This construction prevents the high-density projectile punching through the plastic part of the sabot. After leaving the launch tube, aerodynamic forces separate all sabot parts from the projectile as shown in Fig. 5.21.

5.6 Sabots for Subscale Bomb and Geopenetrator Models

Field tests with large caliber guns or sled tests with real bombs or full-scale geopenetrators are very expensive. The preparation time associated with full-scale tests, the costs for the field crew, test equipment, the ammunition and the target increase as the caliber of the gun increases. Only a few tests can be conducted; parameter studies with variations of the target construction or weapon design are only exceptionally possible. These and other reasons make model testing in this area less expensive than prototype testing.

Application of replica model laws have shown that experiments with small-scale models satisfy many mechanical similarity requirements associated with penetration of full-scale projectiles.

Bombs or geopenetrators impact targets in the low or intermediate transition velocity stage. The low velocity region of impact is characterized by little or no deformation of the projectile. The strength of the bomb is greater than the dynamic pressure of impact and the bomb is only insignificantly deformed. The mechanical properties of the target and the bomb are of importance.

To study the influence of concrete strength, effects of burster slabs, sand and rock-rubble overlays, bar screens, multilayer materials, fiber reinforcement, etc., tests with scaled classical general purpose (GP) bombs are performed at scale factors of 1/4 or 1/5.

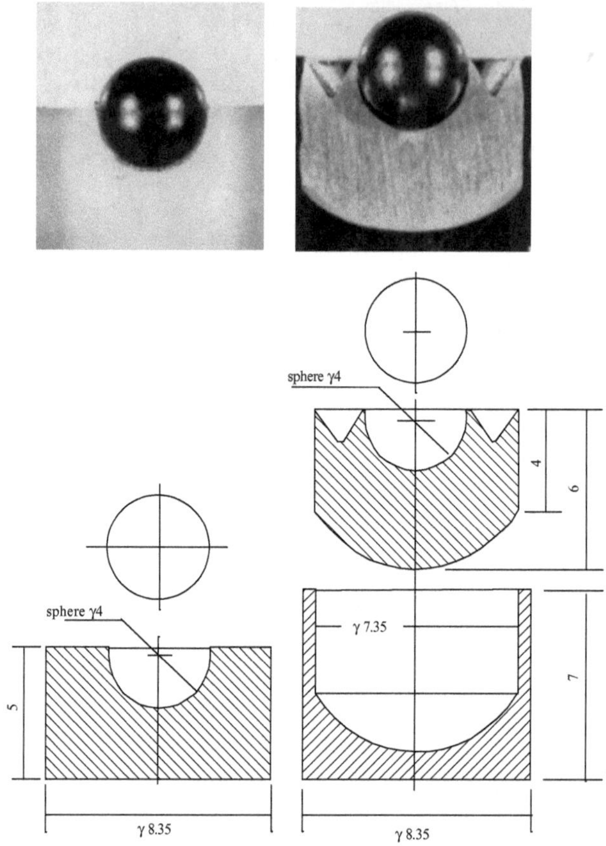

Fig. 5.20. Sabot designs for launching high-density spherical projectiles

Figure 5.22 shows the cross-section of a 500 lb bomb and its 1/4.5 scale model. The sabots for this model are shown in Fig. 5.23. The sabot consists of a four-finger petal part fabricated from hard Styrofoam. These petals are melted in a mold. Differential swelling of the basic Styrofoam material allows for the changing of the density and strength of the Styrofoam petals. Once fabricated, they retain their form and are resistant to deformation. Their surface is compact and smooth. The bomb fits into an obturator in the form of a polyethylene cup. The bomb–sabot assembly is accelerated to velocities as great as 400 m/s with a smooth-bore powder or gas gun of 100–120 mm bore diameter. The same basic sabot construction is used for other GP bomb and MK bomb models, the latter being a prototype of World War II German bomb (Fig. 5.24).

In recent decades crucial changes have taken place in both threats presented and protective measures taken to mitigate these threats. The increase of the protection level is no longer restricted to trivial solutions such as

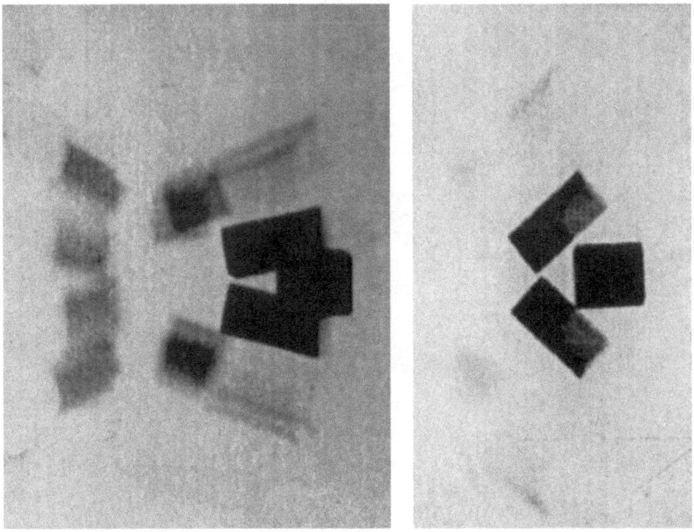

Fig. 5.21. X-ray photograph of the separation of the outer and inner cups

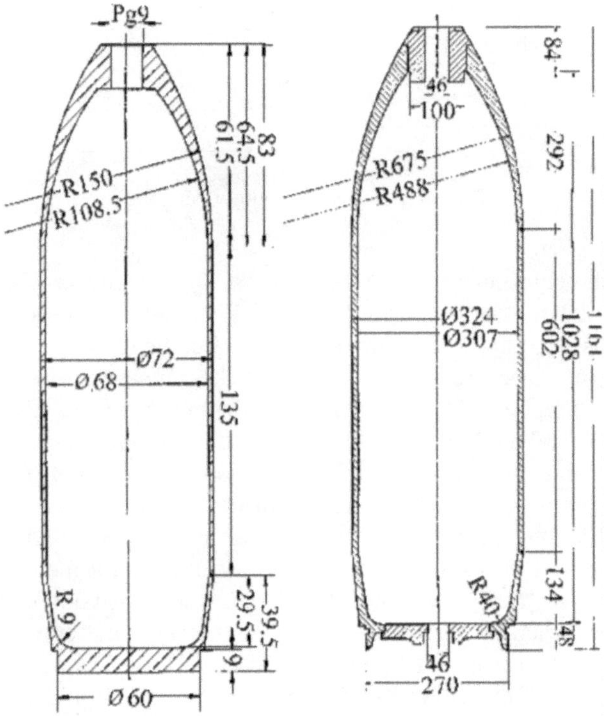

Fig. 5.22. Cross sections of a 500 lb bomb (*left*) and its subscale model (*right*)

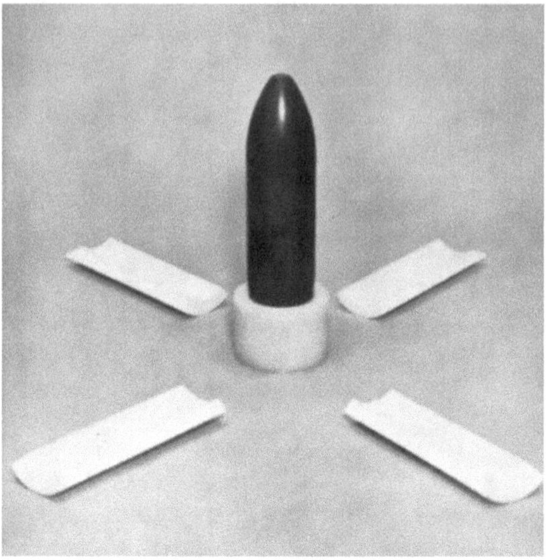

Fig. 5.23. Sabot for the 1/4.5 scale bomb model

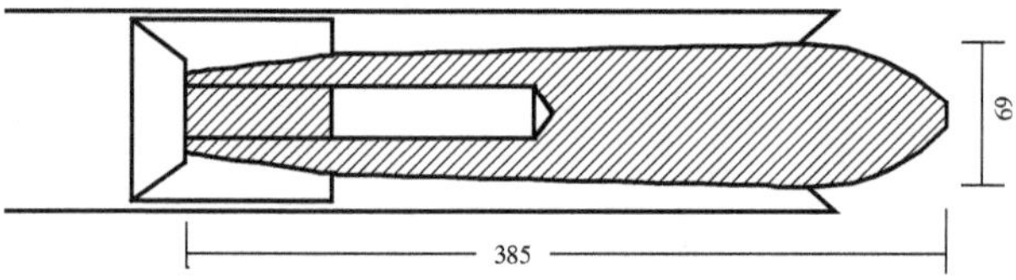

Fig. 5.24. Bomb model with sabot

increasing the thickness of concrete barriers. Buried bunkers and tunnel facilities with a large variety of anti-penetration shields have been developed. Burster slabs with loose boulders or boulders embedded in concrete are installed over the structures. The functions of these measures are diverse. They can deflect a bomb, break the bomb case, destroy the fuse, or cause premature detonation. Yaw-inducing barriers can deflect bombs or cause them to impact at an angle to their line of flight. A strong reduction of the penetration capabilities is the consequence. In response to this new challenge, new generations of bombs and improved geopenetrators have been born. These new types of bombs, which are capable of perforating several meters of concrete and overlays, had their baptism of fire during the First Gulf War.

It was necessary to test high strength penetrator materials, new shapes with large L/D ratios, and highly sophisticated nose configurations to increase the angle of attack and shift the ricochet angle to higher values. The impact velocity of these new bomb types is higher than the drop velocity of classical gravity bombs. Again, tests with scaled models were very effective for parameter studies.

The sabots for these penetrator models must be stronger than those used for GP bombs, and Styrofoam is no longer applicable. Serrated polycarbonate fingers with a resin filler adapted to the fin contour, titanium alloy pusher plates, and polycarbonate obturators are developed, tested, and shown to provide good solutions over a wide range of velocities (250–1700 m/s). Figure 5.25 shows a geopenetrator model in its sabot.

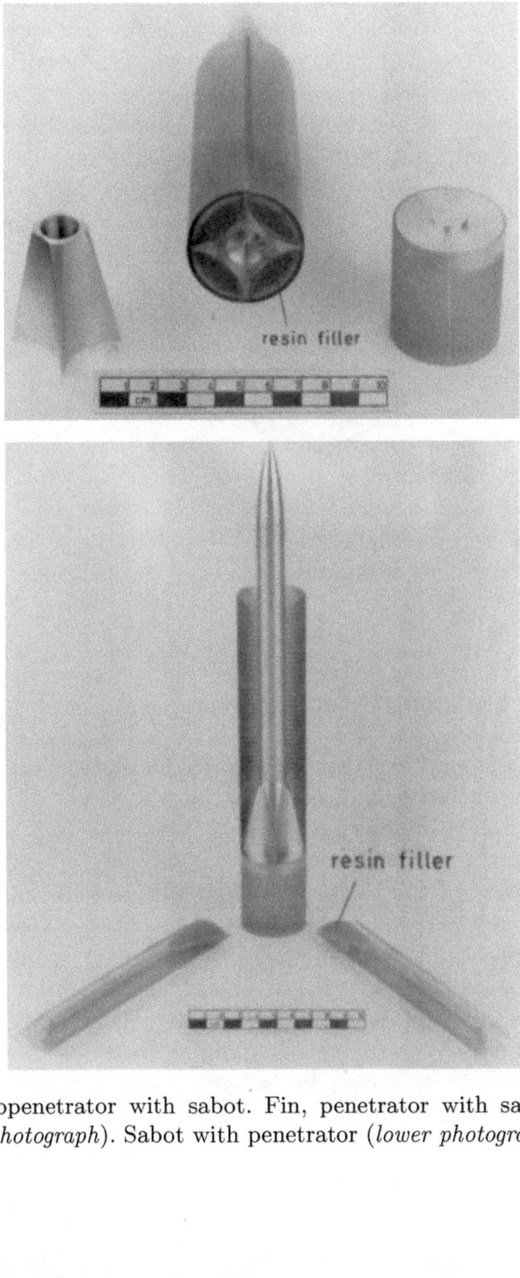

Fig. 5.25. Geopenetrator with sabot. Fin, penetrator with sabot fingers, and pusher (*upper photograph*). Sabot with penetrator (*lower photograph*)

5.7 Sabots for Bird and Hailstone Impact Simulation

The loss of aircraft and repair costs caused by bird and hailstone impact are an important economic factor. Hailstones, birds, or pebbles that impact aircraft engines, windshields, and canopies can produce serious damage or catastrophic failure resulting in total loss of the aircraft, loss of the crew members, or mission abortion. The impact velocities of these objects vary from several meters per second up to several hundred meters per second, depending on the point in the mission at which the impact occurs.

Besides computer simulation, experiments are necessary for the design of aircraft components and to provide understanding of the impact process and the structural response of the aircraft components to impact. Usually free-flight ranges with gas guns equipped with double-diaphragm or other fast opening valves are used for the simulation tests. The acceleration of the objects must be kept at a low level to prevent destruction of the ice balls or the birds being launched. To prevent their breakup during launch, appropriate sabots must be used. The objects must be separated freely from the sabot. The sabots are usually of the cup type. They are often decelerated in a tapered launch tube extension with longitudinal slits to vent the driver gases. The sabot starts to decelerate in this section and the object to be launched begins to move out of the cup. A stopper plate or a sabot cutter at the end of the tube stops the sabot or deflects the sabot parts from the object trajectory. The sabots are usually fabricated from balsa wood or Styrofoam. The density of each of these materials can vary, with the denser and higher strength materials being used for higher velocity tests. Balsa wood and Styrofoam are good thermal isolators suitable for firing ice balls and bird simulants (chickens).

5.8 Sabot Separation Techniques

For all free-flight operations, and especially for impact studies, the sabot parts must be separated before the impact on the target can be carried out. Fragments of the sabot flying close to the projectile disturb the aerodynamic flow field or cause undesired secondary impact damage to the target. Depending on the size of the range and its mission, several methods of sabot separation have been developed. Some of these are discussed briefly in the following paragraphs.

The most common way to separate the sabot parts from the projectile is the use of aerodynamic forces. When the sabot–projectile assembly leaves the muzzle a combination of aerodynamic forces from muzzle blast and the flow field outside of the blast act on the sabot parts and the projectile. Strong asymmetries of the muzzle blast caused by blow-by of the driver gas or defects such as small burrs or indents of the muzzle, begin a chain of causation of undesired separation effects and projectile instabilities. The aerodynamic forces increase with the strength of the muzzle blast and the density of the ambient atmosphere [5].

The sabot parts should separate radially from the projectile without subsequent contact. Beveling of the front of the sabot front increases the surface subjected to aerodynamic pressure and, thus, the rate of radial separation of the sabot from the projectile. The sabot pieces can be stopped by impact into a steel catcher that has a hole in its center that is large enough for the projectile to pass through without touching.

Catching the pusher plate and the obturator is a difficult problem because these parts are follow the same trajectory as the projectile. Double- and triple-plate catchers are most suitable for catching the pusher plate and obturator.

It is necessary that the sabot parts be stopped, deflected, or destroyed with very low probability of interaction with the projectile. For tests at low ambient pressures, the aerodynamic forces are not large enough to separate the sabot parts during a short flight path. For this application a small tank with a higher ambient pressure can be installed between the muzzle and the impact or free-flight area. The entry and the exit of this tank are sealed with very thin Mylar films. This technique is applicable for small and medium size ranges [3].

For medium velocities and robust projectiles, mechanical devices can be used to decelerate the sabot and deflect it from the projectile trajectory. Two devices consisting of a contracting (tapered) section and a deflecting section are shown in Fig. 5.26.

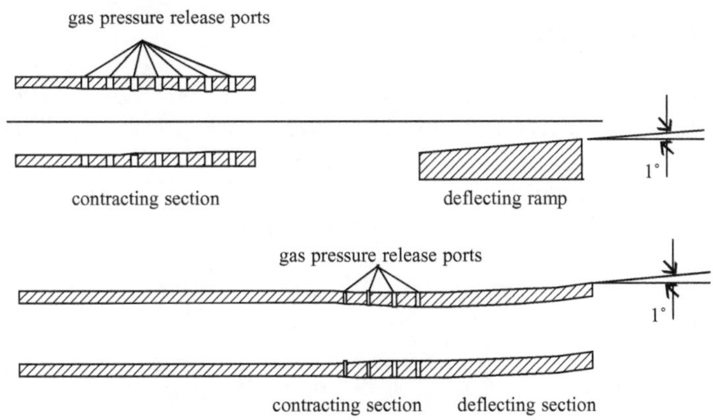

Fig. 5.26. Sabot separation device applicable for medium velocities

The sabot is decelerated in the contracting section. During this process, the projectile starts to separate from the sabot. The base pressure is reduced by gas release ports or slits. These mechanical separation devices are only applicable for cup sabots, they are not applicable for encapsulated projectiles.

The use of tangential forces, induced by a short rifled section at the launch tube entrance can initiate spinning of the sabot–projectile assembly. The spinning is stable and is maintained during both the acceleration and free-flight phases. This method cannot be applied for hypervelocity tests.

As a consequence of the high energy release through the muzzle blast and the sabot catch process, all indoor ranges are enclosed and a blast and projectile catcher tank is an important component of a range with a high safety standard.

The installation of a single- or a double-stereo flash x-ray system close to the gun muzzle gives valuable information about the sabot separation process itself and about the quality of the sabot design and fabrication. Triggering the flash x-ray systems close to the muzzle is accomplished an intense laser and narrow filters on the photocell side.

5.9 Conclusions

Sabot design is highly empirical, it has required a great degree of intuition to fulfil the basic design requirement that a sabot be strong and stable.

Modern sabot technique allows for tests with variety of projectile shapes, masses, and materials over a wide range of velocities up to 10 km/s under exactly defined and controlled test conditions. Sabot design is the secret of light-gas-gun performance.

Acknowledgment

The photographs, flash x-ray pictures, drawings, etc. not specifically attributed to others originate from the former Impact Physics Division of the Ernst-Mach-Institute, Freiburg (1970–1997). Many of the references cited below as private communication are published in the ARA conference proceedings and can be obtained by contacting the authors.

References

1. H.F. Swift and D.E. Strange, "Analyses of Sabot Operation" Published in the proceedings of the 49th ARA Meeting. (Private Communication).
2. A.J. Stilp and V. Hohler, "Sabot and Launch Techniques for Hypervelocity Impact Studies" Published in the proceedings of the 45th ARA Meeting. (Private Communication).
3. D.A. Rabern, "Numerical Simulations of Gun-Launched Kinetic Energy Projectiles Subjected to Asymmetric Projectile Base Pressure," *Int. J. Impact Engng.* **12(3)**, 1992.
4. Orbital Debris–A Technical Assessment, Nat. Acad. Press, Washington D.C., 1995.

5. V. Hohler and A.J. Stilp, "Long Rod Penetration Mechanics," in: *High Velocity Impact Dynamics* (eds. J.A. Zukas, T. Nicholas, H.F. Swift, L.B. Greszczuk, and D.R. Curran) John Wiley, New York 1990. Chapter 5.
A.J. Stilp and V. Hohler, "Experimental Methods for Terminal Ballistics and Impact Physics," in: *High Velocity Impact Dynamics* (eds. J.A. Zukas, T. Nicholas, H.F. Swift, L.B. Greszczuk, and D.R. Curran) John Wiley, New York 1990. Chapter 9.

6 High-Speed Diagnostics for Ballistics and Explosive Studies

Vernon C. Draxler

6.1 Introduction

High-speed data acquisition systems are often a necessity when conducting studies in ballistics and explosion dynamics. This is a result of the high reaction rates occurring in these areas, which leave normal-speed instrumentation being of little value in obtaining meaningful information. Two of the more widely used diagnostics for this research are flash x-ray and high-speed optical photography. Although considerably different in physical characteristics, these technologies compliment each other by sharing one important feature – their ability to record data during rapidly moving events. This chapter will discuss selected experiments in these fields and will show how each technology was applied to obtain the desired test information. This work was performed at the Alliant Techsystems Proving Ground, located near Elk River, Minnesota.

Flash radiography and high speed photography have proven to be versatile diagnostic tools for conducting studies in ballistics and explosive phenomena. High-speed photography can yield information on rapid visual events, but the by-products of smoke and flame associated with explosive and ballistic reactions sometimes precludes the use of photography. In these cases, flash x-ray can be applied to overcome the limitations of photography. With its ability to penetrate not only smoke and explosive debris, but also other optically opaque materials, x-radiography often provides test data not obtainable with standard photographic methods. Thus, together, these two types of instrumentation provide the researcher with useful diagnostics for studies in ballistics and explosives.

6.2 Flash X-Radiography

Flash x-ray instrumentation is a system that produces a high intensity, short-duration burst of radiation in the x-ray wavelength spectrum. With its ability to penetrate optically opaque media and possessing a short exposure time, this technology has a unique capability for stop motion recording of high speed events associated with ballistics and explosive reactions.

The following is a brief description of the x-ray equipment. The components of an x-ray system include a high voltage power supply, an energy storage device called a pulser, an x-ray tube that is the radiation source, and delay generators for triggering the system at the proper time. A system is comprised of one or more "channels", each channel capable of producing one single or stereo x-ray pulse per event. To obtain more than two radiographs requires adding more channels.

In operation, the pulser is charged, via the high voltage power supply, to a level between 12 and 30 kV. An event-initiated trigger signal from a device such as a foil switch or a pin switch, or the fire pulse to a detonator, is input to a delay generator. After a pre-determined time delay, which allows the test item to move into the radiographic area, the delay generator sends a trigger signal to the pulser. The pulser, with its Marx generator design, amplifies the charging voltage and discharges this increased voltage across the electrodes of the x-ray tube, thereby producing the x-ray pulse by a process called cold-cathode emission. Designs of the system allow the tube to be positioned within the pulser or in a remote tubehead that is connected to the pulser via a coaxial cable. The x-ray systems used in our testing were manufactured by Hewlett-Packard Corp. in energy levels of 300 kV and 2.3 MeV, the voltages indicative of the tube input voltage. These systems have a fixed flash duration of approximately 18 ns, measured at the full-width, half-maximum (FWHM) point on the intensity waveform. The pulse duration is independent of the x-ray tube voltage. This short exposure time can freeze the motion of such things as a shaped-charge jet traveling at 10 km/s or reveal the integrity and position of internal parts of an ammunition fuze while in flight. Since the penetration of x-rays into material is dependent on x-ray tube voltage, the lower energy model was used for shadowgraph and penetrating radiography of low-density materials and the higher voltage system for penetrating radiography of thicker, high density media. The experiments discussed below will illustrate the systems capabilities and show how both hard and soft x-rays were used, depending on the test parameters and information desired.

6.2.1 Application of Flash X-Radiography to Explosion Dynamics

Explosively Formed Penetrator

An explosive warhead that underwent extensive research and development at the Alliant Techsystems Proving Ground was known as an explosively formed penetrator (EFP) warhead. A diagram of an EFP warhead is presented in Fig. 6.1.

In operation, the high explosive charge is initiated with an exploding bridgewire (EBW) detonator. The high pressure of the resulting detonation front inverts and collapses the liner into a penetrator that remains coherent at considerable downrange distances. Various shapes and characteristics of the

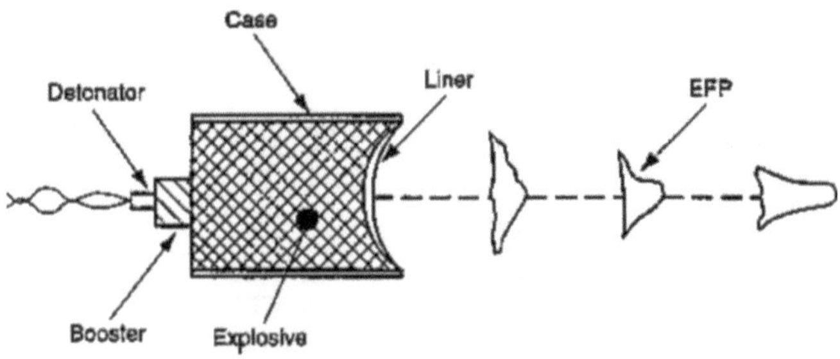

Fig. 6.1. EFP Warhead

penetrator can be obtained by varying the design of the liner and explosive charge.

Flash x-radiography was used to determine the velocity and shaping characteristics of EFPs within the first 20 calibers after being launched from the warhead. The schematic in Fig. 6.2. shows the test arrangement. The x-ray setup for this testing consisted of a fixture called a *ballistic sled*, which contained a six-inch-thick steel blast shield, an angle iron arrangement for securing the orthogonally oriented x-ray films, and a structure for holding the top-view x-ray tubeheads. The side-view tubeheads were encased in protective housings and secured with sandbags. On selected tests, a side view x-ray was placed up-range of the blast shield to obtain an early-time radiograph of the liner. The warhead was mounted to an adjustable wooden stand on the uprange side of the blast shield. A hole in the blast shield allowed the EFP to pass through while preventing the blast and explosive debris from damaging the downrange x-ray films and equipment. Prior to each shot, a precision grid of 1 cm by 1 cm squares, was positioned in the EFP trajectory and statically exposed on the films to provide the magnification factor and interfilm distance correlation. For the film on the uprange side of the blast shield, this grid was replaced with a lead sheet containing a row of coded holes. Exposing the lead sheet prior to the shot resulted in imaging the holes to provide the magnification factor and distance correlation while allowing the data area of the film to be single exposed. The x-ray system for this application was a two or three channel, 300 kV, system in a single or dual tubehead configuration, which produced shadowgraphs of the external shape of the EFP.

Radiographs of an EFP warhead containing a tantalum liner are shown in Fig. 6.3, where the direction of the EFP flight is from left to right. The radiograph on the left is a side view showing early time formation. Orthogonal radiographs obtained further downrange are shown on the right.

Folding of the EFP into a coherent mass generally results in an axial void in the inner part of the EFP. To better understand the mass distribution properties of EFPs, a high energy 2.3 MeV flash x-ray system was used to characterize their internal shape. This high energy radiation was necessary

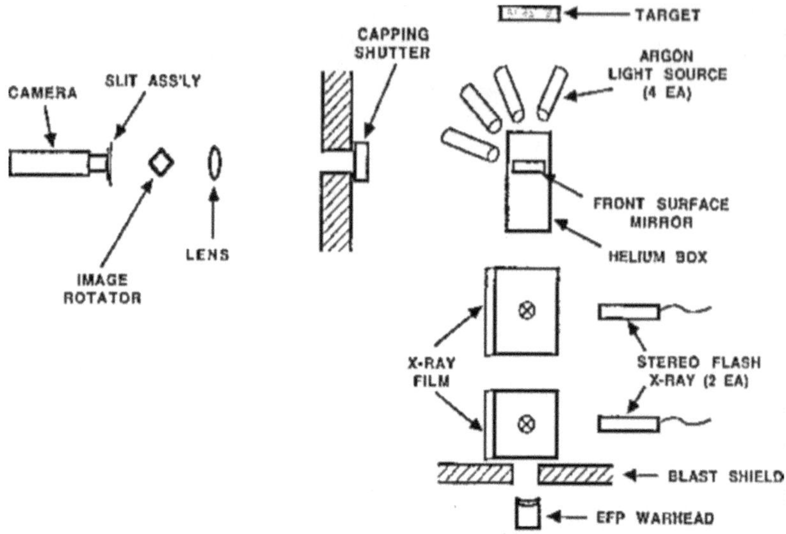

Fig. 6.2. Test Arrangement for an EFP Test

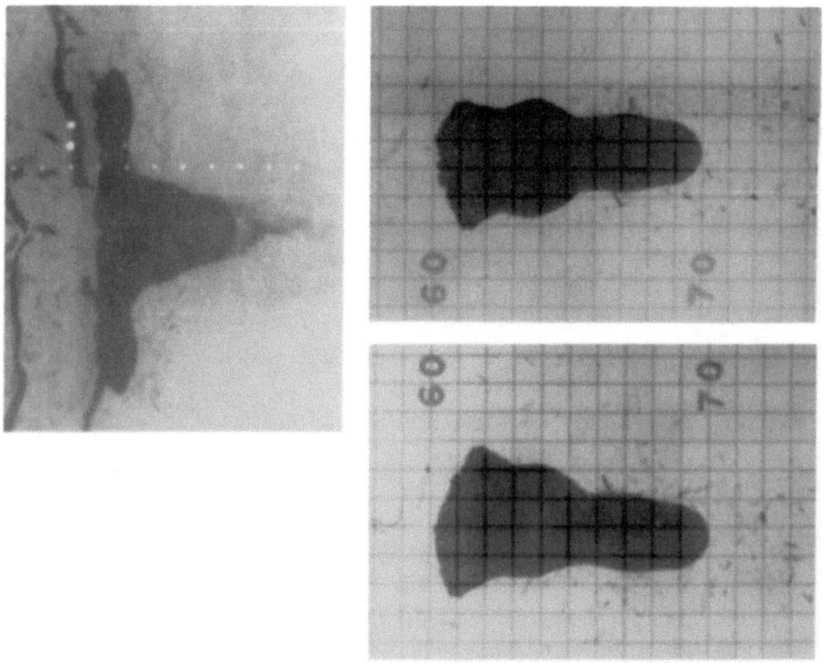

Fig. 6.3. Radiographs of a tantalum EFP

since tantalum, of which many liners are made, is very dense and x-ray absorbent.

Figure 6.4 shows a tantalum EFP radiographed with a 2.3 MeV system. The internal void is evident. Also noteworthy is the mass of tantalum that has formed in the interior and immediately behind the nose of the EFP.

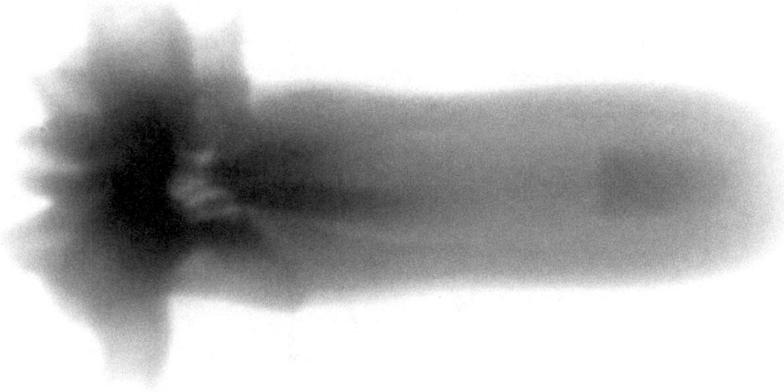

Fig. 6.4. A 2.3 MeV radiograph of a tantalum EFP

Miniature Initiator Experiment

On the opposite end of the explosive spectrum, an investigation was conducted into the functioning characteristics of the M100 detonator, which is a small, electric initiator commonly called a microdetonator (see Fig. 6.5). Although this test was conducted quite a number of years ago, it is presented here to illustrate that, at times, limitations in equipment can be circumvented by using rather simple variations to the standard x-ray method. This work was also of historical significance as it led to the development of the "spit-back" type of fuse that has been widely used in medium caliber shaped-charge munitions.

The detonator for this research was 0.25-inch long and 0.10 inch in diameter with a 0.008-inch thick aluminum case. A stub protruding from the header of the detonator serves as one electrical contact for the bridgewire and the case itself is the other bridgewire contact. For this test, a two-channel 300 kV system producing soft x-radiation was used in a setup shown in Fig. 6.6. For explanation, soft x-rays are those for which the intense radiation occurs in the long wavelength portion of the x-ray spectrum. This type of radiation, along with a complimentary radiographic technique, can better image thin, low density objects. To verify the need for soft x-rays for imaging the small aluminum particles associated with this experiment, hard radiation was attempted with negative results.

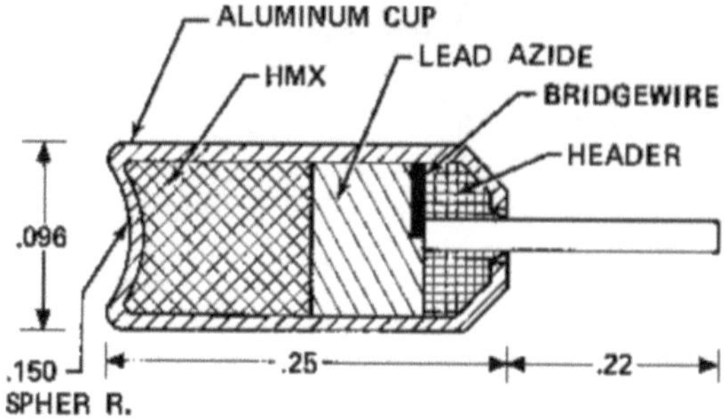

Fig. 6.5. Drawing of a Microdetonator

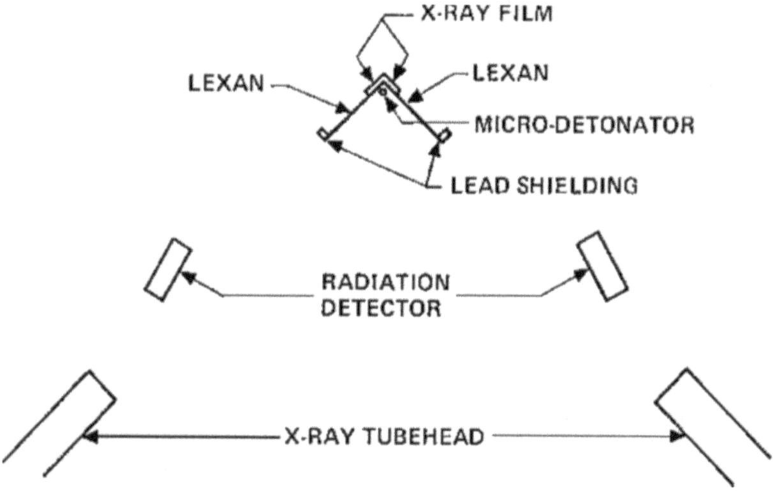

Fig. 6.6. Test Setup for the microdetonator experiment

It was desired to obtain information on the shape and velocity of the end cap of the detonator, called a *flyer plate*, within the first few millimeters of travel. Since explosive equations predicted an approximate velocity of 4 mm/μs, this meant that to acquire data it was necessary to obtain submicrosecond delays between x-ray flashes. However, when this test was performed, the available delay generators possessed a minimum setable delay time of one microsecond, meaning that if they were used in their normal mode, the radiographs would occur too late in time to be of interest. This limitation was overcome by i) setting both delay generators in the "out" position (no delay), ii) triggering both generators with the same trigger signal, and iii) using the jitter (the random, uncontrollable difference in time) in the system to produce the required submicrosecond time intervals. Two dynamic radiation detectors, positioned to sense the x-ray pulse from each tubehead, were input to a high sweep speed oscilloscope to record the x-ray flash times.

Triggering of the x-ray system was accomplished with an ionization type pin switch manufactured by Dynasen, Inc. The pin was positioned against the metal case of the detonator, then retracted until the electrical continuity was broken. Expansion of the metal case, caused by the detonating explosive, shorted the pin and produced the signal to trigger the x-ray system.

Dental x-ray film, in a size of 1-1/4 inches by 1-5/8 inches, was chosen for this test. The films were double exposed, with the static exposure imaging the detonator to provide the magnification factor and distance correlation. Post test radiographs were read on an optical comparator having a resolution of 0.001 inch.

Results of the nine shot test series are given in Table 6.1. Velocities of the flyer plate agreed well with explosive theory. It is also noted that the system jitter for the nine shots produced delay times covering the entire submicrosecond range, from 805 nanoseconds down to 10 ns. By using this jitter technique interframe times corresponding to framing rates as high as 10^8 fps were obtained.

Table 6.1. Results of the microdetonator experiment

Shot Number	Plate Travel mm	Interframe Time ns	Plate Velocity mm/μs
1	1.25	305	4.11
2	2.76	730	3.77
3	3.13	805	3.89
4	0.05	10	5.03*
5	0.93	235	3.94
6	0.52	130	4.04
7	0.45	150	3.00*
8	1.30	360	3.62
8	1.30	360	3.62
9	0.68	175	3.87

*Suspect due to resolution of the time measurement.

Figure 6.7 contains orthogonal radiographs of four of these microdetonators with interframe times of 175 ns, 360 ns, 730 ns, and 805 ns. The rectangular section in the center of each film is an image of the detonator, obtained by a static exposure. These results show that the flyer plate changes from the static concave shape to a convex shape early in flight and only the center of the plate is focused down on axis to provide the energy for impact initiating the next explosive element.

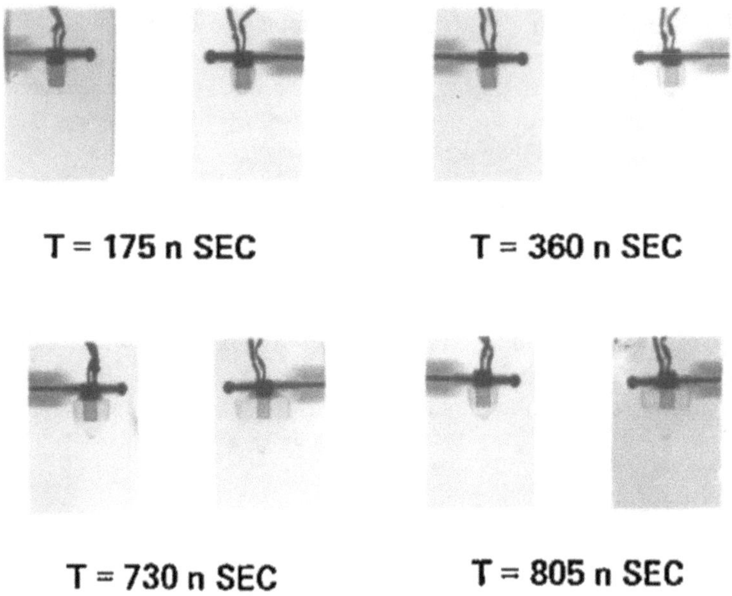

Fig. 6.7. Radiographs of four microdetonators

6.2.2 Application of Flash X-Radiography to Ballistics

Medium Caliber Ammunition

Flash radiography was also effectively applied to ballistic testing of medium caliber ammunition. Requirements in the design of fuzes for medium caliber ammunition usually dictate that the fuze be functionally safe within a certain distance from the gun muzzle and fully armed at a further distance downrange. Flash x-radiography has been used extensively to quantify the fuze arming characteristics. Since the fuze components are completely enclosed within the fuze ogive, data could not be obtained with photographic methods. The radiographs shown below were obtained with a 300 kV x-ray system.

Figure 6.8 contains two flash radiographs of a point detonating, self destruct, 20 mm fuze. The x-radiograph on the left, taken at 15 m from the muzzle, shows the firing pin engaged in a slot on the rotor, resulting in a safe condition. The radiograph on the right, taken at 40 m downrange, shows the piston and firing pin moved forward and the rotor moved to the aligned position, indicating the fuze to be armed at the 40 m distance.

Terminal ballistics effects of medium caliber ammunition were also studied using flash x-ray. One experiment was to determine the sensitivity to oblique impact of a 30 mm fuze similar to that described previously. The test procedure consisted of impacting the fuzed projectiles against a steel target placed at various angles of obliquity.

Fig. 6.8. Radiographs of a 20 mm fuze

The radiograph on the right in Fig. 6.9, taken 50 m downrange, verifies that the fuze is fully armed. On the left is an x-ray of the same fuze taken 20 µs after striking a steel target at 60° obliquity. The reduced film density in the area of the fuze where the explosive booster is located, along with structural deformation of the material below the booster indicate that this explosive component has already detonated at 20 µs after impact. Therefore, the fuze is sensitive to, and normal function has occurred, at 60° oblique impact.

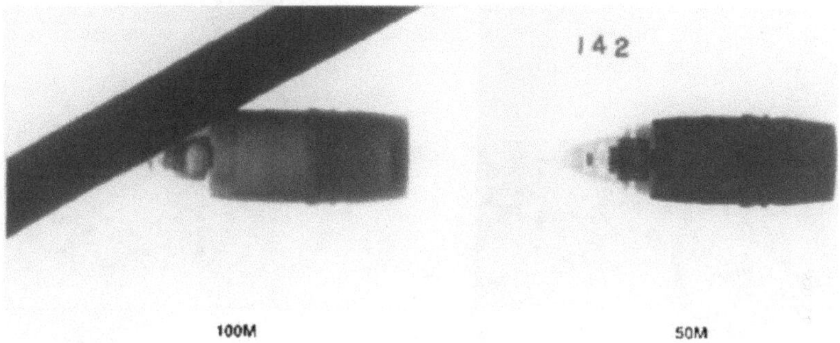

Fig. 6.9. Radiographs of a 30 mm fuze during an oblique impact

6.3 High-Speed Photography

The complimentary instrumentation to flash x-ray for ballistic and explosive studies is that of high-speed photography. Often easier to interpret than radiographs, photographs yield a visual, front view representation of the event. This chapter describes two types of photographic instrumentation that were successfully applied to obtain the desired test information: (i) strobe light photography and (ii) a high-speed rotating mirror camera.

6.3.1 Strobe Photography

As part of a munitions development program, it was desired to measure the velocity and spatial distribution of projectiles launched from a missile bulkhead. The projectiles, called *projectors*, traveled radially outward from the bulkhead, resulting in the need to acquire data over a large circular area. Another important consideration in choosing the data acquisition method was that requirements limited the measurement error in velocity and spatial distribution to less than 0.5% and less than 1°, respectively. Several diagnostic techniques, such and flash x-ray and high speed framing photography were investigated, however, strobe photography was chosen since it appeared to better meet the above constraints.

The missile bulkhead incorporated 210 projectors arranged circumferentially around the warhead perimeter and in five vertical rows (see Fig. 6.10). Note that in rows two through five, the projectors are staggered within each row. The bulkhead was mounted vertically so that the projectors traveled radially outward in a horizontal pattern. Each projector was 2-1/4 inches long by 1-1/4 inches in diameter and contained a cylindrical cavity that was mated to a barrel assembly. The barrels were loaded with a propellant that was initiated by an explosive line cord. Arrangement of the explosive cords was such that the projectors were launched in five vertical rows with a time delay between the initiation of each row.

Fig. 6.10. Photo of Projectors

The strobe system consisted of two xenon-filled flash tube light sources (Xenon Corp.) that could be remotely triggered. To meet the velocity error constraint, the system components were selected to produce a flash duration of less than 10 μs. With this short exposure time, the light intensity was insufficient to photograph the projectors directly. However, this deficiency was overcome with the use of High Gain Reflective Sheeting (3M Corp.), a reflective tape having properties similar to those found on road signs. The tape was applied to the projectors in a coding arrangement so the projector's in-flight orientation could be recorded.

Testing was performed in an enclosed building having an arena diameter of approximately 35 ft. Both the interior of the building and the projectors were painted flat black to minimize light reflections. Two still-frame cameras, one for redundancy, with an 8 inch by 10 inch film format, were positioned approximately 35 ft. above the arena floor (see Fig. 6.11). The primary camera was placed directly above the axis of the warhead. Both cameras were suspended via cables from the roof of the building to isolate them from movement caused by the explosive shock wave.

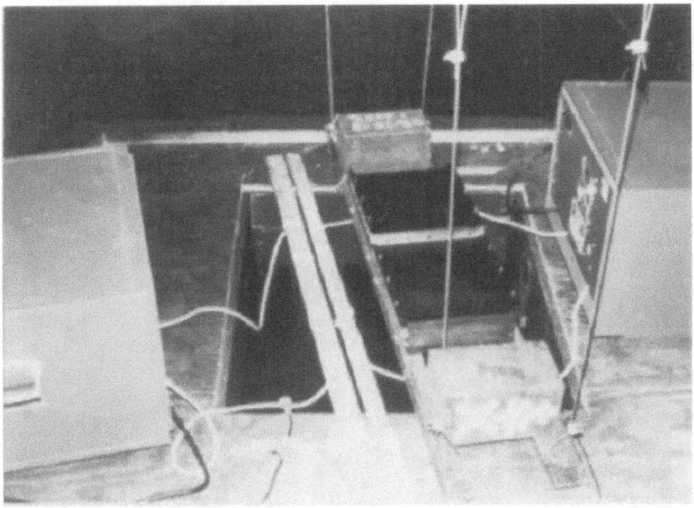

Fig. 6.11. Photo of Camera Instrumentation for the projector test

The two flash tubes were placed as close to the camera lens as possible, (approximately 1° between light, tape, and lens). This was necessary because the reflectivity of the tape is maximized at an angle of zero degrees between the light source and the lens. As this angle increases the reflectivity decreases significantly. Also, early development tests were conducted at night in an outdoor arena. During one such test, it was discovered that even a thin coating of moisture caused by dew condensing on the projectors was sufficient to destroy the reflective properties of the tape. Subsequent tests, in an enclosed building where the environmental elements could be controlled, resulted in all data being acquired.

The camera film was double exposed by the two light sources. The delay time between triggering of the flash tubes was selected to allow the projector pattern to completely separate between the first and second exposures. Subsequent film measurements, done on an optical comparator having a resolution of 0.0001 inch, yielded information from which the velocity of each projector's center of gravity could be calculated.

Since the velocity of the projectors was low (approximately 200 ft/s), it was necessary to account for their gravitational drop in our velocity calculations. This was done by impacting the projectors into Styrofoam panels that were pre-test marked with the theoretical (gravity not considered) impact point and measuring the displacement. Also, pre-test fiducial measurements to determine the magnification factor for each projector row were made by photographing a grid made of reflective tape placed in the projector pattern.

A 6-ft. diameter plywood disk was placed above the warhead to act as a light shield to prevent over-exposing the films by light from warhead initiation.

Results from a typical test are shown in Fig. 6.12. For this test, 209 of the 210 projectors were successfully launched. The strobe system recorded all 209 of these projectors for both flash times, and yielded the necessary data to determine the projector's velocity and spatial distribution within the stringent accuracy requirements.

Fig. 6.12. Results of the projector experiment

6.3.2 Framing and Streak Photography

The camera used for the explosive research discussed in this paper is a Model 116, synchronous type, rotating mirror camera, manufactured by the Cordin Company. The camera is a dual purpose instrument, capable of operation in either the streak or framing mode. The experiments which follow were selected to illustrate the type, quality, and quantity of data obtained in each operating mode. For the framing mode these experiments consisted of: (i) observation of explosive initiators, and (ii) fracturing of the liner of an explosive warhead. For the streak mode the tests were: (i) high-resolution photographs of explosively formed penetrators, and (ii) explosive shock-front simultaneity measurements using fiber optics.

Cordin Camera Parameters and Operation

The camera can be operated in either the streak or the framing mode. Conversion from the streak to the framing mode is accomplished by removal of the streak module from within the camera body and replacing it with a relay lens assembly. External to the camera, the slit assembly is replaced with a framing aperture. Figure 6.13 shows the internal configuration of the camera in both the streak and framing modes.

The beryllium single-surface mirror is turbine driven by a compressed gas such as nitrogen. A rotational speed of up to 5000 rps can be obtained. As implied, the film is stationary in the camera and the image is swept across it by the rotating mirror. At the higher operating speeds, the inside of the camera body is filled with a low density gas such as helium to reduce aerodynamic drag on the mirror.

This camera is of a synchronous type, which means that events to be photographed must be synchronized to the position of the mirror. The photographic record is obtained during only a small portion (10%) of one mirror revolution. Thus, the camera records for only 10% of the mirror period. An optical emitter and sensor are located near the mirror base to detect mirror rotational speed and produce a voltage pulse once each mirror revolution. The emitter and detector are at fixed positions so that detection of the mirror position is constant. Sequencing of explosive firings to the mirror position is done using time delay generators, with delay times chosen so that the event of interest occurs while the mirror is in the portion of its period when recording occurs.

Frame Mode

Black and white or color film, in standard 35 mm format, is used in the framing mode. The film supply can be either a bulk roll or an individual cassette. An electromechanical shutter, located between the aperture and mirror, is automatically opened when the camera reaches its programmed speed and

Fig. 6.13. Photos of a Cordin camera. The *upper* photograph shows the camera in the framing mode and the *lower* photograph shows it in the streak mode

closes after the dynamic record is obtained. Since the closure time of the shutter is about 10 ms, several revolutions of the mirror occur, which can cause rewriting of the film if sufficient light from the experiment is available. A capping shutter, explained later in this chapter, is used to prevent this from happening.

Three objective lenses having focal lengths of 900 mm, 620 mm and 320 mm are available, with selection based on achieving optimum image size on the film. The objective lens is mounted on an optical rail external to the camera.

Although the mirror is sweeping the image along the film track, the internal lens module of 26 relay lenses produces 26 individual frames by a process known as the Miller shuttering effect. The dynamic record consists of 26 frames taken at a rate ranging from 25,000 to 1.25×10^6 frames per second. These parameters are summarized in Table 6.2.

Streak Mode

Operation in the streak mode is quite similar to operation in the framing mode, although some camera modification is necessary. Primarily, this consists of replacing the relay lens assembly inside the camera with a film track module and installing a slit assembly on the external front of the camera. The object is focused on the slit via the objective lens, and the resultant image transferred in a one-to-one size ratio to the film. The film format is 70-mm wide with a data record length of 300 mm. The writing speed can be varied between 0.4 mm/μs. to 20 mm/μs. For a summary of the parameters in the streak mode see Table 6.2. During operation, the rotating mirror sweeps the image of the slit along the stationary film. Thus, the spatial dimension is across the width of the film and the temporal dimension is along the length of the film.

Table 6.2. Camera Parameters

	Streak mode	Framing mode
Writing Speed (max)	20 mm/μsec	1.25×10^6
Record Length	300 mm	26 frames
Film Format	70 mm	35 mm
Mirror Speed (max)	5000 rps	5000 rps
Static Image	Full field-slit removed	Frame no. 5
Fine Focus	Traveling microscope	Traveling microscope
Shutter	Electro-mechanical	Electro mechanical
Slit Width	50–750 μs	N/A
Slit Length	60 mm	N/A

Experiments

Framing Mode

Two experiments are presented to illustrate the framing mode applied to explosion dynamics. The first of these involved a device known as an initiator plate. This plate consisted of a 12-inch diameter by 1/2-inch thick aluminum plate with a track system (grooves) machined into its front face. At the termination of each of the 32 legs of the track was a hole through the thickness of the plate. The track and the hole were filled with extrudable explosive. Behind the plate and centered on the holes were 32 cavities each containing a PBXN-5 booster explosive charge. The objective of this design was to have a single-point detonation on the front of the plate transition to a multipoint detonation at the back of the plate. An equally important criterion was to have the 32 booster outputs occur simultaneously to within one microsecond. The test arrangement is shown in Fig. 6.14.

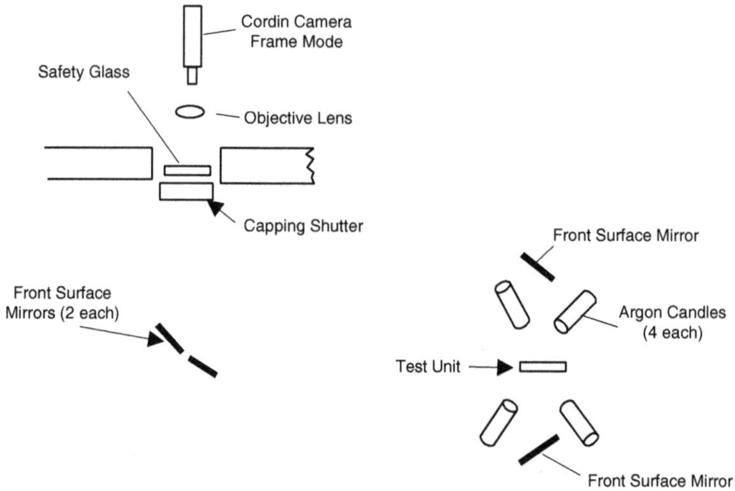

Fig. 6.14. Test Setup for the initiator experiment

Two front-surface mirrors, one front of the plate and inclined at 45° and the other behind the plate and also inclined at 45°, turned the images 90° relative to the optic axis. Two additional mirrors then turned these images 90° into the camera lens. This setup allowed the camera to record both the front and back sides of the initiator plate simultaneously. Two explosively-driven argon light sources illuminated each side of the plate. The camera was operated at 10^6 fps resulting in a 1 μs interframe time.

Three sequential frames from the 26-frame camera record are shown in Fig. 6.15. The first frame shows the detonation after its having progressed through the track system and entering the hole through the plate. The next frame shows that all of the boosters have initiated. Since the boosters have detonated within one elapsed frame, we can conclude that the detonation output of the 32 boosters occurred simultaneously within 1 μs.

This experiment illustrates that the camera's framing mode can record both functioning parameters and timing data of an explosive event. It should be noted that this test was conducted prior to the advent of fiber optics. With a camera speed of 10° fps, the framing mode provided adequate temporal resolution (1 μs) for this experiment. However, a better method would have

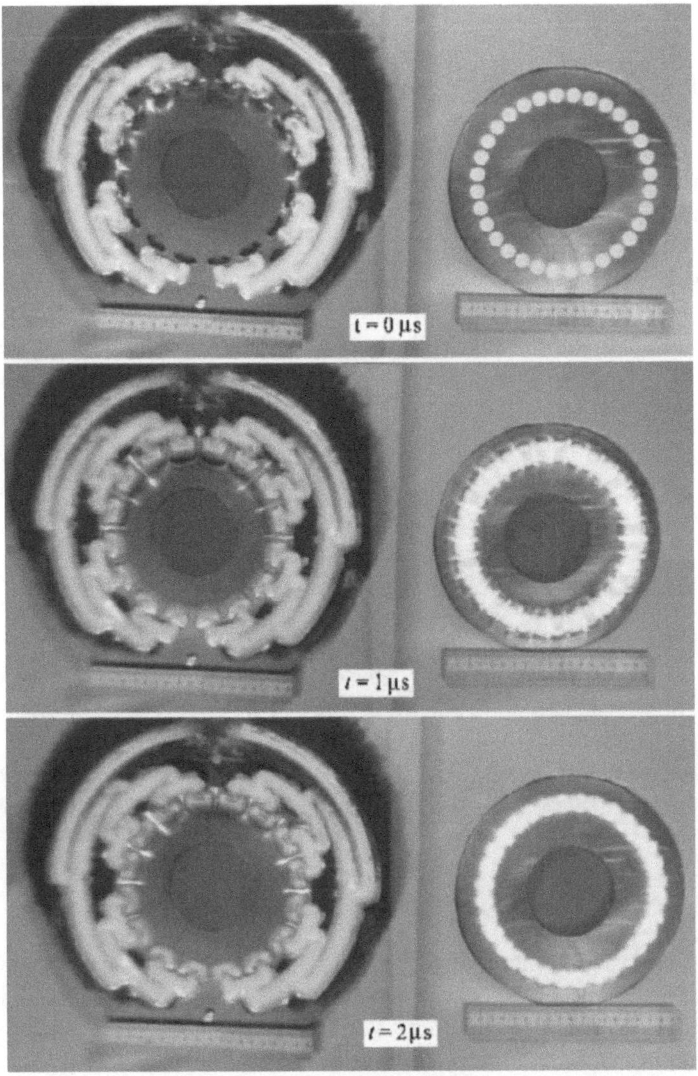

Fig. 6.15. Results of the initiator test

been to use fiber optics in conjunction with the streak mode because, for the same camera speed, an improvement in temporal resolution of about 100× could have been obtained. This technique will be explained more fully in a later section of this chapter.

The final framing-mode experiment that is presented is that of an explosive-filled submunition. This device operates in one of three selectable modes, which produces either an explosively formed penetrator (EFP), a stretched rod, or a fragmenting liner. The framing mode was applied to the fragmenting liner design. The objective of the test was to photograph the surface of the liner as fragmentation is occurring. This technique provided a non-intrusive method of determining the functioning characteristics of a complex, completely enclosed explosive system.

Figure 6.16 shows the test arrangement. The warhead was mounted on a wooden stand that provided both azimuth and elevation adjustments. The stand was positioned such that the warhead was approximately 24 inches up-range from the blast shield of the ballistic sled. The purpose of the ballistic sled was to allow the warhead liner fragments to proceed through the port hole in the shield and on to the target, while protecting down-range flash x-ray equipment from blast and explosive debris. A front-surface mirror located near the down-range end of the sled turned the image of the warhead liner approximately 90° into the camera lens.

Fig. 6.16. Photo of the fragmenting liner test setup

The liner was illuminated with three explosive-driven argon candles. With the setup it was necessary to place the light sources up-range of the warhead and focus the light onto the liner via mirrors, as shown in Fig. 6.16.

A capping shutter, consisting of a bag of powdered charcoal surrounding a short piece of Primacord, was located immediately outside the port hole of the test building. Detonation of the Primacord caused the charcoal to blacken the glass in the porthole, thereby preventing light from entering the camera on post-data mirror revolutions.

Sequencing of the light sources, warhead initiation, and capping shutter was accomplished using time delay generators.

Results of this test are shown in Fig. 6.17, which contains eight selected frames from the 26-frame record.

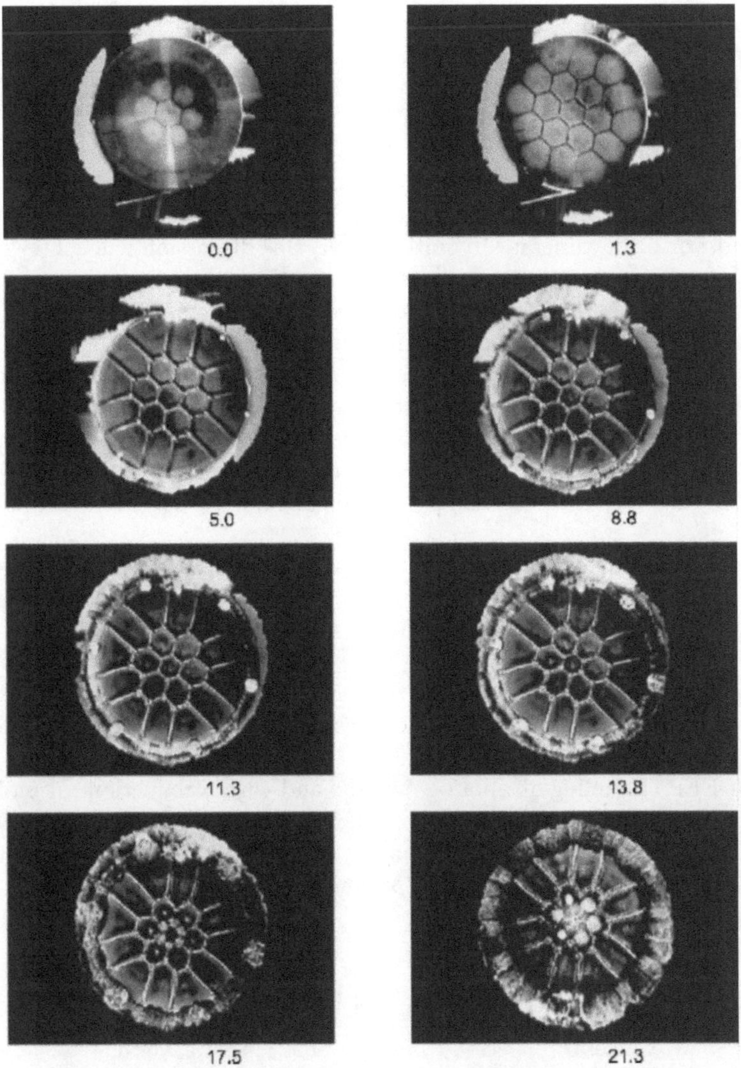

Fig. 6.17. Photographic results of the fragmenting liner test

Streak Mode

Two areas in which the Cordin Streak Camera has seen extensive use are the synchroballistic photography of EFPs and simultaneity measurements of explosive initiators using fiber optics. The following is a description of each of these methods, along with some experimental results.

Front lighted synchroballistic photos of EFPs have been obtained with both the Cordin Model 116 camera operated in the streak mode, and the Model 318 rotating drum streak camera. The primary asset of the Model 116 streak camera is its 70 mm film format, as compared with the rotating drum camera's 35 mm size. Images obtained with the rotating drum camera are adequate, but the larger-format Model 116 camera produces streak photos of exceptional quality. The film size of 70 mm, in conjunction with a long-focal-length lens, produces a large image that allows photographic enlargements to reveal minute features on the surface of the EFP that are normally undetected with lower resolution photographic techniques. Since the photographic methods for both cameras are quite similar, this discussion will center on the rotating mirror camera.

The streak method involves focusing the object, via an objective lens, onto a slit positioned immediately in front of the camera. Relay lenses then focus the slit onto the film plane in a 1:1 ratio, meaning that the size of the image at the slit is the same as the size of the image on the film. As time passes, the rotating mirror sweeps the image along the film. A rotating prism in front of the slit assembly rotates the object (the EFP trajectory) 90° so that the orientation of the image is perpendicular to the horizontal camera slit.

To obtain a synchroballistic photograph, the speed of the image as it is swept along the film must be matched to the speed of the object, taking into account the size reduction (reduction ratio) determined by the geometry of the setup and the lens focal length. Thus, $S = V \times R$, where S is the camera writing speed, V is the velocity of the EFP, and R is the reduction ratio. If this condition exists, dimensions in the spatial and temporal directions are equal.

For EFPs traveling at about 2 mm/μs, and with a reduction ratio of 1/4, the required camera speed is approximately 0.5 mm/μs, which is near the minimum operating speed of the camera.

A summary of instrumentation parameters for the majority of EFP photographs is given in Table 6.3.

Table 6.3. Test Parameters

Camera	Cordin Model 116, Streak Mode
EFP Velocity	1.5–2.6 mm/μs
Reduction Ratio	1/4
Sweep Speed	0.4–0.65 mm/μs
Film Type	70 mm, Tri X, 400 ASA
Slit Width	175 μm
Film Track	30 cm
Light Duration	225 μs

The test setup for EFP photography has been shown previously in Fig. 6.2. After passing through the hole in the blast shield, the EFP is flash x-rayed to record its velocity and shape characteristics. The EFP is allowed to proceed some distance downrange before being photographed. This was found to be necessary as it takes some time for the EFP to emerge from the cloud of smoke and flame associated with the explosive detonation. At the photographic station, the EFP passes through a box constructed of plywood and Plexiglas, and containing a helium atmosphere (see Fig. 6.18). A polymeric film (Saran Wrap) covers the ends of the box. Previous experiments had indicated that many EFPs traveling in air are obscured by a luminous shroud caused by aerodynamic heating. An inert gas (helium) inside the box eliminated this photographically detrimental effect. The box is positioned perpendicular to the optical axis and approximately centered on the plane defined by the camera slit. The EFP is then photographed as it passes the slit plane. For an orthogonal image, a front-surface mirror is positioned in the box along the slit plane to reflect a bottom view of the EFP into the camera lens. The EFP is front lighted with two explosively driven argon light sources, and a similar arrangement illuminates the bottom (reflected) view. After exiting the helium box the EFP impacts a target to determine performance capability.

Fig. 6.18. Photo of the helium box

Figure 6.19 is a single-plane, synchroballistic photograph of a tantalum EFP. As is seen, the photograph reveals folding, buckling, and surface features that would not be visible in an x-ray shadowgraph. Small striations on the surface are also apparent.

The second type of experiment consists of measurements of explosive shock simultaneity. In many sophisticated munitions, precise initiation timing of the explosive components is required to obtain optimum performance. The Cordin Model 116 camera has been used to record the output simultaneity of multipoint explosive initiators using fiber optic cords in conjunction with

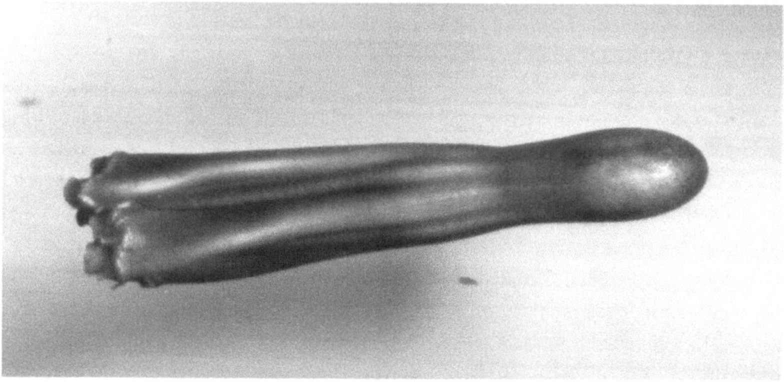

Fig. 6.19. Synchroballistic photo of an EFP

the camera's streak mode. This technique is accomplished by placing the end of a fiber optic cord a short distance from the surface of each initiator pellet, and positioning the opposite end of the cord in a holding fixture called a face plate. The ends of the cords in the face plate are focused on the slit of the streak camera. When the explosive shock front reaches the surface of a pellet, the light from the detonating explosive is transferred to the faceplate via the fiber cord. The camera sweeps each of these light traces along the film as a function of time. To serve as a distance reference line, the slit itself is statically imaged on the film prior to the shot. Subsequently, the distance between the start of each of these comet-like traces is measured on an optical comparator having a resolution of 0.0001 inch. Knowing the writing speed of the camera, these distances are converted into delta-time data of explosive pellet initiation. By operating the camera at a minimum slit width and near maximum writing speed, a time resolution on the order of 5 ns can be realized. This technique lends itself well to geometrically complex-shaped devices, because the optical-fiber cords are flexible and can be bent to accommodate any configuration. Initiators containing as many as thirty eight output pellets have been tested. Thus, the camera can be used as a simple, high speed, multichannel, optical recorder, thereby eliminating the need for expensive electronic signal recording equipment.

Shown below is the photographic record from an integral initiator for a shaped charge. The initiator, loaded with extrudable explosive, had eight output ports and was designed to eliminate the age-old problem of off-center initiation of the high-explosive charge. Simultaneity of the output ports was measured with the above fiber cord/streak camera technique. The plastic sheathed, 1000-μm-core fiber cords were approximately four feet long and were secured against the output ports by a machined receptacle. An RP 2 EBW initiated the process. A photograph of the test hardware is shown in Fig. 6.20 and the streak record of the dynamic test is shown in Fig. 6.21. The data traces are located in the lower right portion of the figure. The long bright streak in the upper part is the flash of a detonator simulator, which was fired for reference timing only.

Fig. 6.20. Photo of test hardware

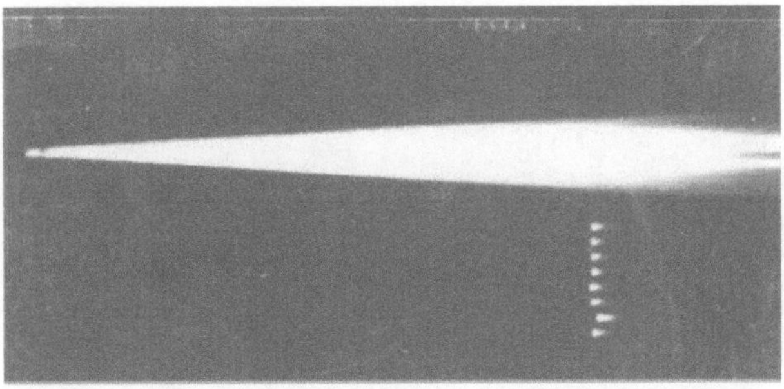

Fig. 6.21. Streak record of the integral initiator test

Results of this test show that the maximum non-simultaneity between the first and last initiator port was 77 ns. Subsequent tests showed this initiator produced a symmetrical, linear, shaped-charge jet.

6.4 Conclusions

High speed photography and flash radiography have proven to be indispensable tools for conducting studies in ballistics and explosion dynamics. The experiments described above highlight some of the uses of this equipment. When conditions of an experiment suggest that one of these technologies may be unsuited, the alternative system can often be applied to acquire the desired information, and do so within stringent accuracy requirements.

7 Some Highlights in the History of High-Speed Photography and Photonics as Applied to Ballistics

Peter W.W. Fuller

7.1 Introduction

This Chapter describes some of the highlights in the history of high-speed photography and photonics (HSPP) as applied to ballistics studies. It describes some of the developments that enabled photographic methods to make a huge contribution to the advancement and development of ballistics research. Indeed, it is true to say that without this particular branch of instrumentation technique, much of what we now know of the detail and data of the behavior and study of ballistic processes would still remain unknown. From the mid 1800s to the present day, photography has played a unique role in ballistics research, and with the rapid advances currently taking place, it promises to continue to do so far into the future. Many of the famous names in scientific research used and developed photographic techniques in their work, which often involved one or more aspects of studies allied to ballistics.

Webster's Dictionary defines ballistics as "The scientific study of projectiles, their launch, flight and impact." Consciously or unconsciously, man has studied ballistics since he first learned to throw stones. Over the centuries as projectiles have become more complex and their method of launch more varied, many of the foremost scientific brains have been engaged in the search for explanations, the production of theories, and the experimental verification of those theories in continuing attempts to fully understand the science. During that long period as scientific methods and new means of measurement have been devised, they have immediately been applied to ballistics studies. In fact, many new measurement methods have emerged in the first instance from attempts to measure specific ballistic phenomena.

As with many other branches of science, the invention of photography offered a new and powerful means of observation and measurement. It had particular advantages for ballistics, bearing in mind that by their very nature, guns and projectiles are not easy or particularly safe to measure and observe. With photography, measurements could be made at some distance from the danger zone, observations could be made on otherwise inaccessible projectiles in flight, and perhaps most valuable of all, unlike most transducers, a picture could give information on more than one parameter at a time.

It is in the nature of ballistics that phenomena are often of short duration, processes proceed at high rates, and objects tend to travel very fast.

It is natural, therefore, that high-speed photography and photonics (HSPP) should be used a great deal in all areas of ballistic studies. Whilst in the present day a large range of high-speed photographic systems are available to the ballistic experimenter, this was not always the case. Ballisticians of the middle and late 19th century were very fortunate that a high-speed photographic method in the form of the electric spark was available to them very soon after the emergence of photography as a scientific tool. Spark photography was to remain their only real means of high-speed photography until the 1890s. This good fortune was recognized by many scientists and much use was made of spark photography, allowing ballistics to be in the forefront in scientific exploitation of the new technology.

7.1.1 Historical Development

The use of photography in ballistics work extends over a period of more than 140 years, it being recorded that a cannon ball was photographed in flight at Plumstead Marshes, near Woolwich Arsenal London, by T. Skaife in 1853. This picture is shown in Fig. 7.1. At this time photography was still in its infancy and somewhat of a mystery to the average person. It is thus not surprising to read that Mr. Skaife was asked in all seriousness "How did you manage to stop the cannon ball in mid flight in order to take its picture?" Skaife took the picture using a camera of his own invention which can still be seen in the Science Museum in London. It has a shutter consisting of two small flaps like miniature barn doors. These doors were driven by elastic strings running round small bobbins on the flap hinges. When triggered, the doors opened for about 0.02 s.

Fig. 7.1. Cannon ball in flight, Woolwich 1851

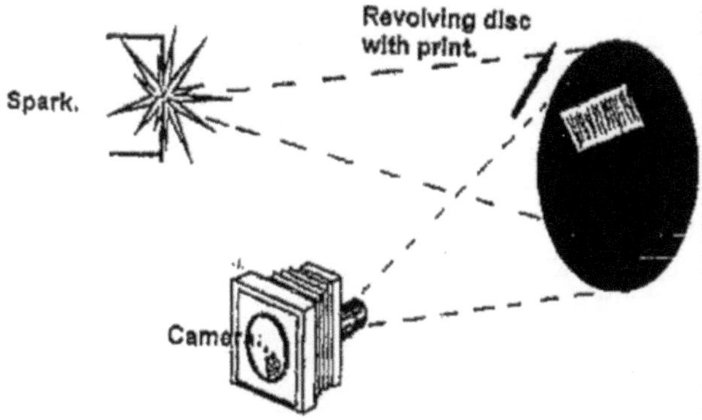

Fig. 7.2. W.H. Fox-Talbot spark light source experiment [1]

Whilst the electric spark had been discovered and the Wimshurst machine had been invented as a laboratory source of high voltages, no one had applied sparks to high-speed photography. The first recorded use was by William Henry Fox Talbot, a famous pioneer in early photography. In 1851 he demonstrated a spark source to the Royal Society. He set up a page of the Times on a revolving disc that could be spun at high speed. Using a spark to illuminate the moving surface, he photographed a small area of print (see Fig. 7.2). On development of the negative the print could be clearly read; the motion of the subject had been effectively frozen (Fig. 7.3) and high-speed photography had been born [1].

By 1883 Mallard and Le Chatelier had made streak records of flames from explosions. Le Chatelier used a photographic plate dropping under gravity upon which the flame image was projected. This could be considered as an early form of streak photography.

In 1884 Ernst Mach in Prague was making use of spark photography to record bullets in flight, which he continued to do for many years. Mach employed a simple Schlieren apparatus recording quite small images and triggered the optical spark by firing the bullet through a series gap in the circuit which fed the main light-producing gap. He later employed an ingenious trigger system, using the bullet shock wave to blow a candle flame into the trigger gap. The ionized gas completed the circuit and fired the spark.

Mach had been led into ballistics studies following his attendance at a lecture on ballistics by the Belgian expert, Meisen. Meisen had proposed that part of the effect of bullet impact was due to a parcel of highly compressed air that the bullet carried ahead of its nose. To explore this theory, Mach attempted to make photographs of bullets in flight.

In his initial attempts he produced no sign of flow in front of the bullets but showed shock waves proceeding from the spark itself. He concluded that the bullets were travelling below the speed of sound and that shock waves would only become visible when they exceeded this velocity. Following help

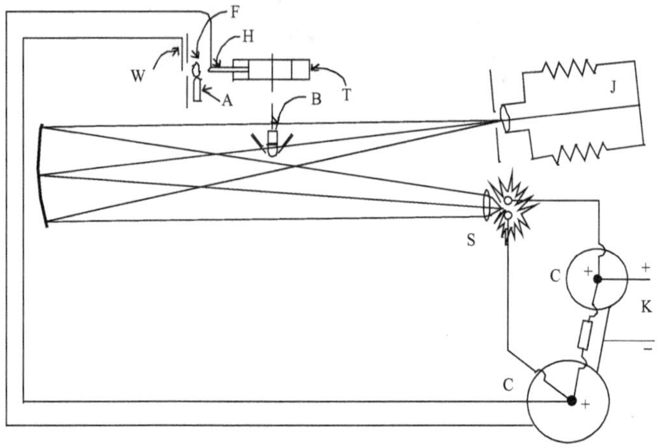

Fig. 7.3. Candle flame trigger. Used by E. Mach for Schlieren photography [2]. K: charging circuit, C: storage capacitor, S: photographic spark, B: bullet, J: camera, A: candle, T: trigger tube, H: side tube, F: blown flame, W: ionization switch

from the Naval Academy at Fiume, he was able to fire bullets at supersonic velocities and obtain pictures of their associated shock waves. By firing bullets through plates, he was able to show that Meisen's theory was incorrect and that the compressive effect ahead of the nose was a continuous process due to motion, with the shock waves reforming again following their passage through the plate.

By 1892 Sir Charles Boys in England was continuing the spark photography begun by Ernst Mach [2]. Boys realized that it was not necessary to use Schlieren methods to visualize flow and used a simple shadowgraph technique to obtain high quality photographs of shock waves and disturbances around projectiles.

Boys also added to the experimental knowledge by measuring the duration of the spark. The flash from the spark was reflected from a mirror revolving at some 60,000 rpm and recorded on film. This was an early use of the high-speed rotating mirror camera principle. The length of the recorded trace was used to determine the flash duration using the known revolution rate and the geometry of the setup. Boys also attempted to determine spin rates by drilling holes in the projectiles at right angles to their axes and taking two consecutive spark photographs at different points along the trajectory. Lights were set up along the trajectory to backlight the holes. By observing the changed positions of the holes at the two photographic stations and knowing the distance between the stations he could calculate the spin rate. It was not a very efficient method but it is interesting as being one of the earliest attempts to measure spin of projectiles in flight.

In 1892 the Prussian Armaments Testing Commission developed a multi-camera system with a revolving disc shutter capable of speeds of 1000 frames/s and exposure times of 100 μs. This had been strongly influenced by the work of Muybridge in California in the 1870s. But, whereas Muybridge had used a line of cameras each one being separately triggered, these later experiments used twelve cameras arranged in a vertical circle and shuttered by a revolving disc with radial slits. By the late 1890s Cranz and Koch began taking spark photographs to study rifle barrel vibrations [3]. Exposures were made at increasing delay times after firing to build up a picture of the whole vibration cycle.

From the 1900s onwards there were steady improvements and innovations. The principles of rotating mirror, disc, and drum cameras became well established and framing rates increased.

By 1905, Kranzfelder and Schwinning had succeeded in discharging 10 capacitors through a single spark gap in rapid succession to give a series of light pulses. The sequential discharges were achieved by using a rotating switch. This was followed in 1909 by Cranz building a multi-spark camera capable of 5000 pictures/s. This used an oscillator feeding a pulse network to discharge a spark gap. With this apparatus he was able to take multiple pictures of projectiles of a wide variety of calibers.

In 1913 Major Franz Duda of Vienna took a series of photographs of projectiles in flight using a camera having a cylindrical slit shutter. Sequential photographs were recorded on a single plate which was moved by clockwork between exposures. Timing marks were introduced by a tuning fork moving another separate shutter. In 1916 in Germany, Ernemann built a cine camera using a rotating polygonal mirror for optical compensation and having a recording rate of 500 frames/s. This camera was used successfully for ballistic studies during the First World War.

In the early 1920s Dr. G. Bull (France) used a drum camera with a synchronized magneto-driven spark to take multiple pictures of bullets in flight. By 1925 Karolus had succeeded in making Kerr's electro-optical polarizing discovery into a practical camera shutter. These shutters were used for photography of atomic bomb explosions in World War II. They are sometimes still used today in ultra fast cameras, but the development of fully electronic cameras has meant that Kerr cells have largely been superceded.

In 1925, P. P. Quayle tried to resolve the controversy as to whether projectiles continued to accelerate after leaving the muzzle. At the United States Bureau of Standards he used improved spark gaps and a shock operated diaphragm trigger switch to take a series of photographs of bullets leaving a rifle barrel. These showed clearly that the bullet continued to accelerate up to 6–8 inches after exit. He also produced much interesting new information about the gas flow in the intermediate area, i.e., the period just after shot exit from the muzzle.

In 1926 Jenkins produced a camera with a ring of rotating lenses capable of 3200 frames/s using 35-mm film. These developments later led to cameras such as the Vinten HS3000 in 1939. In 1938 Western Electric Corp.

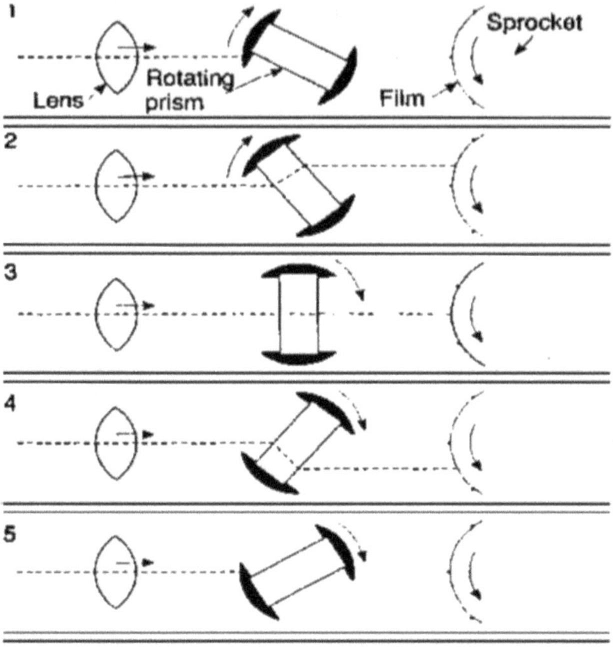

Fig. 7.4. Rotating prism compensation Principle of the Tuttle system using a two-sided rotating prism [4]. This was used in Eastman 3, Fastax T, and Stalex WS3 cameras

in the United States introduced the Fastax camera (Fig. 7.4) which is still in current use in many ballistic research establishments. This camera used the rotating prism system for optical movement compensation [4]. Also in 1938, H. E. Edgerton perfected his high-speed gas discharge lamps that were capable of use as an alternative to spark light sources [5].

By the 1940s most of the current techniques of ballistic photography, including image converter tubes for electronic cine and streak cameras, had been established or were being developed.

The Second World War provided a boost to ballistic and general high-speed photography development, particularly in high-speed streak and framing cameras for the study of explosives and high-speed cine of projectiles and rockets in flight. In the latter part of the war the C4 rotating mirror camera capable of 200,000 frames/s was developed at Aldermaston for the study of explosive reactions. Just after the war, rotating mirror cameras capable of over a million frames/s were developed for the study of atomic bomb explosions.

Since the war probably the main advances in the application of HSPP in the Ballistics area have been in high-speed laser photography, image converter cameras, image intensifiers, the development of cine radiography, improvements in intensifying screens, the use of high-speed video techniques, and improved light sources, all mostly in the electronic/photonic area.

During the development of high-speed video systems, the old high-speed film systems held their position due to their ability to retain high quality resolution at high framing rates. However, digital cameras have now caught up with film for still or lower-framing-rate cine. Due to its many advantages, the use of digital photography will equal or surpass film-based photography in many areas in the near future. It is still likely, however, that film will retain its special advantages at the highest framing rate areas for some time, although new breakthroughs in digital technology may suddenly wrest even that area from it.

Photographic methods have needed to advance to keep pace with advances in ballistics. During the 1960s and early 1970s, light-gas guns developed very quickly and it became necessary to be able to study projectiles travelling at hyper-velocities, i.e., Mach 5 and higher, and to visualize impact processes at similar velocities. Since then the renewed interest in electromagnetic guns, liquid propellant guns, shaped charges, higher velocity long rod projectiles, explosively formed projectiles, and continuous research into new types of propellants has produced new problems for the ballistics photographer. The constant leapfrog progress of new weapons and counter defenses has ensured that there are always more problems to which high-speed photography and photonics (HSPP) can be applied.

Today, ballistics research investigators have a formidable range of techniques to call upon wherever a photographic record is required and there are few areas of ballistics in which photography cannot make a very valuable contribution.

7.2 Application Areas

7.2.1 Introduction

The use of HSPP in the ballistics field is probably one of the most interesting and challenging areas for the photographer. The variety of applications is such that, apart from a few special areas where routine repetitive methods are employed, each project represents a new problem and possibly the development of a new technique. In addition, the exacting requirements are such that a huge range of cameras, lighting, and backup equipment are likely to be employed, including the latest state-of-the-art developments. Add to this the difficulties of often having to operate literally "out in the field," in possibly far from ideal weather conditions and away from the benefits and convenience of the studio, and it can be seen that this area of photography is both rewarding and demanding.

In order to achieve blur-free photographs of the very fast movements encountered in ballistics, much use is made of short exposure shutters or short duration light sources. One way of looking at the problems of blur for a variety of sizes of subject is to consider the time it takes for the subject to move through its own length. If this time is say t, then exposures of the order of $0.001\,t$ will give quality equal to good 35-mm photography of stationary objects. If exposure time rises to $0.002\,t$, then quality falls to the standard of good quality television pictures. Let us consider, for example, a project to record a good quality picture of a 7.62 mm bullet moving at right angles to the camera axis. Let us consider the worst case in which the image is required to nearly fill the frame area. If its velocity is about 2800 feet/s, it will cover its own length in about 30 µs. Thus, for a good quality picture, we shall need to use an exposure in the order of 0.03 µs. This is readily obtainable with modern equipment, but serves to illustrate the high demands of even a relatively slow ballistic photography problem.

High-speed photography is employed directly in all branches of ballistics including, internal, intermediate, external, and terminal ballistics. It is also applied to the study of explosives, hypervelocity ballistics, rockets, flow visualization in wind tunnels and ranges, and in every sphere of research and development which has applications to ballistics. Whilst the term *ballistics* conjures up thoughts of guns and projectiles, many of the projects may involve studies of combustion, the observation of processes, mechanisms, or machinery and, in the research area, subjects which at first glance have little obvious involvement with ballistics at all. Photography is also used extensively in an indirect way as a recording device for information from transducers of various kinds via oscilloscope traces or other means of visualizing waveforms.

7.2.2 The Two Basic Types of Studies

The studies in the main areas of interest are split into two kinds, basic research and development and routine or field testing. Whilst basic photographic techniques will be common to both areas, the general approach and requirements for method and apparatus will be different as shown by the comparisons in Table 7.1.

Thus, it will be seen that whilst a particular research problem may be solved using a new or adapted old technique, the transformation of that technique into one that can be used routinely in the field may take some time and considerable effort.

7.2.3 Five Main Regions of Interest

In the following paragraphs, a selection of five main regions of interest is defined together with the kind of measurement required in each region so that it can be seen in what way and where photographic methods may be applied.

Table 7.1. Comparison of high-speed photography for research and testing

High-speed photography for research	High-speed photography for routine testing
Initially short term	Long term
Can be complex	Preferably simple
Can require careful handling	Preferably rugged
May demand new techniques	Preferably well proven techniques
May need long setup period	Preferably short setup
Initially may need a favorable environment	Operates in a hostile environment
Possibly unreliable	Preferably reliable
Specific application	General application
Can perform at the most favorable time	Should perform on demand
Opportunity for many attempts	Every attempt should count
May be in development	Fully developed
Processing and data extraction may be long and complex	Processing and data extraction should be simple and quick

Internal Ballistics

This region is concerned with the gun and projectile during the period between charge ignition and shot exit. This involves the pressure and temperature in the chamber, the axial and yawing motions of the projectile in the bore, the motion of the gun barrel and carriage, and the general interactions between gun and projectile including the flow of propellant gas. Internal ballistics presents obvious problems of access for photography. However, the projectile can be observed directly down the bore or through the barrel with x-rays. In some experiments, windows or fiber optic paths in the barrel wall have provided alternative means of access. The external behavior of the gun can be followed more easily and direct photography of gun movement and barrel vibration occurring during the in-bore motion of the shot can be carried out with fewer restrictions. Observations of barrel temperature changes can also be made using thermal imaging.

Internal ballistics is also a major area where photography is used as an indirect means for recording information from transducers of many kinds.

Intermediate Ballistics

This region is concerned with the period during which the projectile has just escaped from the physical restraints of the barrel but is still within the influence of the escaping propellant gases. Access for observations is easier but complicated by the presence of the hot, self-luminous propellant gas full of particles, and by the accompanying blast waves. However, these problems can be overcome in some measure by the use of special techniques. In this region it is required to study projectile velocity, physical condition, orientation, spin rate, sabot separation processes, the temperature, composition and flow patterns of the propellant gas, and flow patterns about the projectile itself. X-ray shadowgraphs together with special selective optical techniques using differential focussing or laser light are particularly useful in this region. The need for special imaging techniques is due to the presence of smoke, shock waves, and high temperature gas flows, all of which may prevent the use of conventional photography. Many of the techniques employed in flow visualization can be used here.

External Ballistics

This region is concerned with the projectile during the time from when it escapes the influence of the propellant gas to its arrival at the target. In the case of full range firings in the open, the trajectory of the projectile may well be such as to prevent direct observation. However, in closed range firings or in high velocity flat trajectory firings on open ranges, the projectile will be accessible for the majority of its flight path. Here we are concerned with the orientation of the projectile (yaw), spin rate, its condition, velocity, deceleration (drag), and the surrounding flowfield. If it is possible to set up sufficient photographic stations as in an aeroballistic range, the trajectory can be reconstituted. Data can be produced for many of the observations mentioned, together with the yaw cycle times. In this case spark photography provides a convenient means of illumination and shuttering for shadowgraphs. In addition, ballistic-syncro cameras or flight follower systems together with single or multiple shot cameras can be used for frontal-illumination photography on the open range. Flash bulbs, flash tubes, sparks, or laser light will be used for illumination.

As well as looking at projectiles fired from a variety of launchers, studies can also be made of various stores carried on rocket propelled sleds. In this case high-speed cine may be used as well as the methods mentioned above.

Terminal Ballistics

This region concerns the moment when the projectile strikes the target and the processes which follow. In this case, observations are complicated by the possible presence of self-luminosity from the impact and the production of flying fragments. It is required to observe the way in which the projectile penetrates the target, the condition of the projectile and target with time, the production of debris, and the formation of shock waves. X-ray processes again play a particularly important role in this area, although ordinary photographic methods can also be used. Usually this equires use of long focal length lenses to allow the cameras to be placed at some distance away.

Dynamic Processes

A further region, which does not fit conveniently into any of the others is a very important one which we can call dynamic processes. These processes, whilst highly dynamic, do not involve movement over long distances which require the camera to be tracked to keep the subject in view. In this region we can include the study of explosions, the burning of propellants, the action of primers and ignition devices, the action of mechanisms, rocket motors, studies of models in wind tunnels and shock tubes, shaped charge and explosively formed projectile processes, and a large number of other ballistic related phenomena which do not fit into the four primary regions. Nearly all types of photography will find application in this region but, because of the nature of the phenomena, exposures will usually have to be extremely short.

7.3 Equipment for Ballistics Photography

7.3.1 Lighting Requirements

Lighting requirements are usually demanding due to the short exposure times needed in ballistic events, and the photographer invariably wishes more light were available. Both continuous and short duration lighting is used as appropriate, including strobe lighting synchronized with cine cameras and short duration flash to provide short exposure times for still or multiple-still photography.

7.3.2 Types of Lighting

The important types of lighting used for high-speed photography include those listed in Table 7.2.

Sunlight/Daylight

Sunlight and indirect daylight are used as much as possible but of course have the disadvantage of not being fully under the control of the photographer. In bright sunlight using wide aperture rotating prism cameras, framing rates up to 10–20,000 fps are possible using fast film. Forced processing may also be needed to achieve satisfactory density. For digital cameras, there should be no problems, as the results from a typical test exposure can be checked immediately.

Table 7.2. Types of lighting used for high-speed photography

Source	Typical Durations
Sunlight	Continuous
Tungsten filament lamps of various types	Continuous
Arc Sources	Continuous
Flashbulbs	$0.5–5 \times 10^{-3}$
Electronic Flash	$10^{-3}–10^{-6}$
Argon Bomb	$10^{-6}–10^{-7}$
Electric Spark	$10^{-5}–10^{-8}$
X-ray Flash	$10^{-7}–10^{-9}$
Pulsed Laser	$10^{-6}–10^{-12}$
Super Radiant Light Sources	10^{-12}

Tungsten Filament Lamps

Tungsten filament lamps have a low efficiency and low color temperature and are not very appropriate for ballistic subjects. Tungsten–halogen lamps are used for general lighting purposes including studio type photography of static or slow moving subjects. Projector type lamps which incorporate a reflector are often used where general lighting is required. Low voltage types, such as those which will operate on 6 or 12 volts, are particularly useful in external trials where it may be difficult to connect to ac transmission lines.

In general, continuous lighting lamps of the filament or discharge type that are operated using utility power should be used with caution as the light often fluctuates at the ac frequency. Whilst appearing continuous to the naked eye, the fluctuations may cause problems when the short duration exposures commonly needed in ballistics are used.

Gas Discharge Lamps

Xenon gas discharge lamps are suitable for many applications and the short arc high intensity point source types are very useful where intense collimated beams are required, for example in illuminating down gun barrels or other long narrow apertures.

Metal–halide discharge lamps have a high efficiency but also have a particularly high fluctuation level when powered by ac transmission lines.

Expendable Flashbulbs

Expendable flashbulbs occupy an intermediate position in that their useful operating time of around tens of illiseconds makes them suitable for both cine and single-flash photography. For ballistic applications the exposure time is rather long for any but the slowest moving events but, combined with a separate fast shutter, they can be used successfully. Their light output is roughly equivalent to a 1000 Joule electronic flash tube, and this factor combined with their small size, weight, portability, and simple power requirements make them very efficient light sources. Additionally, their relative cheapness and expendability make them ideal for photographic situations where the light source may be damaged or destroyed. As a consequence, they are extremely valuable as photographic light sources for ballistic purposes.

By the simultaneous firing of a large number of bulbs, large areas can be illuminated to a level equivalent to many kilowatts of light output from other sources. Using modern control devices, bulbs can be fired singly or in groups or can be ripple fired to give almost steady light output over half a second or longer. Because of their long burn-up time appropriate delays must be allowed in synchronization circuits. A major drawback is the need to replace all bulbs after each firing. Flash bulbs are used extensively for cine applications such as synchro-ballistic studies.

Flash Tubes

Electronic flash tubes are used extensively in ballistics photography for many applications. The flash duration may range from 1–2 ms down to one microsecond or less, with output levels going down as exposure time decreases. The flash tube is convenient because it will operate in a single or repetitive mode and can be synchronized with a cine camera if required to flash during the period in which the shutter is open. Synchronization is relatively simple because triggering delays are very short, a useful attribute for ballistic subjects. Several sources can be fired simultaneously from a common trigger or can be fired from built-in triggers that detect the light from another flash unit. Flash tubes are often used in range photography where a short section of the range is enclosed in a temporary light-proof tunnel through which the projectile passes by piercing thin covers at each end of the section. The flash tubes are triggered during the passage through the center of the tunnel and good front-illuminated pictures can be obtained because the lighting direction is controlled by the placement of the flash tubes.

Argon Bombs or Argon Candles

Argon bombs are a light source reserved nearly exclusively for ballistics photography. They are intended primarily to allow the photography of highly self-luminous events such as explosions or shaped charges. They produce sufficient light output to allow the self-luminosity to be effectively ignored; the cameras are stopped down to the point where only the argon bomb output is significant. A typical bomb is made using a cardboard tube that has a transparent diaphragm at one end and a small explosive charge at the other. The tube is filled with argon gas. The light is produced when detonation of the explosive charge sends a shock wave through the container of argon gas. The shock wave ionizes the gas, producing an intense light flash. The intensity is governed by the strength of the shock, and its duration by the length of its path through the gas. Typical durations are of the order of 100 μs and the sources are suitable for use with high-speed rotating mirror cameras which provide the necessary short duration shuttering required. Because of the use of explosives in its construction and its primary application areas, this device is usually only found in government or approved research establishments. Winning and Edgerton showed that the flash duration is equal to about 2 μs for each centimeter of the argon path. Many types of argon explosion driven sources have been made [6–8].

Spark Sources

As was mentioned earlier, spark light sources have an important place in the history of ballistics photography. For many years the spark source was the only available method for effective freezing of ballistic subject motion and much very important work was carried out with its use. Spark light sources can be made in various configurations; however, basically, they consist of a high voltage source, which may be many thousands of volts, that is used to charge a capacitor through a high resistance. When triggered, the charge is allowed to jump across a gap between metal electrodes, producing a bright short duration flash. To allow close control of the flash timing, a trigger electrode is usually incorporated into the spark gap. A trigger pulse applied between this and a main electrode will induce sufficient ionization in the gap to allow the main charge to discharge across the gap.

Shadowgraph photography is an important method in the study of projectile aerodynamics, and spark sources are ideal for this application because they can be of short duration and can be constructed relatively easily as point sources. Durations are normally one microsecond or less and sparks can be fired singly, repetitively, or in groups from multiple positions. A very important application is in aeroballistic ranges where multiple orthogonal spark pictures are taken and analyzed to reconstruct projectile trajectories. A particular feature of point source sparks is that they can be employed to obtain high quality shadowgraph photographs using only the spark source and a sheet of film, i.e., without using a camera.

X-rays

X-rays are used in many areas of ballistics due to their unique ability to penetrate normally opaque objects. Particularly applicable in the internal and terminal areas, both flash and cine x-rays are widely used. Flash x-rays are produced using similar circuitry to spark sources. However, the stored energy is discharged across the electrodes of an x-ray tube. Here the electrons bombard the tube anode and produce an x-ray beam that is emitted from the front of the tube. X-ray systems come in many configurations and penetration capability. The higher the operating voltage, the harder or more highly penetrating the x-rays become. The x-ray film is normally kept in cassettes that are placed on the further side of the object under test. The variation in object absorption and transmission are reproduced on the film as varying shadow density. The x-ray beam effects are often enhanced by the use of screens on one or both sides of the film. The screen material fluoresces and changes the x-radiation into wave-lengths which react more readily with the film stock.

Modern practice is tending toward the use of digital detectors instead of film, with their concomitant advantages.

Laser Sources

Lasers have had a very high impact on ballistics photography methods. Usable in continuous mode, or as a single or multiply-pulsed source, the laser has made possible many new techniques. Its inherent characteristics of monochromatic output, intense light levels, low dispersion, and extremely short pulse duration have brought many benefits to the ballistics photographer. Lasers are used in synchronized mode with cine cameras for short duration exposure cine photography, in conjunction with narrow band filters to observe highly self-luminous subjects, to freeze the motion of extremely fast small objects, and to allow still and cine holography of ballistic events. Another very useful attribute is the ease with which the output can be coupled into an optical fiber and conveyed into areas otherwise inaccessible to an illuminating beam. Another benefit is the wide range of wavelengths available including infrared, which can be used in some areas where visible light may be absorbed or highly attenuated.

Super Radiant Light (SRL)

Super radiant light (SRL) sources represent an alternative to argon bombs or lasers as intense short duration light sources. SRLs are obtained by applying a short intense burst of electrons to certain special crystals. A typical source might be a plate of CdS (cadmium sulfide), backed by a thin aluminum grounding foil. The crystal plate is stimulated much in the same way as a crystal laser, and the individual crystals lase to produce an intense light output during the duration of the electron burst.

SRL sources are not widely known or used as they require an electron beam tube for operation. However, some flash x-ray systems can be used in a dual role for x-rays and as electron beam sources, and can be modified for SRL work. Exposure durations of a few nanoseconds ensure that very fast moving objects can be imaged without blur. For specific applications on self-luminous events or where very fast objects must be frozen in motion, SRLs are very appropriate.

7.3.3 Photographic Recording Devices

Camera Requirements

Camera requirements will range over an extremely wide sweep of capability and complexity and a choice will need to be made bearing in mind the characteristics of the subject.

Some of the techniques and equipment in current use will be described in the following sections. As ballistics photography is such a wide-ranging subject, it is proposed to cover only techniques that are reasonably well established and shown to be capable of use "in the field." To give coherence to a variety of camera descriptions they are ranked roughly in increasing framing speed and decreasing exposure times.

Generally events take place quickly and objects move at high speeds in ballistics work. In order to freeze motion and give clear pictures, exposures are usually short. As object movement becomes faster, exposure times must decrease and illumination must increase or the image brightness must be enhanced electronically.

Subjects may range from a guided missile launch where framing rates of 400 frames/s may be adequate, to the detonation of explosives where a framing rate of 10^{12} frames/s or above may be required.

Exposure times may range from 1/1000th second to tens of nanoseconds or less. In each case the method and apparatus are selected according to the subject characteristics, available light, synchronization, event duration, location, etc. Particularly important will be the form in which the record is required and the purpose of the record, i.e., pictorial or data retrieval.

Slow Speed Cine

In the slow speed cine camera the shutter is mechanical and the film motion is intermittent, the film being advanced by a claw mechanism that engages with the film perforations. The film remains stationary in the gate during the exposure time and is advanced to the next frame whilst the shutter is closed. There are various methods of achieving this motion but all are limited to an intermittent film speed of about 5 m/s to avoid overstressing the film. This results in framing rates limited to about 300–400 frames/s for 35-mm film. Exposures are usually of the order of 1/1000 s. This type of camera is particularly useful for high quality pictorial work where the subject movement is not too fast. Typical applications might be the initial stages of rocket launch, studies of gun recoil motion and generally subjects which have a characteristic time span of seconds rather than milliseconds. Video cameras are finding steadily increasing usage in the range of application of slow speed cine.

Medium Speed Cine

This type of camera, giving cine sequences up to about 10,000 full frames/s and up to 40,000 frames/s with quarter height frames using 16-mm film, is extremely versatile and probably the most widely used in ballistics work. Beyond 300–400 frames/s the cine film must be moved continuously past the lens and compensation must be provided to prevent a blurred image. Most commonly used cameras use a rotating glass block or prism geared to the film drive sprockets to provide movement compensation [4].

The picture image is arranged to sweep in the same direction as the film motion and at the same speed, thus giving a picture that, although moving, is stationary relative to the film. Although compensation is not perfect at the frame edges, definition is adequate for all but the most demanding requirements. Exposure time will be about 1/3 to 1/8 of the frame repetition rate. The framing rate is controlled by the motors driving the sprocketed drive wheel and take-up spool which, in turn, are controlled by applied voltage. In some cameras, exposure time will be variable by the use of different shutter discs. Useful recording time is governed by the time used in accelerating the camera film to operating speed. Use of electronic control systems in current cameras such as the Fastax 11, Hyspeed, or similar types ensures fast acceleration and a long controlled operating speed utilizing 70–80% of the film footage for useful recording. It is common practice to record a time base on the film edge to allow accurate estimation of the event timing. Formerly achieved by small spark gap light sources or pulsed neon lamps, this is now performed by pulsed LEDs. In some systems coded pulses can print information on each frame of elapsed time from event start or the frame number from a given reference.

Event synchronization is usually achieved by triggering the camera from an external detector, and then allowing the built-in control system to take over. In earlier cameras a separate system was used, one well known example being the "Goose" control unit used on Fastax cameras. Most recent camera designs have the control and timer units built into the camera itself, making for a compact portable system.

Alternatively, the camera may be allowed to control the event, sending out a triggering pulse when it has reached operating speed and is ready to record.

Special cameras, such as the Lexander type, which use controlled pinch rollers to carry a short loop of film through the camera at high speed, overcome some of the drawbacks of the normal designs. They are very economical on film, have formats up to 70 mm, and need only a few milliseconds from trigger instant to full speed recording. The main disadvantage is that the top framing rate is limited to 1000 frames/s.

Subject illumination from natural sources is not usually adequate and artificial lighting must be used. This will be from continuous high intensity lamps or a shutter synchronized strobes using flash lamps or lasers. In many ballistic applications, banks of simultaneous or sequentially fired flash bulbs will be used. Continuous high intensity lighting uses a great deal of power and in the field flash bulbs giving a light duration of 50–500 ms can be used to advantage, although synchronization must allow for burn-up time for the lamps to reach full brilliance. Over the last few years more and more applications are being found in ballistics work for metal vapor laser/medium speed cine combinations, where the laser pulses are synchronized with camera shutter opening. These allow much shorter exposure times than the camera alone is capable of, but retains the advantages of a cine record.

High-Speed Cameras

Beyond a certain acceleration rate and velocity, the movement and handling of unsupported film becomes impracticable as the film fails mechanically and higher framing rates must be achieved by alternative means. There are a large number of types of high-speed film cameras, including rotating drum, rotating prism, and rotating mirror, the latter being currently the most widely used. In the rotating drum camera, the film is supported by being fixed outside or inside the periphery of a rotating drum. Thus, although the film may be moving at an extremely high-speed, it is not subjected to destructive forces. These cameras give framing rates into the hundreds of thousands of frames per second.

Alternatively, rotating prism and rotating mirror cameras (Fig. 7.5) allow the film to remain completely stationary and sweep the image across it. They are constructed in two forms, as streak or framing cameras and, according to construction, will be synchronous or continuous access types. They are capable of writing rates of several cm/μs or several million frames/s. These cameras are used very often for the study of explosive events or hypervelocity impact processes as, despite their extremely high framing rate capability, they produce excellent quality pictures. They were initially developed to study early atomic bomb explosion experiments.

Drawbacks for range use are the large amount of light required to illuminate the subject (i.e., they have high f numbers) problems of event and mirror position synchronization, limited number of frames/film length, and general bulk and power requirements. They, thus, tend to be set up in test facilities where they can be left in place for a long time and where the work is brought to them rather than the other way around. Backup lighting will often be supplied by the use of argon bombs. Because of their high framing rates and limited film capacity, at top speed they are only applicable to very short duration events of the order of half a millisecond or less.

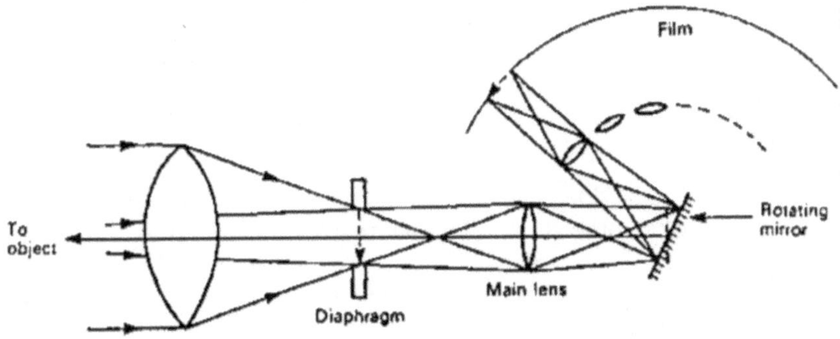

Fig. 7.5. Rotating mirror framing camera with diaphragm

Image Converter Cameras

If still higher framing rates are required than are readily available with high- speed film cameras, the experimenter must turn to electronic recording techniques. Image converter cameras, coupled in some cases with image intensifiers, have two main advantages. They can electronically enhance the brightness of an observed event and, through this property, can give very short exposure pictures which can be transferred to ordinary film, i.e., they have relatively low effective *f* numbers. They can also have extremely high framing rates for cine sequences, although the number of frames is limited. For example, the framing rates of the Imacon type 468 camera can range from 100 frames/s, to 100 million frames/s with exposure times down to 10 ns. The cameras can also be used in streak mode. With the arrival of multiple tube cameras of this type it is now feasible to manipulate framing rate, framing intervals, and exposure times in the same photographic sequence, which offers imaging possibilities unavailable in other types of cameras. Images stored on CCD chips can be fed out directly to frame grabbers and thence to computers for subsequent analysis. For the ballistic experimenter the immediate availability of the image, as with video cameras, offers great advantages in time and money. Decisions on whether to proceed with a firing series can be made very quickly and vital measurements can be available very soon after an event.

It can be argued that resolution and quality may still lag behind film possibilities, but for many requirements this is not a major problem. Also, some of the advantages of electronic cameras are unique to the type. As with all photographic work, equipment is chosen to give the best usage and results for a given project. These cameras are extremely versatile and are used for a large number of purposes such as impact, explosion, and initiator research and, in fact, for any extremely fast events where the very high framing rates are invaluable. Synchronization requirements for these cameras are relatively modest as the response times are almost instantaneous. Also, due to their low light level capabilities, their lighting needs are less onerous than those for film cameras under equivalent lighting conditions. Cameras such as the Imacon 486 and Ultranac can be remotely set up and controlled by a computer.

Video Systems

Over the last decade video cameras have been gaining increasing use in the ballistics area. As in other high-speed photographic work, video cameras are already playing an important role. Whilst still not suitable for cine applications at the very high speed end of the scale, variations on the basic systems are becoming extremely powerful tools for ballistic work.

The immediate playback, on-screen analysis, re-usability of tape, and comparatively low complexity of operation are extremely valuable in the ballistics testing field. The arrival of solid state video storage has also added new features in terms of variable framing rates, no moving parts, and programmed recording so that no memory is wasted during a sequence. Test facilities such as ranges are extremely costly for hiring time and the long wait for conventional film results has always been a problem. Using video, subsequent analysis work, the immediate production of results, and the enhanced possibilities for automation, are very attractive.

For areas where short exposures and fast framing rates are essential, developments such as the still/video cameras have made an enormous difference. In general, the inability to produce fast video framing rates is due to the problem of clearing the CCD chip ready to receive a new image in a very short time. The CCD response itself is very fast and the still/video system takes advantage of this by recording multiple images on the chip, which can then be downloaded at a relatively low rate. Images may be superimposed or placed on different sections of the pixel area. Whilst the number of images is limited, the quality and definition is good and in many instances no other photographic system is required. Currently several developments in image sensor types such as CMOS may bring further increases in performance.

Where the higher framing rates and definition of film cameras are considered really essential, video cameras are often used to provide enough information to enable film cameras to be set up and optimized relatively quickly. As time passes, video capability steadily improves and much research is going on to devise video camera replacements for aeroballistic range cameras. In this area analysis time for film is extremely long and labor intensive, and video coverage could allow extensive automation of these processes.

As the development rate of video cameras is so fast, it is difficult to predict how soon they will equal film or image converter camera performance in the very high framing rate region of imaging. This may come about with the dramatic introduction of newly invented systems which circumvent the problems of pixilated systems.

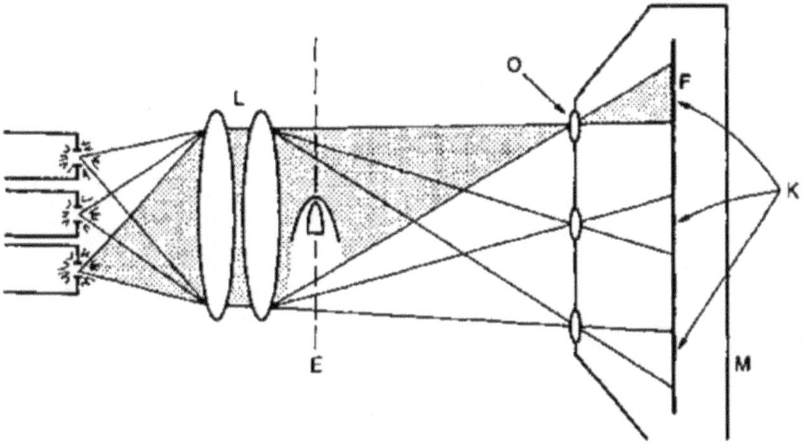

Fig. 7.6. The Cranz–Schardin camera system using multiple spark sources. The spark light sources are fired in a chosen sequence with selected delays. S: spark source; L:condensing lens; E: observed event; O: objective lens; M: multi-lens camera; F: film surface; K: shadowgraph images

7.3.4 High-Speed Still Photography

An alternative method of obtaining short duration exposures is to allow the light source duration to become the means of shuttering the film whilst using an open shutter on the camera. Convenient sources for this are flash tubes, sparks, argon bombs, flash x-rays, and lasers. Any good quality camera is suitable for this kind of photography, but advantages such as interchangeable lenses, interchangeable film backs and variable camera geometry are very worthwhile.

Multiple spark photographs can be made using repetitive triggering of a single gap or multiple gaps. By controlling the spark interval, valuable time related information can be obtained. A typical example is the Cranz–Schardin system shown in Fig. 7.6, where multiple independent source cameras are arranged in a block, all using a common large lens and recording onto a common film.

In a case where a subject is self-luminous, e.g., gas at a gun muzzle or a rocket jet, a spark can be used with special optical arrangements to enhance the spark source light whilst cutting out most of the extraneous light (Fig. 7.7).

7.3.5 Special Techniques

Streak Photography

When a camera is used without image motion compensation, it is called streak photography. There are two types of streak camera where a slit is used, one

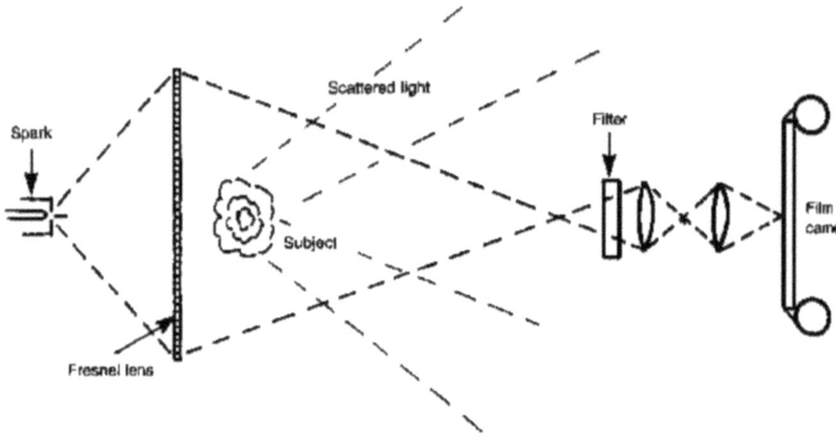

Fig. 7.7. Optical system using a Fresnel lens for spark photography of self-luminous events

in which the slit is parallel with film motion as shown in Fig. 7.8, and the other in which it is perpendicular to film motion. (For the perpendicular case see section Ballistic-Synchro Photography). In the parallel case, it is particularly applicable to explosives studies and the record is taken in a continuous exposure as the film passes the lens. It records boundary and particle motions and intensity changes from the subject image which fall within an area defined by a slit in front of the lens parallel to the direction of subject motion. The film motion is used directly to give movement with time to high resolution without the interframe gaps that would normally be present. As only a fine strip of the subject is seen, it ceases to be recognizable as a discrete object. The record for a solid subject will appear as a diagonal shadow on the film, its angle being dependant on subject velocity and film velocity.

With a knowledge of film velocity and the geometry of the photographic setup, the subject velocity can be calculated. If V_p and V_f are the velocities of projectile and film respectively, M is the magnification of the system, and θ is the angle between the streak image and a stationary image of the slit, then

$$V_\mathrm{p} = \frac{V_\mathrm{f}}{M \tan\theta} .$$

The method is often used to record the outputs from a series of fiber optics placed to observe the arrival time of flame fronts arising from burning processes. Alternatively, fiber optics embedded in an object and illuminated from an external source can show the passage of a shock wave through the subject as their light is sequentially cut off. The streak method is applicable to most cine cameras and these may include medium speed rotating block, high-speed rotating mirror or prism cameras, drum cameras, and image converter

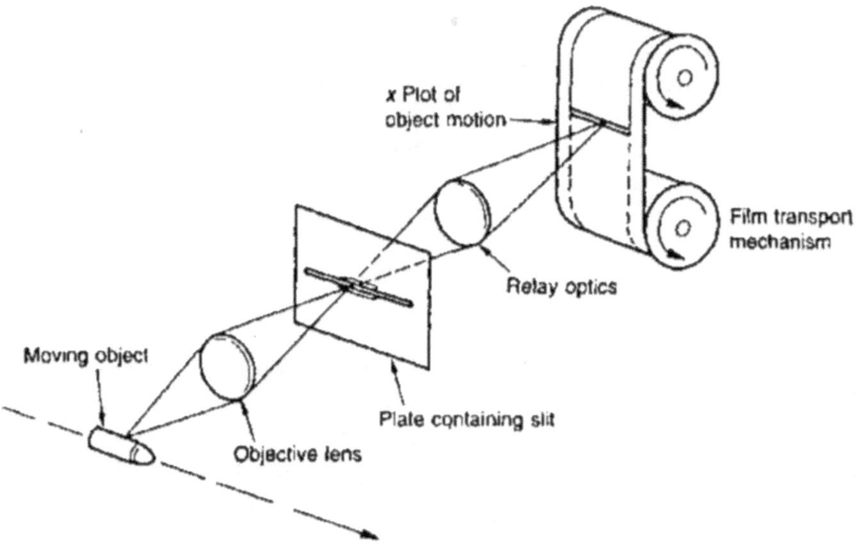

Fig. 7.8. Streak camera system operating in the parallel-slit mode

cameras in which the image motion compensation is temporarily removed. Streak photography is very valuable in the ballistics field due to the very high time resolutions obtainable. Writing rates of several tens of millimeters per microsecond are common, with resolutions of 2–3 ns.

An ingenious method of combining streak and ballistic-synchro photography has been used at Wright Patterson Air Force Base in the United States [9]. As will be seen, the essential difference between the two systems is the orientation of the slit relative to the film travel direction. When using a streak system to obtain projectile velocity a small Dove prism, which turns incoming light beams through 90 degrees, was inserted into the middle of the field of view which covered the path of the projectile. The placement was so arranged that, as the projectile passed the field of view of the prism, the slit orientation was effectively turned through 90 degrees. The resulting picture showed the normal diagonal line with a break in the center in which there was a ballistic-synchro image of the projectile. This effect would not necessarily be possible in all such setups as the optical system and the actual projectile size need to be compatible in scale. Also, the projectile trajectory must be predictable within close limits.

Projectiles in Flight

When there is a requirement to photograph projectiles in flight, several possible methods are available. The usual requirement is to make the view normal to the trajectory so that the projectile is seen from the side. Unfortunately this represents the hardest problem from the point of view of obtaining a blur-free picture and also filling the picture frame with the object. The reasons for this are twofold. First, movement normal to the camera plane requires very short exposures to ensure a blur-free image and, second, we need precise synchronization to ensure that when the exposure is made the subject is within the camera field of view. One way around the problem would be to stand back from the trajectory and take a wide angle view. This will give a chance of registering the projectile at some point in the field of view, even if synchronization is not too well resolved. Also, as the crossing rate is lowered, the exposure time will not need to be so short. Unfortunately, when this approach is adopted the subject image will be so small that usually it will not be of much practical use. Also, the picture will not be very bright.

Another approach would be to somehow follow the projectile with the camera to dramatically cut the crossing rate or to arrange that the projectile image moves in synchronism with the film movement. If this were possible it would then be feasible to once more greatly reduce the image size to frame size ratio. Fortunately there are some special methods which have been devised to overcome these problems.

Syncro-Ballistic Photography

Syncro-ballistic photography is a powerful technique mostly using the medium speed cine camera in a streak mode with its compensating block replaced by a slit perpendicular to the film motion. This technique produces high quality front-illuminated photographs of objects such as projectiles or missiles passing in front of the camera. In the public area the method is more commonly known in the world of sport where it is used to give a photographic record of competitors as they pass the finishing post. Due to loose correlation of film speed and image speed the pictures may be quite strongly distorted in the horizontal dimension. Nevertheless they indicate quite clearly the order of arrival of competitors. This device is used to decide the winner and has led to closely contested results of many kinds being described as “photo finishes.”

In ballistic applications, an appropriate name for the method has been the subject of debate for many years, and it is variously known as ballistic-synchro, synchro-ballistic, smear, ballistic streak, or image motion compensation photography. More recently the term image-sync photography has been suggested. As the method involves the attempted synchronization of image and film movement and in this chapter it is being applied to the study of ballistic subjects, the term synchro-ballistic photography has been chosen. The first ballistic application was by Paul in 1924 when he used a two-slit system to measure projectile velocity.

The basic layout of the system is shown in Fig. 7.9. The objective is to project the image of the subject onto the moving film so that the velocity of image movement and film surface movement are as near as possible the same. This can be achieved by arranging a 1:1 ratio by a suitable choice of lens,

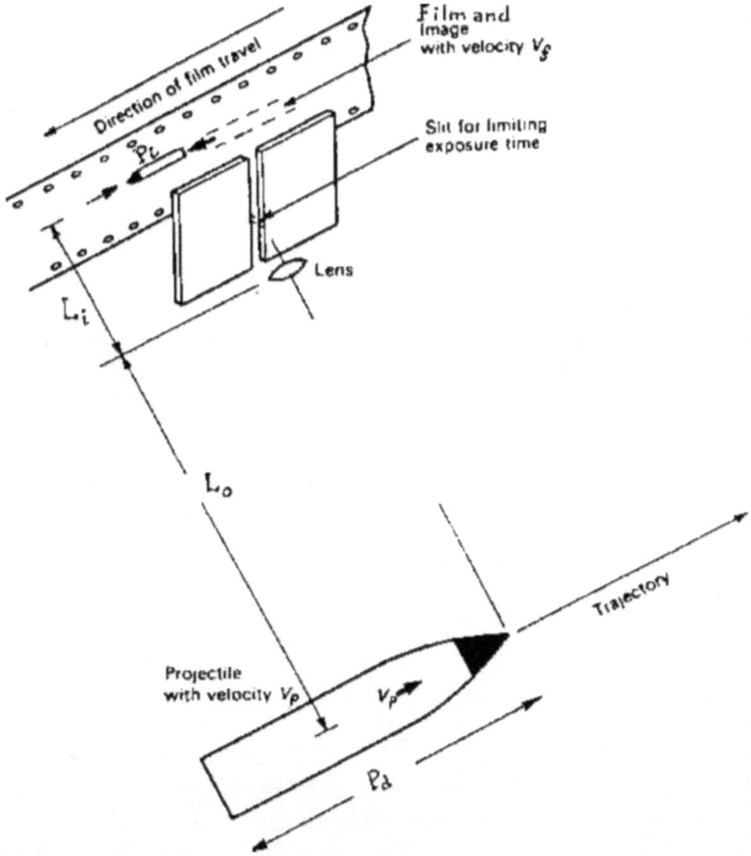

Fig. 7.9. Synchro-Ballistic technique

film velocity and the distances L_0 and L_i. For reasonable success, it must be possible to estimate the velocity of the projectile to quite close limits. In the figure, V_p is projectile velocity, V_f is film velocity, L_0 is the subject to lens distance, L_i is the lens-to-recording-medium distance and F is the focal length of the lens.

To produce a completely blur-free picture the film and image velocities need to be identical. In practice, the subject velocity is seldom predictable to an accurate enough degree to achieve this requirement. It is thus arranged that the moving image is focused onto a narrow slit perpendicular to the film movement. The slit may be placed very close to the film plane or the slit image may be refocused onto the film or recording surface at some distance away. The image is relayed onto the recording surface as a continuous series of narrow strips. The strips form a complete image of the projectile along the direction of motion of the film. By considering Fig. 7.9 it is possible to formulate some of the relationships involving the various parameters involved.

If projectile velocity is V_p and film velocity is V_f, the ratios of distances L_0 and L_i must be arranged such that

$$\frac{V_p}{V_f} = \frac{L_0}{L_i} = \frac{P_d}{I_d}$$

$$T = \frac{P_d}{V_p} = \frac{W}{I_d},$$

where P_d and I_d are projectile length and image length, respectively, with P_d/V_p being the time, T, for the projectile to travel its own length. The quantity T is also the time for the image of the projectile to traverse the edge of the slit. If the slit width is W, then the exposure time for one slit width of the image is T_i, where $T_i = T\,W/I_d$. This will also be the effective exposure time of the whole image as it is composed of multiple slit widths. Image exposure time is thus a function of slit width. If the slit image width could be reduced to be equal to the film grain resolution there would be no blur and the film could even travel in the opposite direction. However, in practical terms this is not feasible and the slit width is normally reduced to a point where blur satisfies the criteria of the experiment. We can also note that exposure time is effectively reduced without the usual penalty of having to increase the lighting level. When L_0 and L_i have been determined, the actual focal length of the lens required can be calculated. From simple lens theory we can say

$$\frac{1}{F} = \frac{1}{L_i} + \frac{1}{L_0} = \frac{L_0 + L_i}{L_0\,L_i}.$$

This is an exact solution for shorter standoff distances. For long standoff distances it can be seen that F will be approximately equal to L_i. Scan rate will be limited by the camera characteristics, i.e., if the film can only be moved through the camera at velocity V_f then the lens focal length and stand-off distance from the trajectory must be arranged such that image scan rate is less or equal to V_f. For example, if the maximum film velocity is, say, 60 m/s and projectile velocity is 600 m/s, then the ratio of stand-off distance to lens focal length would need to be approximately 10:1. Thus, a 250 mm lens would be used at a distance of 2.5 m from the trajectory. If the projectile was 40 mm in diameter and 200 mm long, the image would be 4 mm wide across the film width and 20 mm long. Its total scan time would be 0.2/600 or 1/3000 s. For a slit 1 mm wide the effective exposure time is reduced to $(1/3000) \times (1/20)$ or 1/60000 s.

With a fixed upper limit to camera film velocity, it will be seen that in order to achieve a reasonable image to frame size ratio some problems will arise as the projectile size is reduced and projectile velocity increases. Cameras of the normal rotating prism type have a relatively low top film speed of about 60 m/s. In cameras specially built for ballistic-synchro work, speeds of up to about 85 m/s are possible. For a reasonable picture, such cameras will only be suitable for projectiles of say 30 mm diameter travelling at less than about 600 m/s. If it is required to photograph projectiles that are smaller and faster, the reduction ratio required will reduce the image size to the point where little useful information will be obtained. As projectile size increases, the ability to adequately capture higher velocity subjects will of course also increase. Other types of film camera do have higher possible film velocities. However, they also have relatively short film lengths and thus will need very high resolution for synchronization, particularly in the case where projectile velocity is not accurately predictable. Because of such problems, the only other type of camera that is practicable for regular use with smaller and faster projectiles will be the electronic image converter types which have a very high range of sweep speeds. The sweep speeds available will allow low reduction ratios to be used and as sweep speeds can be finely controlled it will be easier to match sweep speed to projectile velocity [10].

A slight mismatch in relative velocities of film and subject image only causes the recorded image to be slightly longer or shorter than its true proportions, the length mismatch being proportional to the velocity mismatch. The vertical size will remain constant for a velocity mismatch as it is independent of velocity.

A typical result is shown in Fig. 7.10. A valuable attribute of the system is that it records all objects passing the camera during the recording period and relates them in time and sequence. This attribute is particularly useful when a multi-part projectile and sabot is being studied or when a projectile breaks up or when unknown objects fly before or behind the main projectile. When using normal detectors for velocity measurement, incorrect velocities will be measured if objects other than the projectile trigger the detectors. If false readings are suspected, the use of the single or dual synchro-ballistic system will enable such objects to be seen and identified.

Other projectile characteristics such as velocity, angular attitude, and spin rate can be obtained from single synchro-ballistic records but are not often carried out as other independent methods are available which are generally more easily analyzed [11].

Dual Synchro-Ballistic System

The synchro-ballistic method can also be used as a velocity measuring system using two or more stations arranged to record their images on a common film. The earliest record (1924) of a similar system is from Paul. His system used only one incoming image but had two fine slits about 30 cm apart. The ratio of projectile to film velocity was eliberately made unequal so that two separate images were formed on the film. Knowing film velocity and using the system geometry, projectile velocity could be deduced. A system recording multiple separate views from two and four stations on the same film was produced at Fort Halstead, England in the 1960s. In this case the station images were all combined into one imaging lens and slit by the use of mirrors and beam splitters. In the United States the arrangement is sometimes known as the

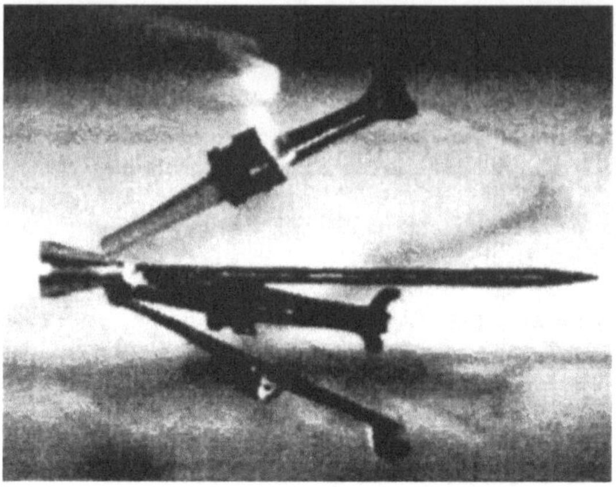

Fig. 7.10. Synchro-ballistic photograph showing 105 mm sabot separation FS-APDS (projectile velocity 5000 feet/s) taken using a Hytax 2 camera at 250 frames/s. Lighting using the Visual Instrumentation continuous xenon high intensity photographic illuminator (HIP!)

Hall system [12]. This system was slightly different because, instead of using a slit placed in front of the film, the effect of the slit is reproduced by the use of fine line light sources placed on the further side of the trajectory or by viewing the backlit image via a slit before it passes onto the objective lens (see Fig. 7.11). All these systems give shadowgraph images at the film plane. Layout geometry will need to take into account that all the images will need to lie within the confines of the film width.

The dual synchro-ballistic system has two major attributes, it allows the measurement of object velocity to high accuracy and it also, in common with the single system, records all objects that pass the camera during the exposure period and relates them sequentially and in time scale with one another.

In some instances the views can be arranged to give orthogonal images if required. Analysis is usually done by reading the film under a travelling microscope. The distance between viewing positions along the trajectory is accurately known and the time of separation of the images can be deduced from a knowledge of the film velocity. Usually the images will be contained within a short length of film and so film speed variations can be ignored. If projectile velocity is V_p, film velocity is V_f, X_t and X_f are the distances between the axes of the viewing stations along the trajectory and the distance between the shadowgraph images on the film in the direction of travel, respectively, then

$$V_\mathrm{p} = \frac{X_\mathrm{t}\,V_\mathrm{f}}{X_\mathrm{f}}\,.$$

As the system is primarily intended to measure projectile velocity and not produce high quality pictures, the film and image velocities do not need to be

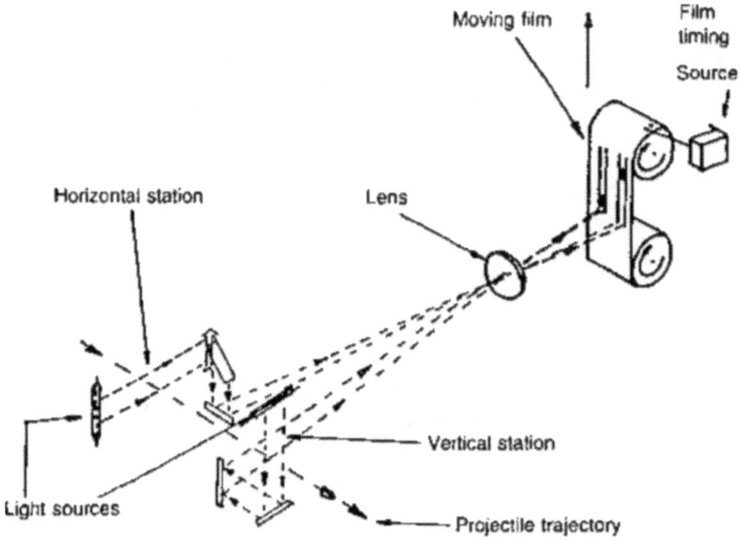

Fig. 7.11. A two station Hall Velocity Measurement system. One station provides a view of the passing projectile from a horizontal viewpoint, and the other provides a view from a vertical viewpoint. Note the displacement of the images on the film

so closely matched. Elongation or shortening of the projectile images will be the same for the several images and will not affect the measuring accuracy.

Disadvantages are the delay during film processing and another delay in making the readings under the travelling microscope. Obviously it is essential to have very good time calibration on the film.

Flight Follower Systems

For a single object the ballistic-synchro method will give a single high quality picture at one point along the trajectory of the object. It will, of course, record a picture of all objects that pass the camera during the recording period. If, however, we wish to obtain a close up cine record of the projectile during a reasonably long section of its trajectory, the ballistic-synchro method is not suitable and an alternative must be found. In subjects such as motor racing, it is feasible for the cars to be photographed for quite long distances by the camera operator panning the camera to keep the cars in view as they pass. However, the translational rate of projectiles is so fast that moving the camera in this way to keep the subject in view is not feasible. An alternative and feasible approach is taken in the flight follower system.

In flight follower systems the camera is kept stationary and the image of the moving subject is relayed into the camera objective via a fast panning mirror as illustrated in Fig. 7.12a. The basic principle is simple, but practical implementation is relatively difficult. The first successful system using the

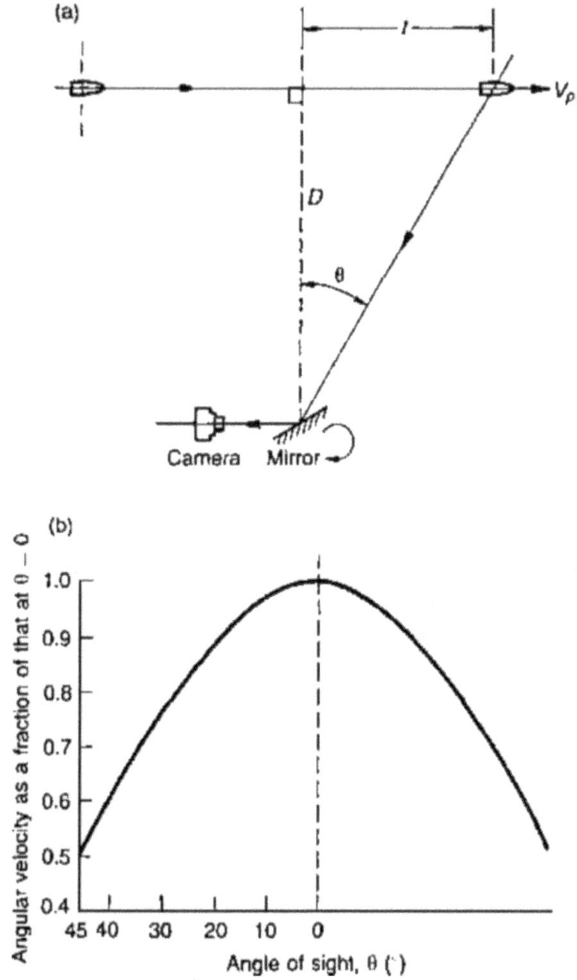

Fig. 7.12. Flight follower system: (**a**) Optical arrangement in basic layout. (**b**) Angular velocity for correct tracking

swinging mirror principle was produced at RARDE UK [13]. In order to follow the trajectory and direct the image into the camera, the mirror must be swung at a varying rate. Initially it must be accelerated to a peak rate when the projectile is normal to the viewing axis and then decelerated to follow a parabolic curve as shown in Fig. 7.12b. For a projectile velocity of V_{p}, the correct angular displacement of the line of sight to give perfect tracking will be when $\tan\theta = V_{\mathrm{p}}\,t/D$, where t is the time for the projectile to cover the distance from a point where the line of sight is normal to the trajectory to the point under observation and D is the stand-off distance of the camera, measured normal to the trajectory. The angular rotation of the line of sight is given by

$$\frac{d\theta}{dt} = \frac{V_p \cos^2\theta}{D} .$$

The mirror angular speed is half this rate. In the RARDE system this was achieved by mounting a mirror on a large galvanometer coil and driving the coil with a series of increasing pulses from storage capacitors. At the point of highest swing rate, the pulses were reversed in polarity and gradually reduced to slow the swing rate again. The pulses were fed in via a wiper arm connected to the coil which swung across a series of contacts. Total swing angle was about 90 degrees. To accommodate various projectile velocities the pulse amplitude could be varied to change the swing rate over a range up to 40 radians /second. By using various stand-off distances and varying the focal length of the camera lens, a variety of lengths of trajectory and image to frame size ratios could be obtained.

The system could be operated from rechargeable lead–acid batteries so that access to an ac transmission line was not required.

Because of the inherent inertia in the mirror mechanism, the mirror swing (panning) must be started before the rojectile comes into view such that the mirror swing (panning) rate has reached synchronous speed by the time the subject first appears in the mirror view as it exits the muzzle. This was achieved by triggering the mirror swing from the gun firing with an appropriate delay or monitoring the projectile movement within the bore. Obviously a close estimate of the subject velocity was required in order to set the apparatus. If a series of experiments of approximately the same projectile velocity was scheduled, the system values could be quickly adjusted to keep the projectile in the middle of the field of view. For bench calibration purposes, several systems could be used. In one, a line of small diodes was set up and pulsed sequentially at varying rates to represent the positions of a passing projectile. This could be recorded on the film and the panning accuracy determined. The possible distance of trajectory over which the projectile can be observed is a function of mirror panning angle and stand-off from the trajectory. The panning angle is limited by the angle subtended from the extremes of trajectory span such that the mirror will be able to reflect the projectile into the camera. A basic system set up is illustrated in Fig. 7.12.

Let us consider the set-up calculation for a perfect situation in which $V_p = 1500\,\text{m/s}$ and required trajectory observation length, L, is 100 m. The projectile is one meter long and a field of view of $2 \times 1.3\,\text{m}$ is needed, i.e., approximately proportional to the 16-mm film frame format. If the full 90 degree swing is used, $\theta = \pi/4$. Then $\tan\theta = (L/2)/D$, where D is camera standoff, and $D = (L/2)/\tan\theta$. Substituting, $D = 50\tan(\pi/4) = 50\,\text{m}$. Let the angular rotation rate of the line of sight be U. Then $U = (V_p/2D)\cos^2\theta$. Maximum value of U is when $\theta = 0$, i.e., when $U = V_p/(2D)$. Then max $U = 1500/(50 \times 2) = 15\,\text{radians/s}$. The lens is required to give a field of view of $2 \times 1.3\,\text{m}$ at 50 m. Considering the horizontal field, the magnification M will be

$$M = \frac{\text{field width on the film}}{\text{field width at the projectile}} .$$

Using 16 mm film, the frame size is 10.4 × 7.4 mm. Then

$$M = 0.0104/2 = 0.0052 .$$

Using the relation $(1/M) + 1 = u/F$, where u is the distance to the object and F is the focal length, we obtain

$$F = \frac{u}{(1/M) + 1} = \frac{50}{193.3} = 258\,\text{mm}$$

for the horizontal field. Then the lens required will have a focal length of 258 mm.

To allow for errors in tracking and settings, the actual lens chosen will be of shorter focal length than ideal to give a slightly wider field of view. In practice image/frame ratios were often 1:5 or greater although, with a series of firings at very close velocities, lower ratios were possible.

Whilst the original system worked adequately, it did have some areas where it could be improved. As a consequence, alternatives to, or developments of, the original system have been made or proposed. In the mid 1980s a system was developed for Aberdeen Proving Ground in the USA. A nickel plated beryllium mirror 300 mm by 500 mm and backed by a honeycomb structure was panned using a servo motor. Mirror acceleration profiles for some 20 velocity variations were stored in a computer and selected according to the particular gun and charge in use. Closed-loop control with mirror position feedback coming from a laser rotary encoder allowed peak accelerations of 100,000 (deg/s)/s. Sensors on the gun tracked the projectile motion in the bore and started the mirror panning motion at the appropriate time to allow for a 3 millisecond start-up delay that would have the mirror pointing at the muzzle at projectile exit time. They also allowed the projectile velocity to be calculated. With this information the computer switched in the appropriate acceleration program for the mirror. Typical mirror swing rates reached 600 deg/s and driving currents for the servo motor were 200 amps at 165 V dc. The system has worked effectively, typically taking pictures at 1 meter intervals along the trajectory [14].

The latest United Kingdom design includes a servo-galvanometer carrying the mirror, which is driven by a current supply controlled by a microprocessor. Initially, mirror and camera are carefully aligned with the gun and the expected trajectory to ensure that the projectile will remain in view over the required distance. After experimental details are fed into a computer keyboard, the processor selects a tangent acceleration curve appropriate to the expected projectile velocity. The mirror is automatically positioned to allow for a few milliseconds delay in reaching the programmed swing rate such that its line of sight will coincide with the muzzle as the projectile exits. For even more precise tracking or where the projectile velocity cannot be closely predicted, a velocity detector system can be placed close to the muzzle. In real time the velocity is determined and the mirror is instantly adjusted through a sight angle to the correct position relative to the projectile velocity. Simultaneously, the corrected new tangent velocity curve is applied to the control of the mirror swing rate. Extra features include the ability to operate on either side of the trajectory as the mirror operation can be bi-directional, and the possibility of sequential operation of several units each programmed with corrected information and triggered to take up the record as the previous unit reaches the end of its sweep coverage. The units are compact, man portable, and capable of operation from either an ac transmission line or a battery supply.

7.4 Flow Visualization

Flow visualization is used extensively in research ballistics to assist in the design of projectiles, missiles, and various stores. Flow visualization enables the shock waves and flow around bodies passing through liquids and gases to be determined. This enables the designer to see where external shapes can be modified to cut drag and generally increase aerodynamic effectiveness.

In recent years flow visualization has become very important in the study of interactions between bodies flying in close proximity to one another. Typical examples include multiple long rod or segmented long rod projectiles, launch of tandem small arms bullets, and interaction between missiles or shells and their sub-munitions during the release phase. In aeronautical research, it is also of use in checking interactions between external stores such as bombs or rockets and the aircraft carrying them. In intermediate ballistics, the way in which sabot segments separate from the main projectile comes under detailed study.

Studies are initially made in wind, shock, or water tunnels and then proceed to model or full scale studies in free flight conditions. Shadowgraph, Schlieren and, to a lesser degree, interferometry and holography techniques can all be used to build up the information required. The various techniques used can be seen in any good book on flow visualization. Unfortunately, there is not space to discuss them here (see [15, 16]). An excellent example of flow around a projectile can be seen in Fig. 7.13.

7.5 Synchronization and Triggering

Synchronization and triggering are extremely important in ballistics photography due to the generally fast movement and fine resolution required for effective triggering of exposures.

Fig. 7.13. Differential Interferometry using YAG laser polarized light. The projectile caliber is 20 mm, its velocity is 500 m/s, its Mach number is 1.46, the Reynolds number is 2.7×10^6, and the exposure time is 15 ns

Triggering techniques are much the same as for general HSP in the sense that similar physical parameters are used. However, in the ballistics area there is a need to make all equipment in an extra robust form to withstand the rigors of blast, overpressure, heat, shock, and vibration. In many cases it may also be necessary to make devices that cannot survive more than one experiment, so considerable ingenuity may be required to produce reliable and cheap equipment.

7.6 Specific Applications in the Five Main Areas

In the following sections more specific details are given of some of the type of problems and high-speed photographic solutions that may be encountered in the main areas of ballistics photography. It must be emphasized that they represent only a very small sample of the huge range of ballistic work that is done with the help of high-speed photography. Unfortunately, although much of the most interesting work is not made public for security reasons, many reports are published and the reader is referred to proceedings of meetings listed in the General References.

7.6.1 Interior Ballistics

Shot Movement From Rest

The measurement of shot acceleration and velocity within the gun barrel from rest to shot exit is currently carried out using microwaves or laser beams using a Doppler system [17]. Before the development of these methods, various other electronic instrumentation methods had been employed. In early attempts in this area, a photographic method had been used. The early movement of the shot was measured by cutting the gun barrel down until it was only a few feet long. A long rod connected to the projectile nose and moving within a centralized guide at the end of the barrel was photographed from the side by a cine camera. The motion of the rod tip could be seen against a reference background and hence the shot movement could be monitored with time as it accelerated from rest.

Gas Blowby

As has been mentioned earlier, direct observation of the projectile can only be carried out by observing down the barrel. This can be difficult if propellant gas escapes past the projectile and obscures the observation path. Gas blowby is also of interest to the gun and ammunition designer since any blowby constitutes a loss of propellant gas which should be assisting the projectile from the gun and thus represents a waste of energy. A method of observing this phenomenon has been devised in which a powerful collimated beam of light is directed down the bore via a sacrificial mirror. The source can be a high intensity xenon beam or a laser beam. The light is reflected from a pattern of retro-reflecting tape that is fixed onto the face of the projectile. The tape reflects the light beam back down the barrel where it is redirected into a cine camera and recorded (see Fig. 7.14).

The gas blowby can be clearly observed as the reflections are cut off and gas leaks in front of the moving shot. Reference to the synchronized and calibrated film record gives a timed history of the process. The system has the added bonus that the spin rate can also be measured from the record [18].

In-Bore Yaw

The yaw of the projectile as it moves up the barrel is of great interest to gun and ammunition designers as it has an effect upon wear and accuracy. Several methods have been devised to measure the angular yaw. Again access is via the muzzle with a powerful collimated beam of light being shone down the barrel onto the front of the projectile. Mirrors have been attached to the projectile nose so that the movements of the reflected beam could be recorded on a cine camera. Whilst successful to a degree they have suffered from complex calibration requirements. A different approach has used a special diffraction grating on the projectile nose [19, 20] (see Fig. 7.15). This grating with two halves inclined at 90 degrees, reflects a double line of three dots that are at right angles and superimposed to give a square of light spots with another spot in the center. The spot pattern is reflected back and recorded on a cine camera. As the projectile spins and yaws during its progression up the barrel, the spot pattern also turns and moves across the film frame. The spot

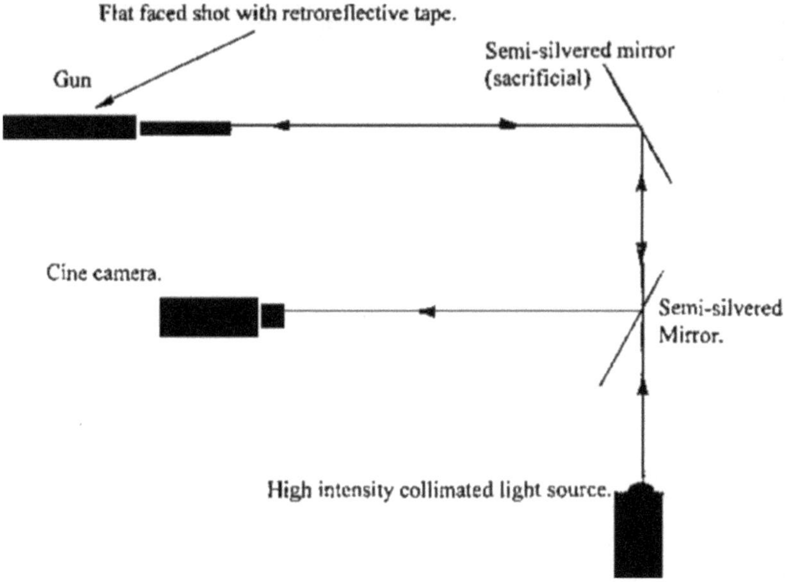

Fig. 7.14. Setup for gas blowby observation

separation distance is a function of the diffraction grating design and thus each film frame carries its own calibration and the yaw angle for each frame and the spin rate can be measured. Static fiber optics are rear illuminated to also appear in each frame and provide an absolute measuring reference. The calibration markings on the film provide a time scale against which the motion can be plotted.

Studies of In-Bore Movement

There is interest in the way the projectile moves within the bore and also in the way that propellant burns and moves within the gun chamber and bore. Flash x-rays have been used to observe both phenomena. Radiographs of internal functions in guns require high energies in order to achieve penetration. Sometimes a temporary barrel may be made of a substance such as fiberglass and resin that is less opaque to x-rays. Often the gun barrel will be machined to give a thinner section through which the x-rays can pass more easily. In the study of charge movement and burning, lead salts are often mixed with the propellant. These salts have a higher absorption of the x-rays and show propellant movement much more clearly.

Alternatively, the barrel may be made of transparent material or have windows fitted along its length. High intensity lighting and ordinary film cameras can then be used to show the internal events [21]. As transparent materials such as plastics will not have the same tensile strengths as steel, the

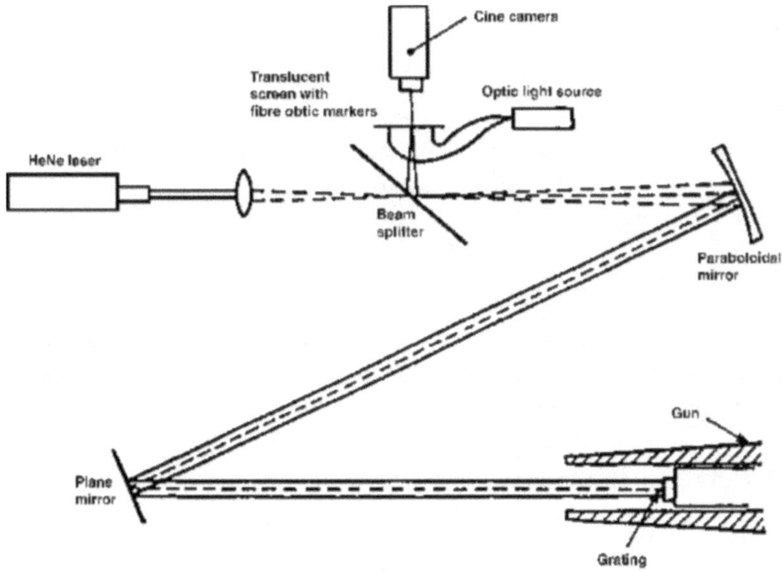

Fig. 7.15. Fuller system for measurement of in-bore yaw

constructions will sometimes fail under the pressures involved, but by then the desired observations will have been made.

7.6.2 Intermediate Ballistics

The region just beyond the muzzle where the projectile is moving through the expanding propellant gas is a particularly difficult region for photographic observations due to the large amount of flash emanating from the burning gases and their high density of entrained particles. Radiographs can be used to study the mechanisms of sabot detachment and the integrity of the projectile itself. By using x-ray photography against a background of preset datum wires, the projectile yaw can also be seen when required. For smaller calibers the smoke and flash problem is not so great and the normal methods of flow visualization can be used [18]. To combat the problem of high flash levels, a different system was used [22]. A Fresnel lens is used to focus the light from a point source, often a spark, into the camera. The event takes place between the Fresnel lens and the camera and the general gun-produced flash is diffused. The camera can be stopped down to receive the focused light and sometimes a narrow band filter is also used to exclude still more of the unwanted illumination (see Fig. 7.7).

To study the growth of expanding shocks from the propellant cloud, multiple spark pictures can be used [22]. The shock velocity can be found in two dimensions from the known intervals between the sparks.

The incandescent gas cloud at the muzzle is used successfully as a light source to take cine photography of the exit of projectiles into the intermediate region. The camera is placed down range and focussed on the muzzle area. The expanding gas cloud provides a luminous background against which the projectile is clearly seen. The illumination lasts sufficiently long for the discard process of saboted projectiles to be observed.

Holography has also been applied in the flowfield [23]. A Q-switched laser with a pulse duration of 20 ns was used in a double pulse mode to make holograms of the flow from the muzzle of a 20 mm cannon. One exposure was made without the disturbance and the other during the firing. The beam was initially split into two parts, one passing straight to the photographic plate and the other passing through the disturbance. Three-dimensional shock formation about the projectile was clearly evident.

7.6.3 External Ballistics

In the case of full range firings the projectile trajectory will usually be so high that normal detailed photographic studies will not be possible. However, photography is still used to provide a record in optoelectronic trackers that follow the trajectory of projectiles and missiles. Once confined to film cameras, this area is now employing video systems as well. As in other applications, the ability to apply digital processing directly to the image carries with it many advantages.

In the case of firings in enclosed ranges or for flat high velocity trajectories on open ranges, the projectile will be photographically accessible over much of its flight path. Several of the methods used in this area are discussed in detail in Sect. 7.3.5.

Considerable use is made of mirrors placed to give views of several dimensions on a single frame. Temporary enclosures also offer the ability to control the type and level of lighting over a short part of the trajectory. Thin walls at each end allow unimpeded passage of the projectile whilst keeping light out.

Flash x-rays are not used so much in external ballistics, as in this area conventional photography is usually more effective. However, x-rays can be usefully employed for observations near the gun muzzle to study the condition and operation of internal mechanisms such as fuzes after the launch process.

Rocket Sled Experiments

As an alternative to free flight, many projectiles and missiles are tested on rocket sleds that are confined to move on rail tracks. The use of rockets enables very high velocities to be obtained. Such tests have many advantages where the pure aerodynamic characteristics of the object are not the paramount study. The specific constraints of movement allow instrumentation to be set up with the certainty that the subject will pass in a certain direction or distance from a measuring device. Thus, cameras and lights can be set close to the rails to give close up detail of events and even placed on the sled to move with the subject.

In some instances synchro-ballistic pictures may be taken or flight follower techniques may be appropriate if a record covering a long distance of travel is needed. Subjects other than guns or missiles will also be studied. These will include items such as the operation of aircraft canopy release and the launch of aircraft ejection seats and parachutes.

Aeroballistic Ranges

A special case of external ballistics is the aeroballistic range. In both the pre- and post-design stages of projectile development as much information as possible is required of the aerodynamic behavior of the device in actual flight. In the early stages this may be gained by firing scale models, but it will eventually have to extend to full scale experiments. The desired measurements can be gained by firing the device in a controlled and instrumented facility called an aeroballistic range. Such ranges may be completely enclosed or have only part of the trajectory enclosed. In both cases the main observations will be made photographically. The experiments are normally carried out in darkness and shadowgraph pictures are recorded on multiple cameras arranged in orthogonal pairs at carefully measured positions along the trajectory. Spark sources are employed to provide lighting and control exposure time. At each station pairs of retro-reflective screens provide a background against which the projectile is imaged.

Suspended catenary wires along the full length of the range carry small markers at carefully measured intervals. These wires appear in all the photographs with at least one wire appearing in both photographs of an orthogonal pair. They provide a means of relating the photo pairs to each other and to the successive photo pairs along the length of the range. The cameras are focussed on the shadowgraph picture on the retro-reflective screen. The front-illuminated image of the projectile is thus out of focus. If both camera and light source axes can be arranged to be coincident, the out of focus image will appear within the confines of the shadowgraph image. This makes for a less cluttered picture and assists analysis. The pairs of photographs thus obtained can provide information on yaw, pitch, precession, nutation, and projectile status. The full projectile trajectories and three dimensional orientations can be reconstructed from these photographs (see Fig. 7.16).

The analysis of the photographs and the reconstruction of the trajectories was a long and tedious manual process, which, with cuts in manpower and funding, led to the closing of several such facilities. However, much of the work has been automated and complex computer programs now calculate the results in a much shorter time. To enable the process to be further automated, much effort is being placed on replacing the film cameras with very high resolution CCD cameras [24]. If this can be done, the electronic images

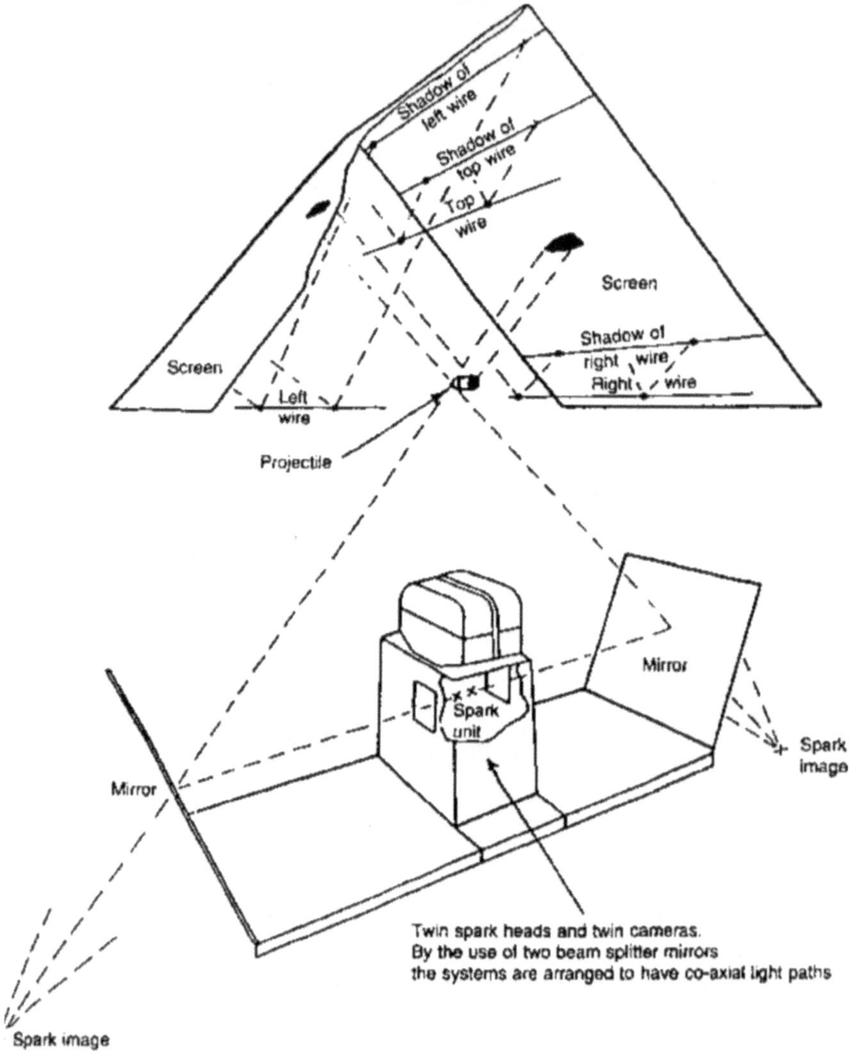

Fig. 7.16. Layout of one orthogonal photographic station in a ballistic range

will be fed directly into computers, scanned, and analyzed in a relatively short time. As well as basic design studies, the aeroballistic range is used to study other phenomena such as the interactions between multiply-launched projectiles and the launch process of sub-munitions. The enclosed, permanent, and controlled nature of the aeroballistic range allows the use of complex photographic methods such as Schlieren, shadowgraph, and interferometer systems to be set up on a long-term basis.

7.6.4 Terminal Ballistics

Terminal ballistics is an area in which high-speed photography can be widely applied. By the nature of the subject, many of the processes involve penetration of objects and thus this is an area where x-rays are widely used. Many target materials of great interest are metal or other materials that are opaque to normal photographic methods. Flash radiography, thus, offers a unique method of observing the interactive process between projectile and target during penetration. Another advantage of radiography is the ability to photograph through the smoke and luminosity often associated with impact. In 1938, Steenbeck was already demonstrating the usefulness of radiography in records of bullets penetrating wood targets.

In order to adequately observe this kind of phenomena, multiple exposures are desirable. Multiple exposures enable the speed of growth of craters and the rate of material flow to be observed. Much use is made of cine x-rays to allow the history of the process to be recorded. In addition, the penetration of projectiles through plates and the production of ejecta and projectile break-up, can be seen as a continuing process. Flash x-rays are also widely used in the study of explosive processes where the insensitivity to the self luminosity of the phenomena is a valuable asset. Although explosives are fairly transparent to x-rays the progress of detonation waves can be followed due to the local increase in absorption. If a calibrated wedge is simultaneously photographed, calculations of the detonation wave density can be made.

The progress of the penetration with time and the evolution, velocity, size, and direction of behind-target debris is all of great interest. This can also be studied with x-rays, but other methods such as still video, high-speed rotating mirror cameras, and multiple pulsed lasers are all used successfully. Some very interesting observations of the growth of the jet from a shaped charge have been obtained using short duration pulses from ruby lasers (see Fig. 7.17). Previously, the jet has been observed from the side with x-rays. The changing shape of the copper cone as it turns inside out and transforms into a thin metallic jet had been modeled on computers but had not previously been observed experimentally in real firings [25].

Work has also been done using a ruby laser stroboscope to study impact phenomena. The ruby stroboscope offers long recording times (1 ms) at frequencies up to 500 kHz with exposure times down to 20 ns [26].

There are many other terminal ballistic phenomena not necessarily concerned with the impact of projectiles on steel armor.

Wound Ballistics

During the development of small arms rounds, there is great interest in what happens when the bullets penetrate into human tissues. The penetration is often associated with the formation of temporary cavities that often produce extreme trauma. In order to study such phenomena, specially prepared

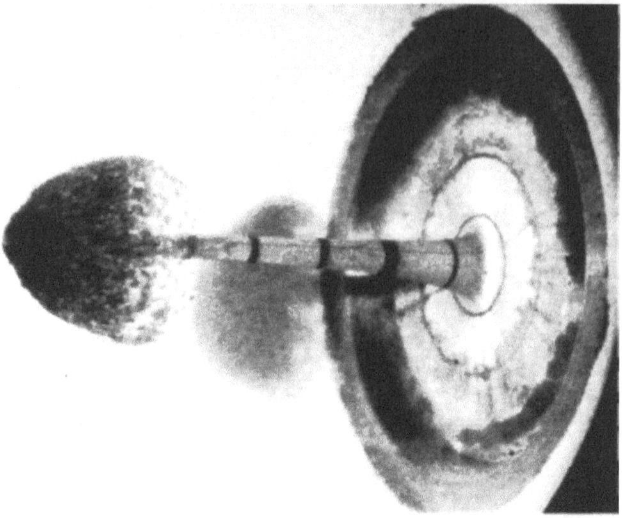

Fig. 7.17. Photograph of a shaped charge jet taken using a ruby laser and an image converter camera at 50 million frames per second

gelatin blocks are used which are translucent and also have similar properties to tissue. High-speed cine cameras are used in conjunction with strong backlighting to observe the growth history of the cavities produced. The records can later be analyzed to make plots of the shape and size of the cavities. Similar studies are made in the development of bullet-proof vest designs. Although it is hoped that bullets will not penetrate the vest there is distortion of the area immediately behind the impact region that can be relayed to the wearer via the protective layers. In this case the studies can be made from the side using cine x-rays or high-speed cameras looking at a sectioned sample. An image converter camera can be used in conjunction with flash lamps [27].

Bird Strike

In the civilian area, much work is done to study the effects of bird strikes on aircraft cockpit covers and their effects on turbine blades if ingested into jet engines. Here the strike velocities are rather low and extensive use is made of medium speed cine and video cameras.

The target subjects are real full-scale objects and in the case of engine tests, individual fan blades may be impacted in the laboratory, but tests on running engines are also made in open air facilities. The birds are sometimes dummy objects made from materials with similar properties to real birds. However, most realistic tests are made with dead chickens of appropriate sizes that are fired from air or propellant guns using sabots.

Bullet-Proof Vests

Other types of test might be made on bullet-proof vests or helmets and the fibers used in their construction. High-speed rotating mirror cameras or image converter types will be used to obtain the framing rates that are needed. Multiple single plate studies have also been made using high repetition rate spark sources such as the Strobokin system. Having all the images on one frame can be very convenient, particularly in the study of the behavior of single-fiber breakage [27, 28].

Similar tests are made to observe the mode of breakup of ceramic tiles often used as facings for vests. Here there is interest in the rate at which the penetration takes place, the rate of crack propagation in the ceramic, and the breakup or distortion of the projectile itself. These processes occur on a very short time scale so very high-speed film cameras or image converter cameras will be required, and may be used together with high intensity lighting.

7.6.5 Dynamic Processes

This area includes processes such as explosions that are dynamic but remain in the cameras field of view for the total event time. Much work is done using flow visualization techniques to observe the blast waves coming from large scale explosions [29, 30]. Typical of processes in the area will be the study of small full scale samples or models of projectiles and missiles in wind, water, or shock tunnels. Here the device will be suspended on a sting in the tunnel and air or water will flow around it to simulate flight conditions. Schlieren and/or shadowgraph pictures will be taken to study the flow patterns. For some subjects that may have an oscillatory flight, medium speed cine sequences will be used to record the motions. (see also Sect. 7.4, Flow Visualization).

Gun Recoil and Barrel Vibration

The motion of the total gun or particular parts of the moving mechanisms such as the recoil system or loading system is slow enough to be recorded by medium speed cine or video cameras. Sometimes particular elements of the subject may be painted with a bright color or have a piece of retro-reflective material fixed to it in order to increase its visibility in the pictures. As the subject is basically static, lighting is relatively easy and some time can be spent to place sources to the best advantage.

The study of barrel vibration is often required when analyzing new designs or when attempts are being made to optimize gun performance and accuracy. In large guns vibration before shot exit, although significant ballistically, is very small in actual amount and must usually be measured with very sensitive electronic or electro-optical devices. However, after shot exit the movement can be of quite large dimension and can be studied by high-speed film cameras or high-speed video. The view will normally be from the side, sometimes in orthogonal arrangements and normally incorporating a scaled grid background to assist analysis. In some experiments a line of bright reflectors has been placed centrally along the length of the barrel and illuminated with a high power strobe source. Observations have been made with high-speed cine cameras located far enough away to include the complete barrel in the field of view. Firing at night to reduce background interference has resulted in the film showing a line of bright dots undulating with time and analysis has given a history of the motion.

Combustion

A large part of the work in this area will be concerned with combustion in all forms. The action of firing devices such as squibs, detonators, caps, and primers takes place in very short time scales, ranging from microseconds to milliseconds. High-speed rotating drum, rotating mirror, or electronic cameras will be used to record their action in framing or streak mode. Lighting will be supplied by flash bulbs, flash tubes, or high intensity continuous sources [31].

For explosions, similar types of camera will be required but, because there will be a large amount of self luminosity, lighting will be by argon bomb or high power laser pulse. Filters will also be used to selectively cut out the light from the combustion process. However, in some cases, it will be required to study the progress of flame fronts and in this case color photography will be employed to assist the definition of various areas. In addition, in some cases fiber optics will be placed to transmit the arrival time of the flame front from various positions on the explosive charge. These arrival times will be recorded by streak cameras. For these experiments cameras will have to be well protected from blast effects and will either view the scene through thick windows in a safe enclosure or will be behind blast walls and set to view the event via relay mirrors.

Flash x-rays are also used to study explosive processes and, in this case, the problems of self luminosity do not apply as the x-ray cassettes can be completely enclosed to visible light. X-rays are particularly useful when the motion of metallic parts of an exploding device is needed. In this case cine x-rays or multiple sequentially fired x-ray units surrounding the area will be used.

When x-rays are used, the film and intensifying screens will usually have to be placed quite close to the explosion site. Over a long period of experiment many ingenious systems have evolved to allow the film to be recovered intact. Film cassettes include internal foam padding and are sheathed externally in plywood covers. They are also often not rigidly fixed in place but are allowed to be blown away by the blast to be recovered after the event. This resilient placement avoids much of the direct shock damage that would be experienced by a rigid attachment.

Arena Firings

In some cases the primary objective will be to find out the general action of an exploding store and its fragmentation performance. It will be set up in a temporary enclosure often known as an arena. The store will be placed in the center and around one half of a surrounding enclosure will be placed sheets of straw board that will be used to trap fragments. Around the other half will be placed light support sheets faced on the inside with thin metallic foil. Under the semi-circular foil wall, mirrors will be placed to show the wall surface to a series of cine cameras arranged around the outside of the wall in protective enclosures. Bright continuous lighting or synchronized flash bulbs will be arranged to illuminate the outer surface of the wall. When the store is fired, fragments from one half of the device are trapped in the strawboard wall for later recovery, with their location, weight, and shape being recorded. When the flying fragments from the other half reach the foil screen wall they pass through and their passage is marked by the appearance of a star-like spot on the camera record due to the external light sources now becoming visible through the hole. Their time of appearance can be deduced from the cine timing record. This will provide data on the velocity of the fragments as their flight path distance will be known. The temporary arena structure will usually be destroyed, but not before the data have been recorded.

Gun Temperature

The temperature history of gun barrels during firing is of great interest as it is closely linked with wear rate. Measurements are often made with thermocouples or other devices placed on the gun structure. A method has been developed [30] in which the output from a thermal imaging device is recorded to give a history of temperature change over a section of the barrel. Thermal imagers have a relatively long response time if a recognizable image is required. This is due to the scanning action used that gradually builds up the image from a series of horizontal scans which are rastered down the viewing screen. However, the time of one scan can be quite short, 5 ms, and if the picture is limited to a narrow strip equal to one scan width it becomes possible to produce a history of temperature change over the area under observation. The scanner produces a false color picture in which color represents a temperature band. In the experiment, these picture were recorded on film and also plotted out onto a color printer from a digital store.

7.7 Conclusions

High-speed photography has been a valuable tool in ballistics research for well over 100 years. The applications and usage are constantly expanding. New methods of applying established techniques are regularly discovered and there is a steady growth of new methods, particularly in the electronic/photonics area. Video capabilities pose a constantly increasing challenge to film systems. However, long established film methods such as medium and high-speed cine, flash, and spark photography continue to form a large part of general day-to-day usage, particularly for work requiring the highest resolution. Means of data recording of waveforms using electronic methods continue to grow in number and complexity and in most areas have taken over from conventional photography of oscilloscope displays. It is certain that, in the future, ballistic work will make increasing use of video and other types of electronic recording. Whatever the methods used, the ballistics experimenter is likely to affirm in the future, as in the past, "that a good image is worth a thousand words," or more likely "several megabytes of explanation."

References

1. W.H. Fox-Talbot, "Sketches of eminent photographers. William Henry Fox Talbot," *British Journal of Photography* **11**, pp. 340–341 (1852), Anon. (1864).
2. C.V. Boys, "Photography of flying bullets," *Nature* **47**, pp. 415–421 and pp. 440–446, (1893).
3. C. Cranz and K.R. Koch, "Investigations on the vibration of rifle barrels," *K gl Bayerische Academie der Wissenschaften Munich Abhandlungen* **19–21**, (1899–1901).
4. H.D. Taylor, "On the use of rotating parallel plate glass blocks for cinematography," *Proc. Phys. Soc. (London)* **49**, pp. 663–670, (1937).
5. H.E. Edgerton, "Electronic flash photography," *J. Soc. Motion Picture Eng.* **53**, (1949).
6. A. Michel-Levy and H. Muraour, "Photographs of phenomena accompanying explosions of brissant explosives," *Compt. Rend.* **204**, pp. 576–579, (1937).
7. J. Sewel et al., "High-speed explosive argon flash photography system," *J. SMPTE* **66** (1957).
8. C.H. Bagley, *Rev. Sci. Inst.* **30(2)**, pp. 103–104, (1959).
9. H.F. Swift, "The Airforce Materials Laboratory hypervelocity ballistic range," Air Force Materials Laboratory Tech. report AMFL-TR-67-2, Wright Patterson AFB, Dayton, USA, (1967).
10. R. Hadland, "Design and application of a ballistic-syncro system using the Imacon image converter camera," Proc. 9th International Congress on High-Speed Photography and Photonics, Society of Photo-optical Instrumentation Engineers, (SPIE), Bellingham, NY. (1970).
11. M. McDowell, H. Klee, and D. Griffith,. A new approach to synchro-ballistic photography. Proc. 19th International Congress on High Speed Photography and Photonics Vol. **1358**, Society of Photo-optical Instrumentation Engineers, pp. 227–236, (1990).
12. D.A. Hall and W.W. Atkins, "New technique for the measurement of the velocity of high speed objects," *Proc. 5th Int. Conf. High Speed Photography SMPTE*, Washington, DC, (1960).

13. E. Damant and F. Knowles,. "The flight follower – a device for the cinematography of objects in rapid translational motion, "Proc. 6th Int. Conf. High Speed Photography SMPTE, Netherlands, (1962).
14. F Yeaple, "High speed mirror tracks projectile", *Design News* 5/9/88, pp. 122–123, (1988).
15. P.W.W. Fuller, "Flow visualisation," in *High Speed Photography and Photonics* (Ed. S. Ray), SPIE Press Monograph, (2002).
16. G.S. Settles, *Schlieren and shadowgraph techniques*, Springer-Verlag, Berlin, (1949).
17. P.W.W. Fuller and J. O'Conner. "A Michelson interferometer for high resolution of shot start movement using a CO_2 laser" Proceedings of the International Congress on Instrumentation in Aerospace Simulation Facilities, Institution of Electrical and Electronic Engineers, USA 81 Record pp. 169–178, (1981).
18. P.W.W. Fuller, "High speed photography in ballistics," 8th Int. Conf. Instr. Aero. Sim. Fac. Record 1979. IEEE. pp. 112–123, (1979).
19. P.W.W. Fuller, "Measurement of yaw in bore," *Proc. 5th Int. Symp. on Ballistics*, Toulouse. Sponsor ADPA USA., (1980).
20. P.W.W. Fuller, "Aspects of yaw in bore measurement," *Proc. 3rd US Army Symp. on Gun Dynamics*, USA, (1982).
21. P.W.W. Fuller, "Instrumented investigation of the internal ballistics of a low velocity launcher," *Proc. 8th Int. Symp. on Ballistics*, Sponsor ADPA. pp.1B5–1B12, (1984).
22. Schmidt E. and Shear D, 1975. *Optical measurements of muzzle blast.* AIAA, **13**, pp. 1086–1091.
23. Bettinger, 1964. *Application of holographic interferometry in the external flow field in muzzle environment of a 20 mm cannon.* Report no AD-A013-465 Case Institute of Technology, USA.
24. D.R. Snyder and F.M. Kosel, "Application of high resolution still video cameras to ballistic imaging," Ultra high and high speed photography, videography, photonics, and velocimetry, Proc. SPIE **1346**, pp. 216–225, (1990).
25. L.L. Shaw et al., "Electro-optic frame photography with pulsed ruby illumination," Proc. 20th Int. Conf. High Speed Photography, SPIE **1801**, pp. 92–105, (1992).
26. K. Kleinschnitger, V. Hohler, and A.J. Stilp, "The EMI ruby stroboscope and its application in terminal ballistics," Proc. 20th Int. Conf. High Speed Photography and Photonics," *Proc. SPIE* **1801**, pp. 278–287, (1992).
27. J.L. van Brie and E.J. van Riet, "Use of an image converter camera for the analysis of ballistic resistance of lightweight armour materials," Proc. 19th Int. Conf. High Speed Photography SPIE **1358**, p. 692, (1990).
28. J.E. Field and Q. Sun, "High speed photographic study of impact on fibres and woven fabrics," Proc. 19th Int. conf. High Speed Photography SPIE **1358**, p. 703, (1990).
29. T.J. Kinsey et al., "Hycam camera study of the features of a deflagrating munition," Proc. 19th Int. Conf. High Speed Photography SMPTE **1358**, p. 914, (1990).
30. J.M. Dewey and D.J. McMillin, "Analysis of results from high speed photogrammetry of flow tracers in blast waves," Proc. 19th Int. Conf. High Speed Photography SMPTE **1358**, p. 246, (1990).
31. M.M. Chaudhri, "The working of a small arms cartridge," Proc. 20th Int. Conf. High Speed Photography, SPIE, **1801**, pp. 108–114, (1992).

32. P.W.W. Fuller, "Gun barrel temperature studies using short time scan thermal imaging," ICIASF Record 1989 Pub IEEE, pp. 159–167, (1989).

General References

R.J. Bracher and A.E. Huston, "High Speed Radiography of projectiles," Proc. 12th International Congress on High Speed Photography and Photonics, Society of Photo-optical Instrumentation Engineers, (SPIE) Bellingham, NY. (1976).

P.H. Dugger and R.E. Hendrix, "Laser photography at AEDC," *Optical Spectra* **9**, (1975).

H.E. Edgerton, "Electrical flash photography," *J. Soc. Motion Picture Eng.* **53**, (1949).

P.W.W. Fuller, "Instrumentation for the determination of the shot exit time in guns," Proceedings of The 7th International Congress on Instrumentation in Aerospace Simulation Facilities, Institution of Electrical and Electronic Engineers, USA, pp. 182–194, (1977).

P.W.W. Fuller, *J. Photo. Sci.* **42(1,2,3,4,5,6)**, (1994).

F. Jamet and G. Thomer, *Flash Radiography*, Elsevier, Amsterdam, (1976).

G.A. Jones, *High Speed Photography*, Chapman & Hall, London, (1952).

R.F. Saxe, *High Speed Photography*, Focal Press, London (1966).

The proceedings of The International Congress on Instrumentation in Aerospace Simulation Facilities. Pub. IEEE Aerospace and Electronic Systems Society. USA. Meetings every two years.

The Proceedings of The International Congress on High Speed Photography and Photonics. Pub SPIE USA. Meetings every two years.

The proceedings of The International Symposium on Ballistics. Sponsored by the American Defense reparedness Association. USA. Meetings each year.

8 Ballistic Holography

Gary Hough

If a picture is worth a thousand words, then a hologram is worth a thousand pictures.

8.1 Introduction

Improvements in the visualization of ballistic events have long occupied the attention of researchers in this field. Conventional methodology includes streak cameras, shadowgraph and Schlieren images, high-speed still and cine cameras, high-speed video, and flash x-radiography. Their use has been described by Swift [1]. However, these approaches provide only two-dimensional images (although a limited degree of three dimensionality may be obtained by using, for example, orthogonal flash x-rays), a serious drawback for study of what is almost always a three-dimensional phenomenon. Recent needs for improving both the quantity and the quality of event data, as well as the opportunity of obtaining high-fidelity information for the validation of hydrocodes, have led to the exploration of pulsed-laser holography as a practical approach for providing such answers.

Pulsed-laser holography has many potential advantages. In addition to providing a true three-dimensional picture, these include a large depth of field and high spatial resolution. Furthermore, two separate and independent holograms can be exposed on the same film by double pulsing the laser with the intervening time precisely controlled. From this double image, vector velocities of bodies of interest can be determined.

Applications for which the promise of holography appears particularly attractive include behind-target debris formation and dispersal, the formation and breakup of hypervelocity jets, and impact crater ejecta.

This chapter introduces the holographic principle, provides a brief history of early work in the field, some subsequent successes, data reduction efforts, and an assessment of the current state of the art.

8.2 Holography

A hologram, literally *whole image*, is a way to retrieve the likeness of a subject from a record of its diffraction pattern. This now well-known concept originated in the studies of Gabor in 1947. It became a practical reality with the work of Leith and Upatnieks in the early 1960s. Their key contribution was that the use of laser light.

The basic technique of hologram formation is to divide the coherent light from a laser into two beams: one to illuminate the subject and one to act as a reference (see Fig. 8.1a.) The reference beam is directed so as to intersect the light transmitted or reflected by the subject. An interference pattern is formed in the region where the beams overlap. A photosensitive medium, placed in this region, undergoes chemical or physical changes due to exposure to the light intensity. After processing, the medium (a glass plate or flexible film) becomes the hologram. When the hologram is illuminated by the original reference beam, the original subject is reproduced (see Fig. 8.1b.) The hologram thus acts as both a recording and a projection system.

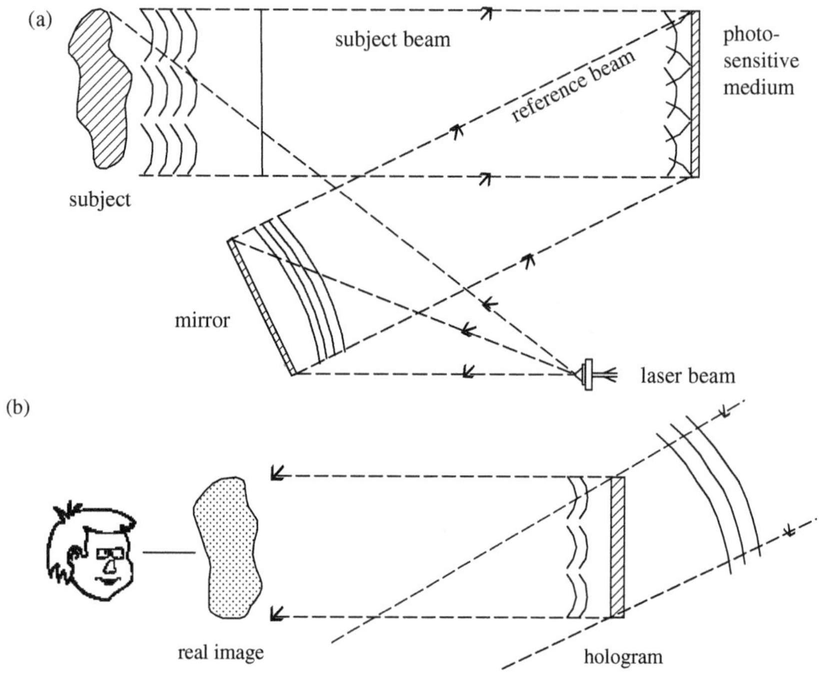

Fig. 8.1. Making and viewing a hologram (**a**) Making the hologram. (**b**)Viewing the hologram

Conventional holograms are made of static subjects and, in fact, motion of the subject (and of the optical components) is strictly limited to less than a small fraction of the laser source wavelength. This restriction is necessary for the proper formation of the interference fringes and appears to preclude application to ballistic events. Fortunately, it can be bypassed if we instead limit the subject travel to be a suitably small fraction of its length during the duration of the laser pulse, as is done in conventional photography. The result is a *shadowgraphic hologram image* that reveals the essential three-dimensional shape of the subject, but not surface details.

An important consideration in making ballistic holograms is the choice of the laser. Two possibilities exist: a pulsed ruby laser or a pulsed, frequency-doubled, Nd-YAG laser. The ruby laser has the advantages of low cost, relatively high power output, ease of triggering, and reliable beam coherence. Offsetting these advantages is the fact that it is not possible to operate a ruby laser in a continuous-wave mode for the reconstruction system. This introduces a scale shift in the image that complicates the required data analysis. The Nd-YAG laser has a key advantage in that Nd-YAG systems are available to provide a continuous monochromatic light source for reconstructing the image system. This simplifies the data analysis considerably. Also, they are available with the shorter pulse durations that may be needed to capture hypervelocity events. On the other hand, these lasers are costly and require a quite complex triggering system to generate a reliable pulse output at the desired time. Table 8.1 expands and summarizes the features of these two lasers.

Table 8.1. Comparison of Pulsed Lasers for Ballistic Applications

Feature	Ruby Laser	Nd-YAG Laser
Wavelength	694 nm	532 nm
Pulse duration	Nanoseconds	Less than Ruby
Output Power	Joules	Less than Ruby
Cost	Modest	More Expensive
Triggering	Simple	Complex
Image Analysis	Complex	Straightforward

Each of these lasers has been employed for ballistic holography. For reconstruction of holograms made using a ruby laser, an inexpensive continuous-pulsed He-Ne laser (633 nm) can be used because these two laser wavelengths differ by less than 10%. Later in the chapter, alternate approaches to this problem are suggested.

8.3 Early Ballistic Holography

Applications of holography to ballistics first appeared in the 1970s. The problems of primary interest were those of characterizing behind-target debris fields and of measuring surface deflections of the impacted target. A paper related to debris characterization was given at an Aeroballistic Range

Association (ARA) meeting in 1973 [2]. Royer and Giraud at Institut Franco-Allemand de Recherches de Saint-Louis (ISL) in France reported on several tests to measure debris particles sizes [3, 4]. Some work was also undertaken at SAI in California [5] to look at both the debris field and the target surface deflection problems. This latter problem was also the focus of work by Albe in France [6].

The various approaches were marginally successful, but apparently were not pursued or extended. Possible reasons range from image interference by impact-generated flash or obscuration by smoke and/or dust to perceived difficulties in data analysis.

8.4 Breakthrough

In the mid 1980s, research was initiated at the LTV Missiles Division in Dallas with the goal of developing a workable system that would overcome the deficiencies of the earlier approaches, provide a capability to capture data equal or superior to that of flash x-radiography, and be cost effective. Here too, the problem of interest was analysis of behind-target debris.

Preliminary proof-of-principle tabletop tests were carried out in 1987, and ballistic-range testing with a 30-mm powder gun was initiated in 1988. A pulsed ruby laser was used, and it and associated optical components were mounted on massive optical benches adjacent to the gun. The laser had a pulse width of 30 ns and a peak energy output of about 350 mJ. The laser was pulsed at a pre-selected time after impact of the projectile on the target plate. A sketch of the experimental setup is shown in Fig. 8.2.

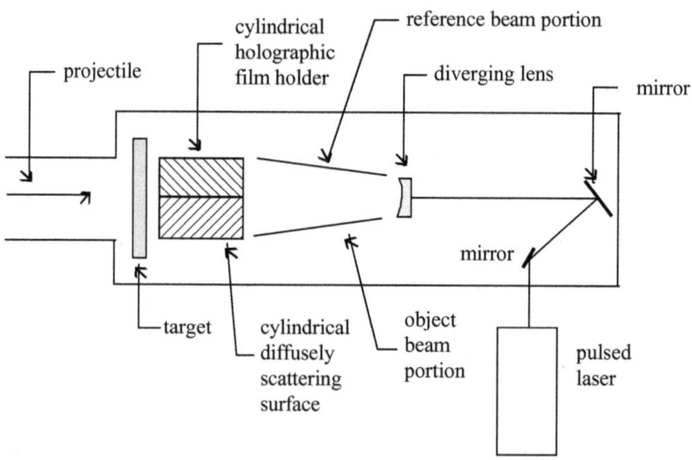

Fig. 8.2. Holographic System Schematic

The holographic film holder (interior diameter 15 cm) is shown in Fig. 8.3. One of the key features of the approach was to make the holograms on flexible film, in a semi-cylindrical configuration, filling one-half of the holder. The other half was filled with a diffusely scattering surface, similar to that of a movie screen, to provide the laser illumination of the desired subject. The advantage of the semi-cylindrical arrangement is that it enables the user to select optimal, well over 180 degree, views of the scene after the hologram is made (holograms made on a flat-plate film have a range of viewing angles less than about 20 degrees). For example, it is quite difficult to observe a common target impact phenomenon, the generation of spall rings, without this technique.

Fig. 8.3. Holographic Film Holder

Although the holographic film placed as described above is vulnerable to being struck by debris particles and perforated, this is not a serious drawback. Any undamaged portion of the film contains all the information needed. However, for some applications, catastrophic perforations could possibly occur. In such cases, the film can be protected behind a thin curved Lexan sheet, with the space in which the film is inserted being filled with a refractive index-matching liquid. The properties of the liquid can be selected to virtually eliminate the effects of spurious reflections from the enclosure. After some experimentation, glycerin was selected as the best compromise.

Another novel feature of this system was the inclusion of a fiducial grid into the hologram to provide linear and angular scaling capability. This was done by inserting a polycarbonate rod (on which a cylindrical co-ordinate grid was scribed) along the film axis and pulsing the laser to record the grid image on the film. The rod is then removed and the gun and laser fired to superpose the debris field on the grid. When viewed, the debris particles appear to "float" in the grid system. An example photograph is shown in Fig. 8.4. This arrangement provides an approximate means of determining the true scale of the image.

Typically, the holograms could be developed within a few minutes after test completion and were available for viewing in less than half an hour. This rapid turnaround provides close to a real time capability for examining the test results. A point worth noting is that a considerable amount of time and effort was spent on the holographic film development process. Bleaching was found to improve the clarity of the image.

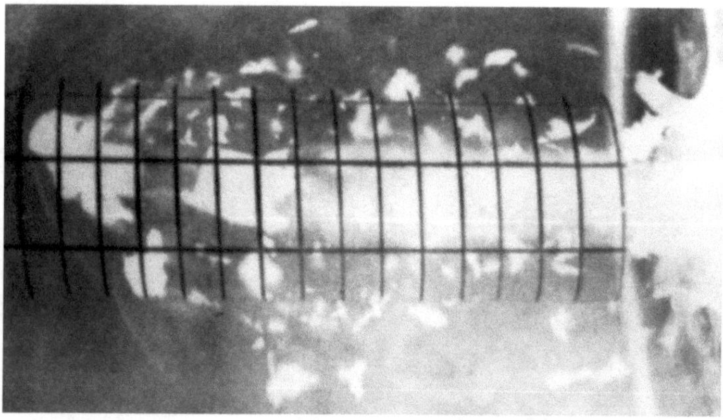

Fig. 8.4. Superposition of a reference grid on a hologram

A fiber-optic delivery system for the pulsed laser beam was also investigated. However, the fiber optic was unable to sustain the peak pulse energies needed to ensure holograms of sufficient brightness and so this approach was abandoned.

Numerous holograms were made at impact velocities from about 850 to 1500 m/s (the ordnance regime). Figure 8.5 illustrates the difference in the behind-target debris patterns for steel and aluminum plates providing roughly equivalent ballistic protection.

Double-pulsed holograms, useful for determining velocities, were also made. There are two important considerations for making these successfully. The first is selection of the time interval between laser pulses, which must be chosen to ensure that all objects of interest appear in the hologram. This interval depends on both the size of the film available and the velocity of

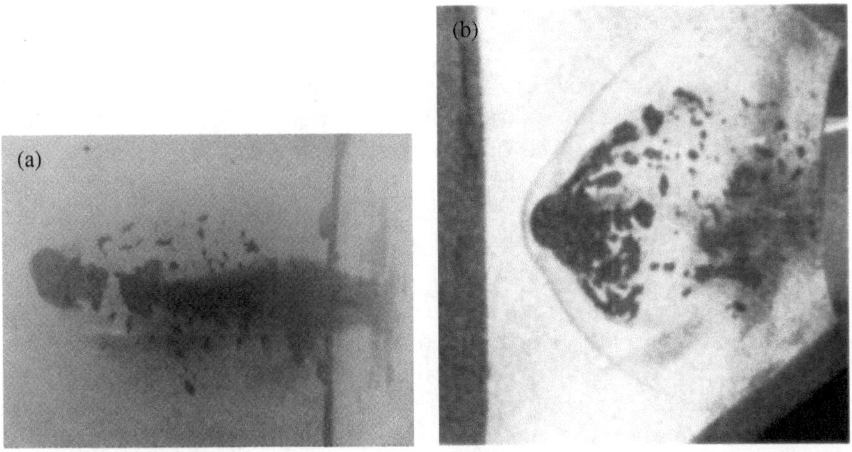

Fig. 8.5. Debris patterns behind aluminum and steel targets (**a**) Aluminum Target. (**b**) Steel Target

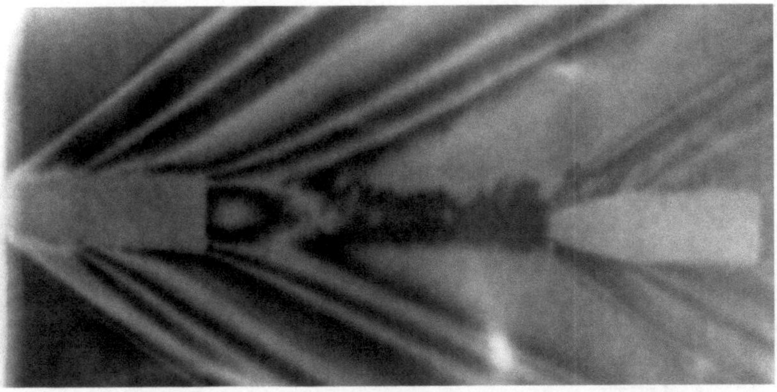

Fig. 8.6. Double-pulsed hologram of 30-cal bullet

the objects. For the LTV studies, time intervals of about 100 to 300 μs were typical. The second consideration is the allocation of the total available laser energy between the two pulses. Experimentation established that the ratio of the energy contained in the first pulse to that in the second should be slightly less than unity. A higher value leads to film overexposure by the first pulse, whereas a value that is too low leads to underexposure of the first image. Figure 8.6 is a photograph of a double-pulsed hologram of a 30-cal. rifle bullet in flight. Using the known laser interpulse time, the bullet velocity was calculated to be 876 m/s. This is in excellent agreement with the value of 881 m/s measured by a laser-beam break signal. The actual hologram shows clearly the flow details such as shock waves and turbulent boundary layers. Summaries of the LTV work are found in [7, 8].

8.5 Later Studies

The LTV efforts led to offshoots and extensions at several other organizations. In the early 1990s, Sandia National Laboratories carried out studies along these lines, with the emphasis on hypervelocity impacts. Sandia chose to use a Nd-YAG laser (5 ns pulse, 750 mJ) as the light source and carried out the testing using their two-stage light-gas gun. A key engineering accomplishment was the integration of the difficult laser triggering subsystem into the gun control circuits. Sandia also used a flat glass plate holographic medium that has better dimensional stability than does flexible film. This is important when more accurate measurement of object size and location is required as, for example, when extremely small objects are of primary interest. On the other hand, this also generally requires some form of plate protection, and the Sandia system placed the holographic plate in a liquid gate with an index-of-refraction matching fluid. A schematic of their system is shown in Fig. 8.7, wherein the perspective is from the impact chamber looking back towards the gun.

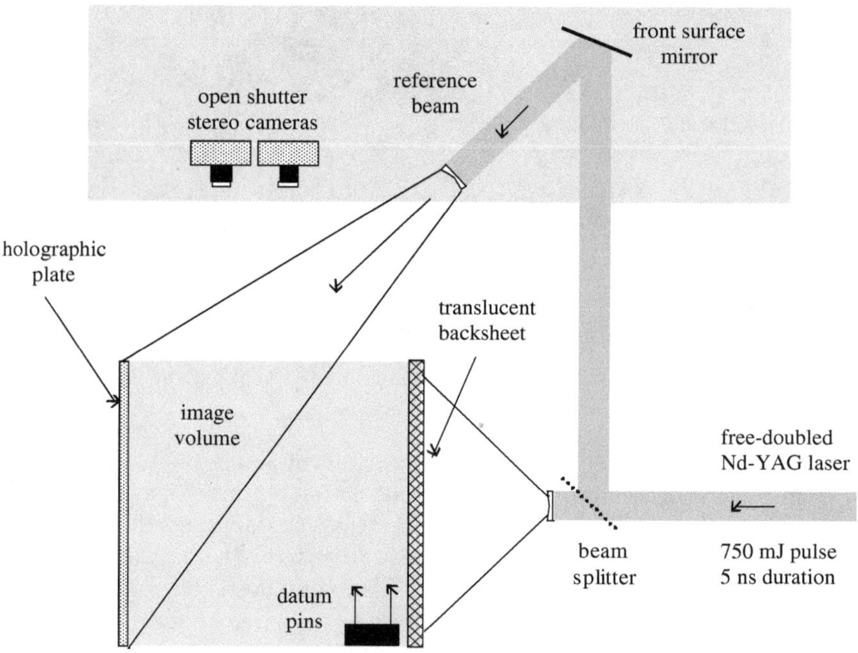

Fig. 8.7. Schematic of Sandia Holography Layout

Sandia made holograms of behind-target debris resulting from impact velocities up to nearly 4.4 km/s. The results, and more details of their holographic system, are reported in [9].

Later efforts have included studies at the University of Alabama in Huntsville using a Nd-YAG laser having a 130 ps pulse width and extensive studies with a powerful ruby laser at the United States Air Force Armament Laboratory, Eglin Air Force Base. The latter was directed towards increasing the useful size of holographic images that could be made.

8.6 Hologram Analysis

Ballistic holograms are a source of valuable information. On the one hand, having for the first time a true three-dimensional visualization provides key insights into the dynamics of these impact events. On the other hand, by extracting specific information such as particle masses and velocities from the holograms, valuable quantitative data can be obtained. The ability to extract these data in a timely and economic manner is the key to the ultimate usefulness of ballistic holography.

The holographic image can be reconstructed by illuminating the plate or film with a continuous laser light source. This source does not have to be of the same wavelength as that used to make the hologram. However, using a source of different wavelength introduces nonlinear amplitude changes into the image. Although this is not bothersome for qualitative viewing, it is important for quantitative analysis. There are several ways to alleviate this problem, including introduction of a fiducial reference into the hologram or by comparison with a static hologram made of objects of known size, but these still raise questions about attainable accuracy (as well as increasing the data analysis time). It is obviously better to use the same wavelength for both making and analyzing the hologram, all other aspects being equal.

Both LTV and Sandia used essentially the same approach for data analysis. This involved converting the information contained in the hologram into two-dimensional form, digitizing it, and conducting subsequent analysis on a computer. The basic LTV arrangement is sketched in Fig. 8.8; a photograph of the Sandia image processing system is given in Fig. 8.9.

The hologram can be placed on a precision rotary turntable, illuminated, and viewed with a suitable CCD camera. The desired image is captured by using a commercially-available image processor board on a personal computer. This software permits the user to locate debris pieces relative to a common reference point and hence to one another. This establishes two of the coordinates necessary to specify the position of the piece. The third coordinate is obtained by a 90° rotation of the turntable and repeating the operation. Some trials at LTV suggested that the center of mass locations of the major debris pieces could be found to within an accuracy of about 5% of their largest dimension. Velocities could be determined from successive individual center of mass positions tracked in double-pulsed holograms.

In principle, the area of each piece can be determined at incremental angles and used to estimate its volume. In practice, areas are more difficult

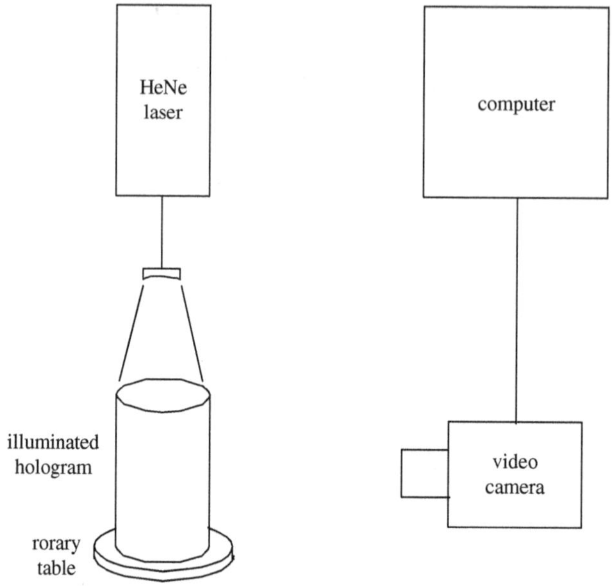

Fig. 8.8. Digital analysis schematic

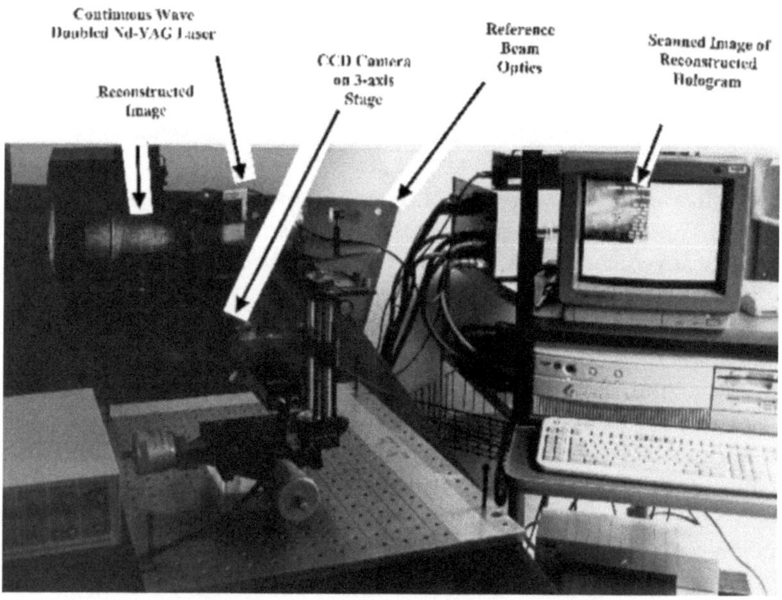

Fig. 8.9. Sandia image processing system

to determine accurately because of subjective user selection of edge-defining contrasts. Accuracy also suffers as areas decrease in that the digital representation must contain a sufficient number of pixels so that this does not become a limiting factor. An additional issue complicating mass determination is that one cannot distinguish among different materials in the hologram. Hence, densities have to be assigned to establish particle masses.

Overall, the data extraction process has not yet been sufficiently well developed for behind-target debris to provide holography with any realistic advantage over the processes used, for example, in orthogonal flash x-radiography. Myriad pieces "clog" the region of interest, and it is not always possible to separate every individual piece from another.

One feature that data analysis has been able to establish is that of the minimum size of particles that can be resolved. In the LTV tests, a diameter of 125 μm was estimated. In the Sandia studies, particles as small as 5 μm were identified.

8.7 Conclusions

The practical application of holography to visualization of high-speed ballistic events has been amply demonstrated. The holograms provide a unique viewing perspective and are potential sources of valuable information relating to event dynamics. However, although the capability to make holograms is now in hand, the task of analyzing the data contained therein remains in relative infancy. Until considerable progress has been made in this area, the value of holography will likely remain in that of qualitative viewing of impact events.

References

1. H.F. Swift, "Image Forming Instruments," in *Impact Dynamics*, John Wiley and Sons, Inc., 1982.
2. J.D. DiBattista et al., "A Ballistic Range Application for Far Field Holography in Determining the Ejected Particulate Velocity Distribution of Impacted Dirty Plates", Paper presented at the 24th Annual Meeting of the Aeroballistic Range Association, Tullahoma, TN, September 1973.
3. H. Royer and M. Giraud, "The Use of Holography for Kinetic Energy Measurements of Target Splinters", Proceedings of the Fourth International Symposium on Ballistics, Monterey, CA, October 1978.
4. H. Royer, "A System for Recording High-Speed Holograms of Very Far Objects," Paper presented at the 16th International Congress on High-Speed Photography and Photonics, Strasbourg, France, August 1984.
5. Personal communication from George Burghart 1985. Also, "Examination of Impact Using Lasers", SAI, Santa Ana, CA, 1 July 1977.

6. F. Albe, *Interférometrie Holographique par Double Exposition, Appliquée a l'Étude du Début de la Déformation d'un Matériau Dur Soumis a l'Impact d'un Projectile*, Thèse, Université de Besançon, France, June 1976.
7. G.R. Hough and D.M. Gustafson, "Ballistic Applications of Lasers," *Proceedings of the SPIE Ultrahigh Speed and High Speed Photography, Photonics, and Videography '89 Conference*, SPIE Volume 1155, (ed. P.A. Jaanimagi), pp. 181–188 (1990).
8. G.R. Hough, D.M. Gustafson, and W.R. Thursby, "Enhanced Holographic Recording Capabilities for Dynamic Applications," *Proceedings of the SPIE Ultrahigh- and High-Speed Photography, Photonics, and Velocimetry '90 Conference* (ed. L.L. Shaw, P.A. Jaanimagi, and B.T. Neyer), pp. 194–199 (1991).
9. J.A. Ang et al., "Pulsed Holography for Hypervelocity Impact Diagnostics", *Int. J. Impact Engng.* **14**, pp. 13–24, 1993.

9 Time-Resolved Gauges for Measurements of Shock Waves in Solids

William M. Isbell

9.1 Introduction

To a large extent, the development of physical models to describe shock wave phenomena has closely followed the development of shock wave instrumentation and the production of physical data. As instrumentation has become more capable, it has been possible to move from measuring simple, one-dimensional geometries in the laboratory to the measurement of "real-world" phenomena, which frequently involve complex, multi-dimensional waves. This chapter describes instrumentation for shock wave research that provides the experimenter with capabilities for measurement of complex phenomena. The intent of the chapter is to inform the reader of selected capabilities and to direct him or her to the appropriate references.

It has been said that, "One good piece of data is worth a thousand expert opinions." This is particularly true in the field of shock waves, where the event is far too brief for the eye to observe and the release of energy which creates the shock waves tends to overwhelm the senses. For such studies, researchers are placed in the role of indirect observers, with only their instruments to record the phenomena. It is a curious fact that we researchers into shock wave phenomena have never actually seen a shock wave – only its effects on gases and solids.

Thus, instrumentation is at the very heart of the science of shock waves and the quality of the available instrumentation substantially influences our understanding of shock waves and material behavior. New developments in instrumentation are welcomed by researchers, as they lead to improved understanding of the mechanical, thermodynamical, and electromagnetic response of materials to high intensity waves.

As the field of shock wave physics matures, the problems that can be addressed grow more complicated. One-dimensional shock waves produced in the laboratory for the measurement of basic material parameters are extended to real-world applications in which multi-dimensional shock waves are customary. To a large extent, the "easy" problems have been solved and the researcher is challenged with more complicated questions – which lead to more stringent instrumentation requirements.

This chapter summarizes several techniques for measuring shock waves in solids

- In-material gauges: Strain-compensated stress gauges for measurements of multi-dimensional stress fields within a specimen
- Rear-surface gauges: X-Cut quartz gauges for measurement of equation-of-state (EOS) parameters and transmitted wave profiles
- Surface velocity: Velocity interferometers for measurement of surface motion (including VISARs and Fabry–Perot interferometers).

Note that the references in the chapter have been augmented by additional entries not quoted in the text. Since neither time nor space in this chapter is sufficient to cover the field of time-resolved gauges completely, the author has included many other references than just those quoted in the paper. In particular, the reader is encouraged to examine papers that describe the other diagnostic techniques listed in the references under "Recommended Reading on other Time-Resolved Diagnostics."

9.2 Instrumentation Techniques

9.2.1 In-Material Gauges: Strain-Compensated Stress Gauges for Non-Planar Waves

Development of gauges for in-situ measurements of planar shock waves in solids began in the 1960s, using thin foils of piezoresistive materials inserted between layers of specimen materials. Manganin, carbon, ytterbium, lithium, and other materials were calibrated for use over a range of pressures of 0.1 GPa to over 100 GPa. Since the late 1970s, thin films of piezoelectric materials such as PVDF (poly-vinyl difluoride, often referred to PVF_2) have been utilized as the stress-sensitive element [2, 3].

Whereas these gauges have proven useful for measuring one-dimensional shock waves, the real world usually involves more complicated structures, producing waves of multi-dimensions. For such waves, standard stress gauges can produce quite misleading results. The problem arises from the fact the gauge materials are sensitive to both stress and lateral (in-plane) strain effects. Thus, as a multi-dimensional stress wave passes over the gauge, the response is the sum of two independent effects, the normal stress and the lateral strain.

A solution has been to provide a simultaneous and independent measurement of the strain in the gauge [2, 3, 8, 9] and then to subtract the strain contribution from the stress record. This is accomplished by placing a material that is sensitive only to strain (typically constantan) in close proximity to the stress gauge itself. Figure 9.1 shows an example of such a gauge. Two dual-element packages are shown, consisting of one stress gauge (typically manganin, carbon, or PVF_2) and one strain gauge.

Details of the calibration and construction of strain-compensated stress gauges are given in [6, 8–10]. Calibration test configurations are shown in Fig. 9.2 for different combinations of stress and strain.

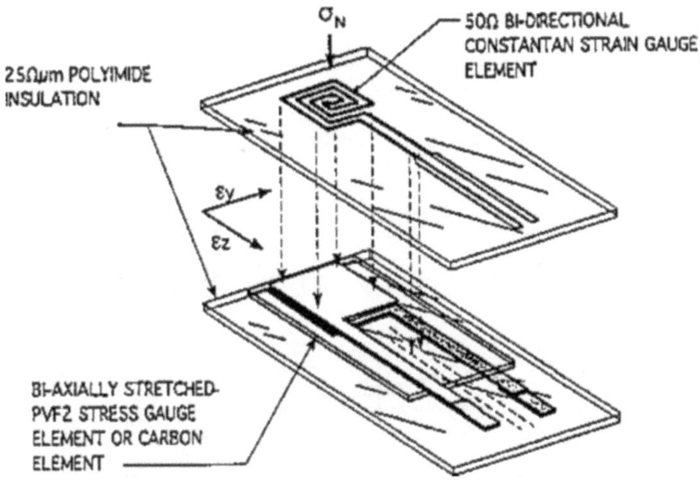

Fig. 9.1. Superimposed PVF_2 and Constantan or Carbon–Constantan gauges

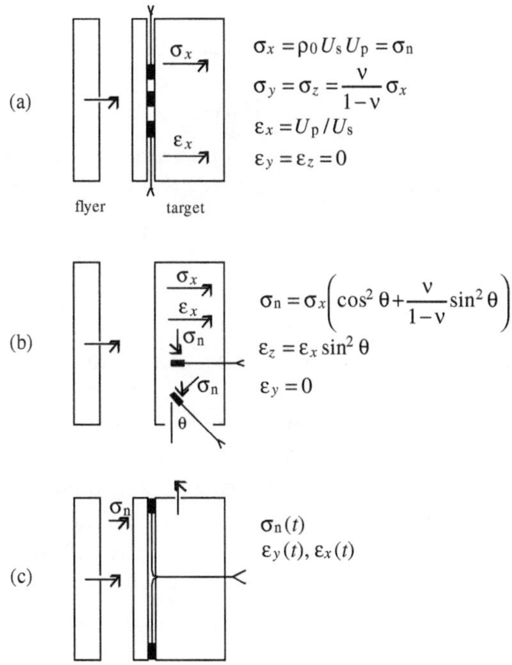

Fig. 9.2. Experimental approach to measuring stress and in-plane strain. (**a**) Plane wave compression test. (**b**) Plane wave compression stress/compression strain test. (**c**) Time-varying compression stress/tension strain test

- Compressive plane wave stress
- Compression plane wave stress – compressive strain
- Multi-dimensional wave stress and strain

The applicable equations are shown beside each configuration.

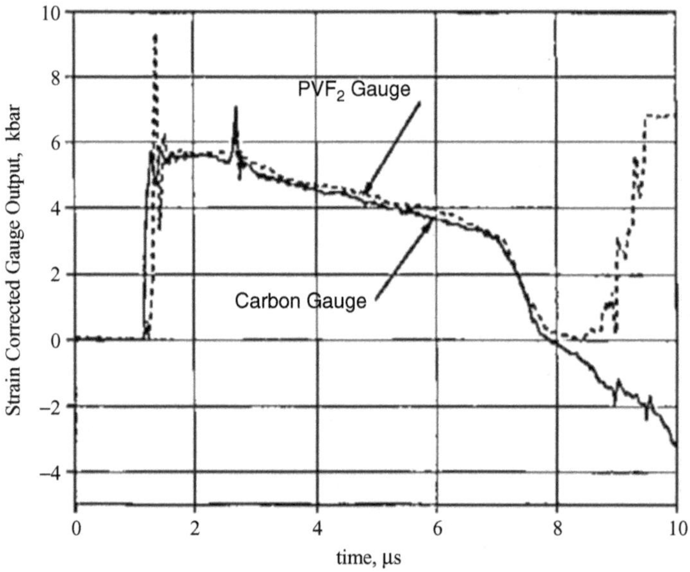

Fig. 9.3. PVF_2 and carbon gauge records after correcting for the effects of lateral strain. These results demonstrate the effectiveness of the analysis and the ability of such gauges to measure multidimensional fields

Figure 9.3 shows the strain-corrected outputs from the PVF_2 and carbon gauges, demonstrating the effectiveness of the analysis and of the gauge concept. Calibration curves utilized in the analysis of the PVF_2 and the carbon gauges are shown in Fig. 9.4.

The development of stress gauges for multi-dimensional testing allows the experimenter to explore a much wider field [1]. One-dimensional tests of basic material parameters are augmented by two-dimensional configurations such as spherically-diverging waves, and by multidimensional configurations such as the impact of an irregular fragment into a spacecraft shield.

9.2.2 Rear Surface Gauges: X-Cut Quartz Gauges for Measurement of Equations of State and Transmitted Wave Profiles

In contrast to the in-material, piezoresistive gauges previously described, a very large amount of valuable test data has been taken with piezoelectric

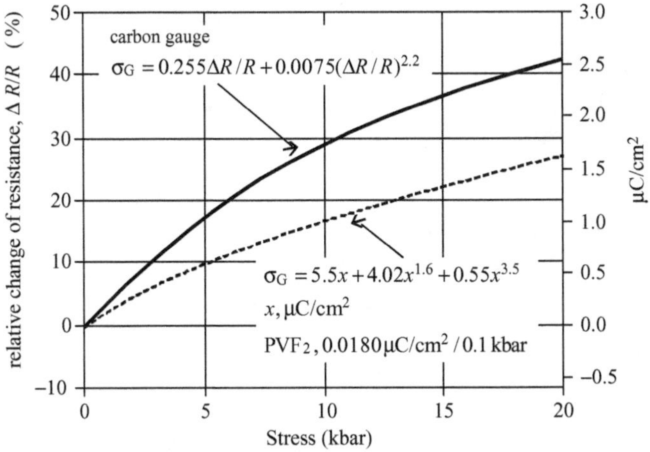

Fig. 9.4. Calibration curves used in the analysis of impact tests

crystals, either mounted at the rear of the specimen or impacted directly on the specimen surface.

X-cut quartz crystals generate a current when opposing faces are at different stress levels. In shock wave applications, one face remains at zero stress while the face in contact with the specimen is subjected to a time-varying stress pulse. This condition lasts until the shock wave reaches the back face of the gauge, at which time the useful recording ends.

Measurement of the generated current as a function of time permits circulation of the stress history. The quartz gauge is generally used in two configurations: direct impact or transmitted wave, as discussed below.

Direct Impact

The direct impact technique is shown in Fig. 9.5. At impact, one stress wave propagates into the quartz crystal and one into the impactor or specimen. The current generated by the stress difference between the impact face and the rear face of the quartz is collected from an electrode in the center of the rear face of the crystal, with the impact face at ground potential. The current is passed through low-loss, foam-dielectric cable to digitizing oscilloscpes, and the piezoelectric current vs. time signal is recorded.

Since quartz calibration data become non-linear above about 3 GPa, a variation in the above technique is used to obtain data at higher stresses. An elastic, high impedance buffer material (e.g., tungsten carbide) is bonded to the front of the quartz gauge and is impacted by the specimen. Because of the large difference in impedance of the buffer and the quartz, high stresses are generated in the specimen and the buffer without exceeding the well-calibrated stress range of the quartz. Tungsten carbide buffers, for example,

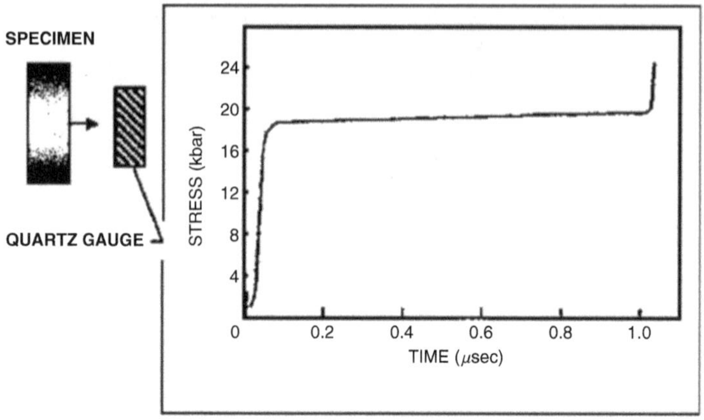

Fig. 9.5. Quartz direct impact technique

reduce the stress induced in the quartz to about one-quarter that in the specimen, allowing measurements to ~12 GPa.

Transmitted Wave

The technique for measuring the profile of a shock wave transmitted through a specimen is shown in Fig. 9.6. An x-cut quartz crystal is placed on the back face of the specimen, and the assembly is impacted with an impactor either of the same material as the specimen (where, because of symmetry, $u_{\mathrm{p}} = v_{\mathrm{I}}/2$) or an impactor of different material with a known Hugoniot. A stress wave is generated at impact and propagates through the specimen to the interface with the quartz gauge. Here the wave is partially reflected back into the specimen and partially transmitted into the quartz, as determined by stress and particle velocity continuity requirements. Data recording is essentially the same as with direct impact.

Elevated Temperature Testing

For testing at elevated temperatures, the target may be radiantly heated by a nichrome-wire heater element placed between the target and the launch tube, as shown in Fig. 9.7. The system is shown both with the heater in position before firing and retracted for firing (with the target removed to show heater and projectile detail). The target is heated from the front face only, to permit instrumentation or optical access to the rear, and temperature is monitored by a thermocouple. Heater element voltage is adjusted to bring the target to temperature in ~10 minutes, and the target is then allowed to stabilize at test temperature for 5–10 minutes.

The target and impactor are aligned at room temperature for planar impact. As the target is heated its alignment is monitored by an optical-lever system in which a laser beam is reflected off the rear face of the target and

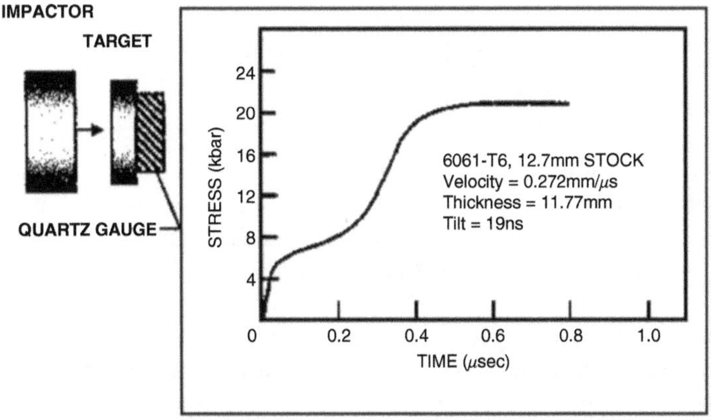

Fig. 9.6. Quartz transmitted wave technique

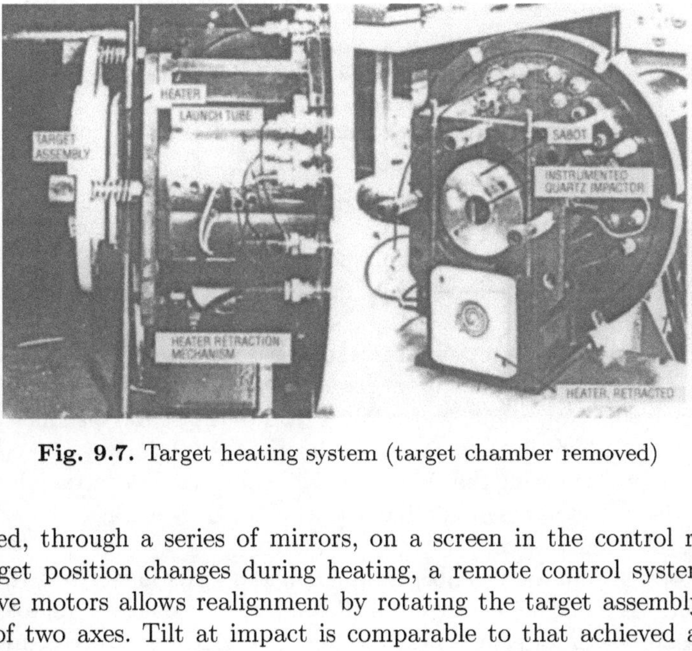

Fig. 9.7. Target heating system (target chamber removed)

displayed, through a series of mirrors, on a screen in the control room. If the target position changes during heating, a remote control system using DC-drive motors allows realignment by rotating the target assembly about either of two axes. Tilt at impact is comparable to that achieved at room temperature.

Although designed primarily for quartz gauge testing, the heating system can be used with any type of instrumentation, provided the transducer, if any, can withstand exposure to elevated temperature, or if the rear surface remains reflective, when optical techniques are being used. Velocity interferometer measurements at elevated temperatures can be performed with this technique.

9.2.3 Surface Velocity Measurements: VISAR and Fabry–Perot Interferometers

VISAR Interferometry

Among the several methods used for measuring the motion of surfaces under the action of shock waves, laser interferometry has become a standard technique for laboratory and field testing. In such instruments, a laser beam is focused on the specimen surface and the reflected beam is returned to the interferometer.

Within the interferometer, the beam is divided into two paths, one path proceeding through an optical delay line (the *delay leg*) and the other through a shorter path (the *reference leg*). Upon recombination of the beams, interference fringes are formed.

As the specimen surface moves under the action of the shock wave, the light in the return beam is shifted by the Doppler effect by an amount proportional to the instantaneous velocity of the surface. The result is a movement of the fringe pattern, which is subsequently recorded by photomultiplier tubes or by streak cameras.

Development of velocity interferometers has taken two forms, a Fabry–Perot interferometer, in which the light is multiply reflected in the delay leg, and the VISAR interferometer [16], in which the light passes through the delay leg but one time before recombination. Each technique has its advantages; the Fabry–Perot technique for the simplicity if its design and use and the VISAR technique for its greater velocity range and higher time resolution.

The use of velocity interferometers in tests involving complex, non-planar shock waves is made possible by the use of diffuse surfaces on the specimen to reflect the beam. Non-planar shock waves tend to rotate the specimen surface. For specular (mirror-like) surfaces, the beam is rotated away the collecting optics, rapidly decreasing the amount of light collected. In contrast, for diffuse surfaces the reflected beam is relatively unaffected by surface tilt for tilts in the range of a few degrees.

The wealth of data available from a single interferometer test is demonstrated in Fig. 9.8. The test involved the impact of two aluminum alloy 6061-T6 plates, one (the target) being twice as thick as the other (the impactor). Upon impact, a complex wave consisting of an elastic wave and a plastic wave is created and propagates into each plate. When the waves reach the free rear surface of the impactor, they are reflected as decompression waves which, having higher velocities, can overtake the compressive waves in the target.

As the waves in the target reach its free surface, they too are reflected as decompression waves. Superposition of the decompression waves within the target results in development of a region of tension, and in this case, formation of a plane fracture surface (spallation) within the target. The remaining

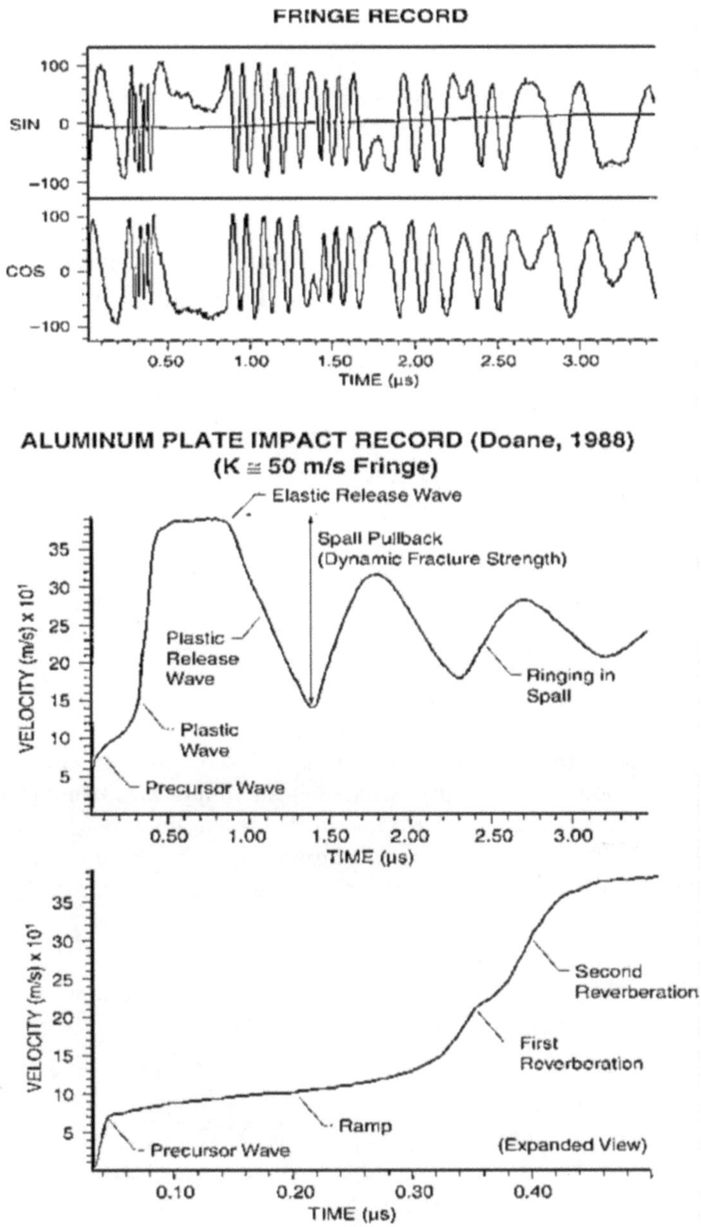

Fig. 9.8. Demonstration of the wide variety of parameters that can be measured in a single VISAR test. Details are described in the text, below

wave, entrapped between the target rear surface and the newly-formed spall surface, reverberates between the two surfaces, producing the damped sinusoidal "ringing" seen at the target rear surface.

From this single test the following parameters can be obtained.

- Elastic wave and plastic wave velocities
- Hugoniot elastic limit (HEL) wave amplitude
- A point on the Hugoniot of the specimen
- Elastic and plastic release wave velocities
- A measure of the spallation strength, from the amount of "pullback" before spallation occurs
- An indication of the number of dislocations created in the target material, from the damping of the spallation signal.

Figure 9.9 shows a fixed cavity VISAR system, developed at Sandia National Laboratories, in which the fringe adjustments are permanently aligned within an optical delay leg [24, 48]. Changing of sensitivities is accomplished by removing one permanently aligned delay unit and replacing it with another. More recent developments in velocity interferometers have concentrated on increasing the utility of the instruments and on their use in more complicated geometries (see Fig. 9.10).

With VISAR Interferometry becoming an accepted standard for measuring the shock loading of materials and providing information for the development of new explosives and pyrotechnics, it has become necessary to extend the lower limit of the velocity range of standard VISARs to tens of meters per second, with velocity changes as small as 1 m/s being resolvable. This was accomplished by developing a stable air delay leg for the fixed cavity VISAR [20]. The new delay module has a maximum sensitivity of 44 m/s/fringe, a factor of two–four times lower than that of standard glass etalon delay units.

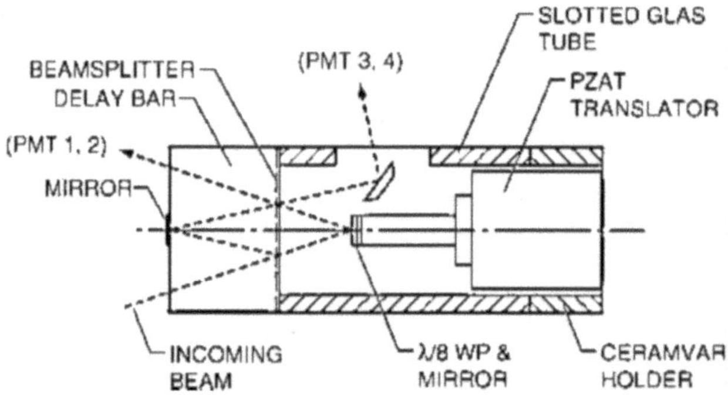

Fig. 9.9. A schematic of the fixed cavity VISAR configuration developed at the Sandia National Laboratories

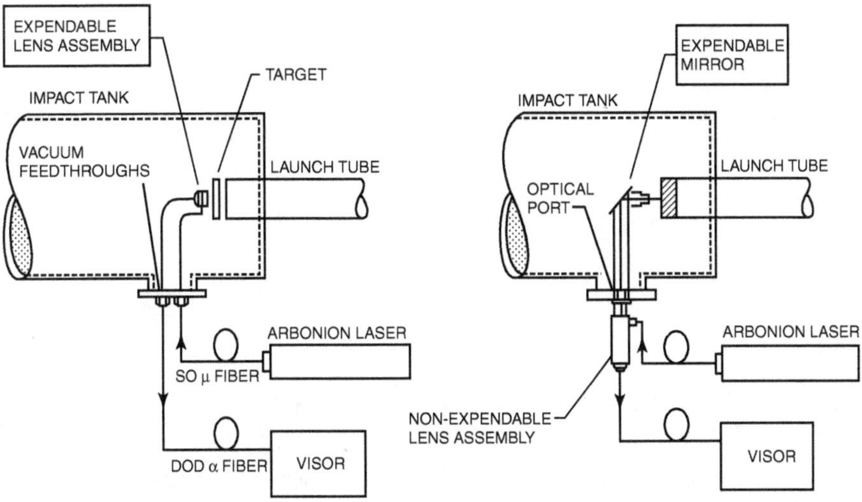

Fig. 9.10. Applications of VISAR systems to impact tests and to measurements of projectile motion. The laser beam is transmitted by fiber optic cable from the laser to the target and from the target back to the VISAR, allowing the laser and interferometer to be placed remote from the test itself

Conversely, new advances in laser-driven foils require very insensitive delays of ~20,000 m/s/fringe, so that fringes will not be lost in the rapid acceleration of the foil. For these tests, time resolutions of tens of picoseconds are required. This was accomplished by a modification to the fixed cavity VISAR design that allows fringe constants nearly a factor of eight greater than previous etalon designs. Combined with electronic streak cameras, theoretical time resolutions of 10–50 ps are possible.

Figure 9.11 shows a modified design, in which a *balance bar* has been added to the reference leg side of the optical delay leg. In effect, this bar subtracts time from the delay bar, making the delay leg much less sensitive and allowing tests at extremely high velocities (to 20,000–30,00 m/s). The fringes are focused on a slit and a streak camera is used to record the fringe motion [25].

For VISARs, the range of velocities is covered by a series of delay units, as shown in Fig. 9.12, which demonstrates the relationship between time and velocity resolutions. Time resolutions can be as short as tens of picoseconds for high velocity events.

Fabry–Perot Interferometers

Concurrent with the VISAR, Fabry–Perot interferometers have also been developed by several organizations (see for instance [18, 23]). This technique also utilizes the Doppler-shift of laser light, reflected from a target surface, to measure surface motion. The light is passed through a Fabry–Perot etalon

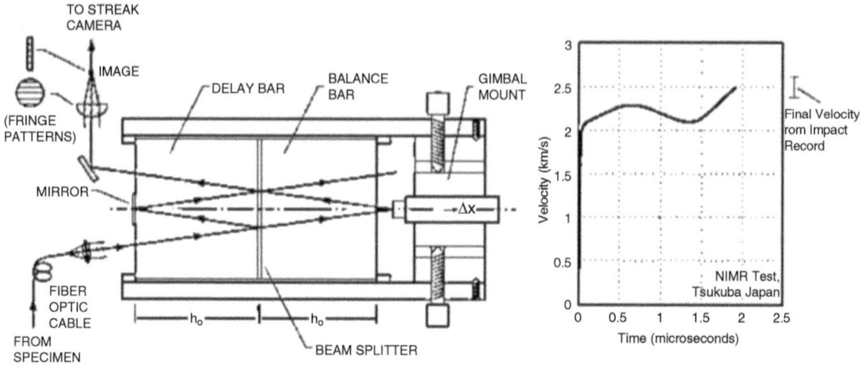

Fig. 9.11. The differential etalon technique provides measurements at very high velocities

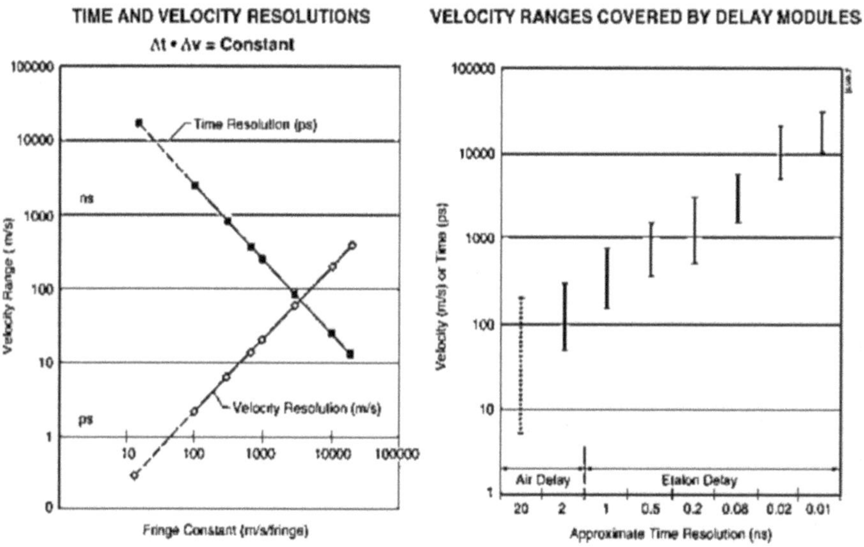

Fig. 9.12. (**a**) Time resolution to tens of picoseconds is possible at high velocity and using streak cameras as a recording medium. (**b**) The velocity range from 30–30,000 m/s is covered with individual delay modules

(for instance, a Burleigh Instruments Type RC150 etalon) in which fringes are created for recording by a streak camera.

Figure 9.13 shows the test arrangement and test results for the acceleration of a thin (250 μm) Kapton flyer from an exploding foil test [23]. The laser beam is led to and from the test fixture by an optical fiber and is focused on the surface of the foil by a small diameter lens.

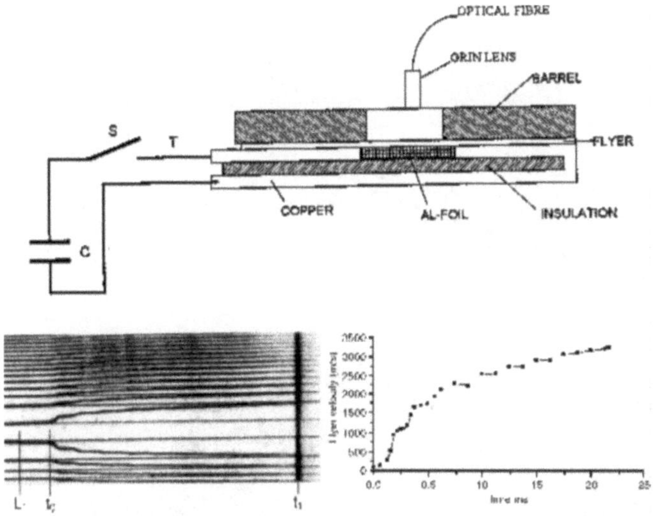

Fig. 9.13. The top part of the illustration is a schematic of a test conducted to measure the acceleration of an exploding-foil-driven plate. Test results are shown in the bottom part of the figure. On the left is the streak camera record and on the right is the analyzed result

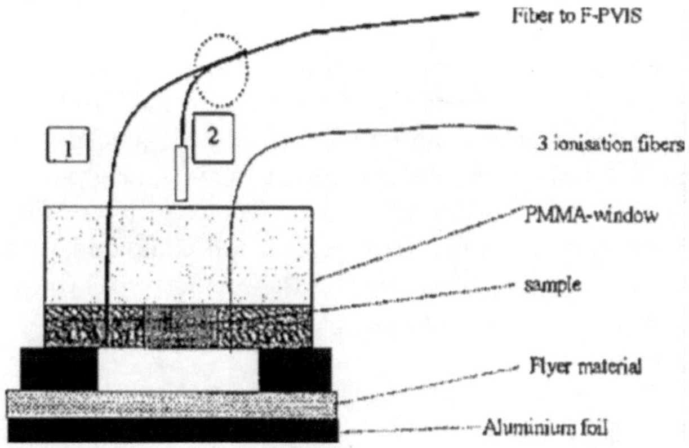

Fig. 9.14. An advantage of the Fabry–Perot interferometer illustrated in this figure is that one can measure (1) the velocity of the Kapton flyer, (2) the particle velocity of the interface, and (3) the shock wave velocity in the sample, measured by both the interferometer and the ionization fibers

The amount of displacement of the fringes from their initial position on the left of the record is a direct measure of the change of velocity. The analyzed record is seen at the right of the figure.

An arrangement to measure simultaneously both the velocity of the flyer and that of the interface between a specimen and a transparent "window" of PMMA is shown in Fig. 9.14. In this configuration, the interferometer has

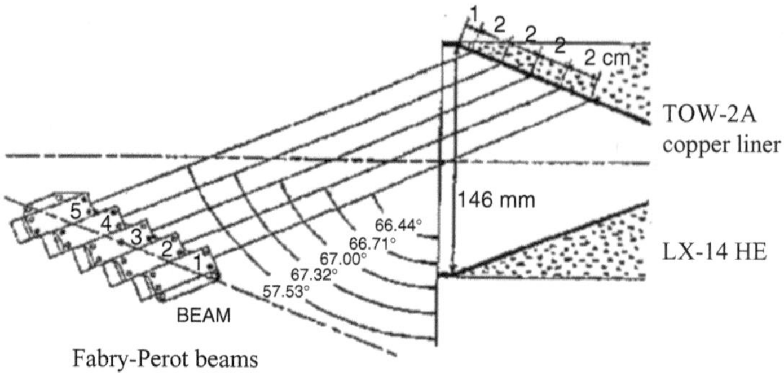

Fig. 9.15. LLNL five-beam Fabry–Perot interferometer, focused on one side of an imploding shaped-charge warhead, provides substantial information and insight into the mechanisms of liner collapse

been modified with a multi-mode coupler that allows the output from both fibers to be recorded on the same streak camera record. Such an arrangement provides an unambiguous measurement of a point on the Hugoniot of the specimen, by recording the flyer velocity, the shock wave velocity in the specimen, and the velocity of the interface of the specimen and the PMMA window. The particle velocity in the specimen can be inferred from these data.

A five-beam Fabry–Perot system has also been built [18] at Lawrence Livermore National Laboratory. Using several of these interferometers, data from as many as 20 positions can be recorded during a single test.

The experimental configuration for a five-beam system is shown in Fig. 9.15. In this test, five beams, spaced 2 cm apart, are focused on the inside of a shaped charge liner. Test results are compared with hydrocode calculations to determine the accuracy of the predictive routine and to aid in optimizing the shaped charge design.

9.3 Summary

The instruments described in this chapter are examples of the types of new instruments that are becoming available to researchers for advanced studies of the interaction of shock waves in materials. As the research moves from the laboratory to the field, the problems faced place greater requirements on the instrumentation. Planar shock waves are replaced by multi-dimensional shocks, requiring measurements at several places in the specimen simultaneously in order to record the complex motion of the materials.

Similar advances in measurement techniques for the bulk motion of targets and specimens are also underway. Examples include laser-illuminated image converter cameras and holographic techniques, which provide new insights into the complex world of shocks in solids. More advances are expected.

Acknowledgments

The author would like to acknowledge his fellow researchers in the Aeroballistic Range Association. Their expert advice and their friendship have inspired him throughout his career.

References

In-Material and Rear-Surface Gauges

1. Barbee, T.W., "Some Aspects of Dislocation Dynamics in Metals", Stanford University, ONR-SU Report No. 65-33, 1965 (D635023).
2. Bauer, F., "Properties and Shock Loading Response of Poled Ferroelectric PVF_2 polymer gauges," The 1987 ASME Applied Mechanics, Bioengineering, and Fluids Engineering Conference. Cincinnati, Ohio, June 14–17, (1987).
3. Charest, J.A., "Development of a Strain Compensated Shock Pressure Gauge," Dynasen Inc, report TR 005, (1979).
4. Charest, J.A., "Development of a Carbon Shock Pressure Gauge," Defense Nuclear Agency (U.S.) report TR DNA 3101 F, (1973).
5. Charest, J.A. and C.S. Lynch, "The Response of PVF_2 Stress Gauges to Shock Wave Loading." in: *Shock Compression of Condensed Matter – 1989* (eds. S.C. Schmidt, J.N. Johnson, and L.W. Davison) North-Holland, Amsterdam, pp. 797–800, (1990).
6. Charest, J.A. and C.S. Lynch, "A Simple Approach to Piezofilm Stress Gauges." in: *Shock Compression of Condensed Matter – 1991* (eds. S.C. Schmidt, R.D. Dick, J.W. Forbes, and D.G. Tasker), Elsevier, Amsterdam, pp. 897, (1992).
7. Charest, J.A. and C.S. Lynch, "Practical Considerations of the Piezofilm Stress Gauge Technique," 41st ARA meeting, San Diego CA, (1990).
8. Charest, J.A. and C.S. Lynch, "Effects of Lateral Strains on PVF_2 Stress Gauges," 42nd ARA meeting, Adelaide, Australia, (1991).
9. Charest, J.A. and M.D. Lilly, "PVF_2 Stress Gauges for Non-Planar Wave Applications, Part I," in: *High-Pressure Science and Technology – 1993* (eds. S.C. Schmidt, J.W. Shaner, G.A. Samara, and M. Ross) American Institute of Physics, New York, pp. 1731–1734, (1994).
10. Charest, J.A. and M.D. Lilly, "Effects of Large Strains on PVDF Gauges," 46th ARA meeting, St Louis France, (1996).
11. Chen D.Y., Y.M. Gupta and M.H. Miles. "Quasi-Static Experiments to Determine Material Constants for the Piezoresistance Foils Used in Shock Wave Experiments," *J. Appl. Phys.* **55(1)**, pp. 3984, (1984).
12. Gran, J.K. and L. Seaman, "Analysis of Piezoresistance Gauge for Stress in Divergent Flow Field," *J. Eng. Mech.*, (1997).
13. Lynch, C.S., "Strain Compensated Thin Film Stress Gauges for Stress Wave Measurements in the Presence of Lateral Strains," *Rev. of Sci. Inst.* **66(12)**, pp. 5582–5589, (1995).
14. Rosenberg, Z., Y. Partom, and D. Yaziv, "The Response of Manganin Gauges Shock Loaded in the 2-D Straining Mode," *J. Appl. Phys.* **52(2)**, pp. 4610, (1981).

Interferometry

15. Amory, B.T., "Wide Range Velocity Interferometer", in: *Proc. Sixth Symp. (International) on Detonation*, Report ACR-221, Office of Naval Research, Arlington, VA, pp. 673–681, (1976).
16. Barker L.M., and R.E. Hollenbach, "Laser Interferometer for Measuring High Velocities of Any Reflecting Surface", *J. Appl. Phys.* **43(11)**, p. 4669, (1972).
17. Barker, L.M., "The New Valyn Multi-Beam VISARs," Proceedings, 49th Meeting of the Aeroballistic Range Association, (1998).
18. Baum, D., "Testing and Diagnostic Capabilities at LLNL", Proceedings, 49th Meeting of the Aeroballistic Range Association, (1998).
19. Hemsing, W.F., "Velocity Sensing Interferometer (VISAR) Modification," *Rev. Sci. Instr.* **50(1)**, pp. 73, (1979).
20. Isbell, W.M., "A Simplified, Compact VISAR: Concept and Construction," Proceedings, 42nd Meeting of the Aeroballistic Range Association, (1991).
21. Isbell, W.M., "An Infrared VISAR for Remote Measurement of Projectile Motion," Proceedings, 39th Meeting of the Aeroballistic Range Association, (1988).
22. Isbell, W.M., "Extending the Range of the Third-Generation VISAR from 30 m/s to 30,000 m/s," Proceedings, 47th Meeting of the Aeroballistic Range Association, (1996).
23. Prins, W.C., R.J. van Esveld, L.K. Cheng, H.J. Verbeek, and A.C. v.d. Steen, "Measurements of Shock Wave Properties with Fabry-Perot Interferometer," Proceedings, 49th Meeting of the Aeroballistic Range Association, (1998).
24. Sweatt, W.C., P.L. Stanton, and O.B. Crump, Jr., "Simplified VISAR System," Sandia National Laboratories report SAND90-2419C, and *Society of Photographic Instr. and Engng.* Vol. 1346, (1990).
25. Yoshida, K., National Institute for Material and Chemical Research, Tsukuba, Japan, private communication, (1995).

Recommended Reading on Additional Interferometric Methods

26. Asay, J.R., (1975) "Shock and Release Behavior in Porous 1100 Aluminum", *J. Appl. Phys.* **46**.
27. Barker, L.M., and R.E. Hollenbach. *Rev. Sc. Instr.* 36, 4208 (1965).
28. Christman, D.R., W.M. Isbell, and S.G. Babcock,"Measurements of Dynamic Properties of Materials, Vol. V: OFHC Copper", General Motors Materials and Structures Laboratory, report DASA-2501, July 1971 (AD728846) (1970).
29. Clifton, R.J. "Analysis of the Laser Velocity Interferometer." *J. Appl. Phys* **41**, p. 3535 (1970).
30. Erskine, D.J. and N.C. Holmes, "Imaging White Light VISAR", 22nd International Congress on High Speed Photonics and Photography, Santa Fe, New Mexico, (1996).
31. Froeschner, K.E., et al, "Subnanosecond Velocimetry with a New Kind of VISAR," 22nd International Congress on High Speed Photonics and Photography, Santa Fe, New Mexico (1996).
32. Gidon, S. and G. Behar, "Multiple-Line Laser Doppler Velocimetry", *Appl. Optics* **27**, pp. 2315–2319, (1988).
33. Gillard, C.W., G.S. Ishikawa, J.F. Peterson, J.L. Rapier, J.C. Stover, and N.L. Thomas, Lockheed Report No. N-25-67-1, (unpublished) (1968).
34. Gooseman, D.R., *J. Appl. Phys.* **45**, p. 3516, (1975).

35. Hemsing, W.F., A.R. Mathews, R.H. Warnes, M.J. George and G.R Whittemore, "VISAR: Line-Imaging Interferometer", American Physical Society Topical Conference, Williamsburg, VA, June 17–21 (1991).
36. Isbell, W.M., "The Versatile VISAR: An Interferometer for Shock Wave and Gas Gun Diagnostics", Proceedings, 26th Annual Meeting of the Aeroballistic Range Association, (1976).
37. Isbell, W.M., and P.W.W. Fuller, "Wide Range, High Resolution Measurements of Projectile Motion Using Laser Interferometry," 27th Annual Meeting, SPIE and High Speed Photonics and Videography Conference, TR-16-83, (1983).
38. Isbell, W.M., "Laser Interferometry for Accurate Measurements of Projectile Motion", Proceedings, 34th Meeting of the Aeroballistic Range Association (1983).
39. Isbell, W.M., "Initial Tests of VISAR Interferometry to Measure E.M. Launcher Projectile Motion", Proceedings, 38th Meeting of the Aeroballistic Range Association, (1987).
40. Isbell, W.M., "Interferometric In-Bore Velocity Measurements of Electromagnetically-Launched Projectiles", Proceedings, 41st Meeting of the Aeroballistic Range Association, (1990).
41. Isbell, W.M., and J.R. Christman, "Shock Propagation and Fracture in 6061-T6 Aluminum from Wave Profile Measurements", General Motors Materials and Structures Laboratory, report DASA-2419, (AD705536), (1970).
42. Isbell, W.M., *Measurements of the Dynamic Response of Materials to Impact Loading*, Doctoral Thesis, Shock Wave Research Center, Tohoku University, Sendai, Japan (1993).
43. Isbell, W.M., "A Combined Displacement/Velocity Interferometer for Impact Measurements at 0.1 to 100 m/s", Proceedings, 32nd Meeting of the Aeroballistic Range Association, (1981).
44. Isbell, W.M., "An Infrared VISAR for Remote Measurement of Projectile Motion", Proceedings, 39th Meeting of the Aeroballistic Range Association, (1988).
45. Isbell, W.M., "Extending the Range of the Third-Generation VISAR from 30 m/s to 30,000 m/s," Proceedings, 47th Meeting of the Aeroballistic Range Association, (1996).
46. Isbell, W.M., "Modern Instrumentation for Measurements of Shock Waves in Solids", Proceedings, Japanese Shock Wave Symposium, Tokyo, Japan, (1999).
47. Johnson, J.N., and L.M. Barker, "Dislocation Dynamics and Steady Plastic Wave Profiles in 6061-T6 Aluminum." *J. Appl. Phys.* **40**, pp. 4321–4334, (1969).
48. Sweatt, W.C., P.L. Stanton, and O.B. Crump, Jr., "Simplified VISAR System", Sandia National Laboratories report SAND90-2419C, and SPIE, Vol. 1346, July. (1990).
49. Yoshida, K., National Institute for Material and Chemical Research for Material and Chemical Research, Tsukuba Japan, private communication, (1996).

Recommended Reading on other Time-Resolved Diagnostics

50. Chhabildas, L.C., and R. A. Graham, in *Techniques and theory of stress measurements for shock wave applications,* ed. By R. R. Stout, E. R. Norwood, and M. E. Fourney, *Amer. Soc. of Mech. Eng.* NY, AMD **83**, 1–18 (1987).
51. d'Almeida, T. and Y.M. Gupta, "X-ray Diffraction Measurements in KCl Shocked Along [100]," in: *Shock Compression of Condensed Matter – 1999*

(eds. M.D. Furnish, L.C. Chhabildas, and R.S. Hixson) American Institute of Physics, New York, (2000).
52. Graham, R. A., and J. R. Asay, "Measurements of wave profiles in shock-loaded solids" *High-Temperatures-High Pressures* **10**, 355–390 (1978).
53. Gruzdkov, C.S. and Y.M. Gupta, "Optical Measurements to Probe Inelastic Deformation in Shocked, Brittle Materials," in: *Shock Compression of Condensed Matter–1999* (eds. M.D. Furnish, L.C. Chhabildas, and R.S. Hixson) American Institute of Physics, New York, (2000).
54. Gustavsen, R. and Y.M. Gupta, "Time-Resolved Spectroscopic Reflection Measurements in Shock-Loaded Materials," *J. Appl. Phys.* **69**, p. 918, (1991).
55. Horn, P.D., and Y.M. Gupta, "Wavelength Shift of the Ruby Luminescence *R* lines under Shock Compression," *Appl. Phys. Lett.* **49**, p. 856, (1986).
56. Knudson, M.D., and Y.M. Gupta, Stimulated Emission to Measure R Shifts in Shocked Ruby," *J. Appl. Phys.* **85**, p. 6425, (1999).
57. Knudson, M.D., "Picosecond Time Resolved Electronic Spectroscopy in Shock," *Rev. Sci. Inst.* **70**, p. 1743, (1999).
58. Kwiatkowski, C.S., and Y.M. Gupta, "Optical Measurements to Probe Inelastic Deformation," in: *Shock Compression of Condensed Matter–1999* (eds. M.D. Furnish, L.C. Chhabildas, and R.S. Hixson) American Institute of Physics, New York, p. 641, (2000).
59. Rigg, P. and Y.M. Gupta, X-ray Diffraction, in: *Shock Compression of Condensed Matter – 1999* (eds. M.D. Furnish, L.C. Chhabildas, and R.S. Hixson) American Institute of Physics, New York, p. 1051, (2000).
60. Winey, J.M. and Y.M. Gupta, "Raman Spectroscopy," *J. Phys. Chem. B* **101**, p. 10733, (1997).
61. d'Almeida, T. and Y.M. Gupta, "X-ray Diffraction Measurements in KCl Shocked Along [100]," in: *Shock Compression of Condensed Matter – 1999* (eds. M.D. Furnish, L.C. Chhabildas, and R.S. Hixson) American Institute of Physics, New York, (2000).
62. Yuan, G., R. Fong, and Y.M. Gupta., "Compression and Shear Wave Measurements to Characterize the Shocked State in Silicon Carbide," *J. Appl. Phys.* **89**, p. 5372, (2001).
63. Gruzdkov, C.S. and Y.M. Gupta, "Optical Measurements to Probe Inelastic Deformation in Shocked, Brittle Materials," in: *Shock Compression of Condensed Matter – 1999* (eds. M.D. Furnish, L.C. Chhabildas, and R.S. Hixson) American Institute of Physics, New York, (2000).

10 Use of Pulsed Magnetic Fields for Quasi-Isentropic Compression Experiments

James R. Asay and Marcus D. Knudson

10.1 Introduction

An understanding of material response to dynamic loading is required in a large number of applications ranging from planetary physics to hypervelocity impacts on space assets. In order to describe these phenomena under a variety of impact conditions, it is necessary to know the physical, mechanical, and thermodynamic properties of materials over the appropriate pressure range. The complete equation of state (EOS) plays a fundamental role in these studies by specifying the thermodynamic states of a material occurring in these events, including solid, liquid, vapor states, polymorphic phase transitions, or chemical reactions. In addition, a wide range of compressed mechanical states are produced, including elastic behavior, plastic flow, and tensile failure.

The high-pressure EOS of materials is usually determined by shock compression experiments [8,29] that provide access to the thermodynamic regions of interest. In the large majority of these experiments, flat cylindrical specimens are subjected to planar loading for time durations of a few hundred nanoseconds to several microseconds and the kinematic properties of the steady shock wave produced in the experiments are measured; this is usually the shock velocity and the particle velocity. The steady shock wave produced by this process is a transition from an initial equilibrium state of pressure, specific volume, and specific internal energy to a final equilibrium thermodynamic state. The data obtained from these experiments can then be used to determine the pressure-volume-energy states produced by the steady shock wave for a specific experiment; results from several experiments determine the Hugoniot curve, which is the locus of end states produced by shock compression. The principal Hugoniot curve initiating at standard thermodynamic conditions is a unique path on the equation-of-state surface and valuable for developing and validating thermodynamic models. The vast majority of dynamic experiments fall in this category.

Although shock wave techniques are extremely useful for generating high-pressure EOS properties, there are many applications where the dynamic loading paths differ significantly from shock loading and are closer to shockless or quasi-isentropic loading. A specific example is the path followed by cryogenic fuels in an inertial confinement experiment to compress hydrogen isotopes to fusion ignition conditions [66]. In these applications, it is desirable to obtain dynamic compression data for conditions emulating the loading path, which is quasi-isentropic compression in this case. Thus, it is necessary to develop experimental techniques that allow EOS information to be obtained for both shock and isentropic compression. In some cases the region of the EOS surface that is of most interest lies away from the Hugoniot curve. A specific example relates to our understanding of the interiors of the Jovian planets. The inferred structure of the Jovian planets, i.e., whether the planets have a solid core, and if so the size of the core, is quite sensitive to the EOS of hydrogen at several hundred GPa pressures and a few thousand degree temperatures, which is much closer to the principal isentrope of hydrogen. The inferred structure provides important clues into the formation process of these giant planets [80]. Finally, Hugoniot and isentropic compression data complement EOS data acquired by other methods, such as isothermal compression, that together provide a more comprehensive description of the complete EOS [8].

The relationship between these thermodynamic responses is shown in Fig. 10.1. The isotherm represents the pressure–volume response obtained by compressing a material at constant temperature. The cold curve is the isotherm obtained at a temperature of absolute zero, in which the compression behavior is controlled by elastic stresses. The Hugoniot represents an adiabatic compression process by steady shock transitions to various final states, but is highly irreversible, resulting in a higher temperature in the shocked state, and a pressure offset from the isotherm. The isentrope, which lies between the isotherm and the Hugoniot, represents the response obtained for continuous, adiabatic, and reversible compression. In this case there is only a modest temperature increase from the isotherm, resulting in a compression curve that generally lies close to the isotherm. At low pressure where the temperature increase produced by the shock wave is small, the isentrope and the Hugoniot are similar, as illustrated in the figure. It is well known that the isentrope and Hugoniot are second-order tangent at the initial volume state and differ from each other at modest pressures as the third order in volume compression [21].

It is extremely difficult to produce states of isentropic compression with static methods since this requires simultaneous control of pressure, volume, and temperature, but it is possible to approach these states in inviscid liquids using dynamic methods. In the case of liquids, dynamic loading on microsecond time scales is continuous, adiabatic, and presumed to be reversible if there are no viscous effects. Dynamic methods can also be used to produce shockless, adiabatic loading in solids with rate-independent mechanical response, but the loading process in this case is irreversible because of elastic–plastic processes. In solids the longitudinal stress differs from the hydrostatic pressure because of resolved shear stresses that produce an entropy increase from the irreversible work done by deviator stresses. For this reason, ramp loading in solids is usually assumed to be quasi-isentropic although detailed studies

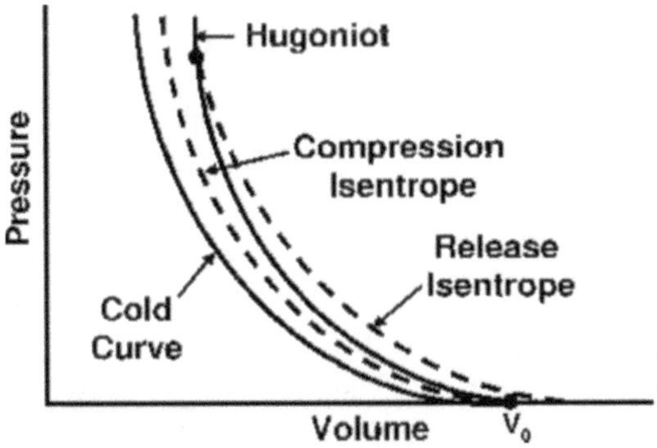

Fig. 10.1. Notional relationship between the zero-temperature isotherm (cold curve), the compression isentrope, and the Hugoniot. Also shown is a release isentrope

are needed on a specific material to determine the extent of departure from isentropic response [33]. For the quasi-isentropic response of aluminum, Ding has shown that the temperature increase at 100 GPa is about 200 K and the resulting stress offset from the theoretical isentrope is within experimental error for an assumed yield stress of 1.6 GPa. He has also explored the effects of strain rate on the departure from isentropic response. Preliminary results assuming a specific dissipation model for aluminum exhibit minimal effects for strain rates two to three orders of magnitude below the shock loading rate.

Flat-plate projectile impact is the presently accepted procedure for obtaining precision loading of condensed matter at high pressures. The basic concept is shown in Fig. 10.2a. Planar impact of a flat-plate projectile results in abrupt loading and propagation of a planar shock wave into the specimen. Each experiment produces a unique state of pressure, volume, and internal energy, which can be determined from the shock jump conditions, as we will illustrate. It is important to note that only a single datum representing a final equilibrium state is obtained in each steady shock wave experiment.

Barker and Hollenbach were the first to produce well-controlled continuous, or quasi-isentropic, loading of solids using projectile impact techniques [14]. The first measurements of this nature were made using a fused silica ramp wave generator, which has a longitudinal stress–uniaxial strain loading path with concave curvature to about 3 GPa. A steady shock is unstable for this compression behavior and will evolve into a compression "fan" that spreads in time with propagation distance. If a planar sample is mounted behind a planar fused silica buffer, as shown in Fig. 10.2b, a continuous loading history is produced at the impact surface of the sample. Generally, this ramp wave will evolve into a shock wave because of convex non-linear response of the specimen. Fused silica ramp generators are limited to pressures of about 3 GPa since the mechanical response above this stress becomes

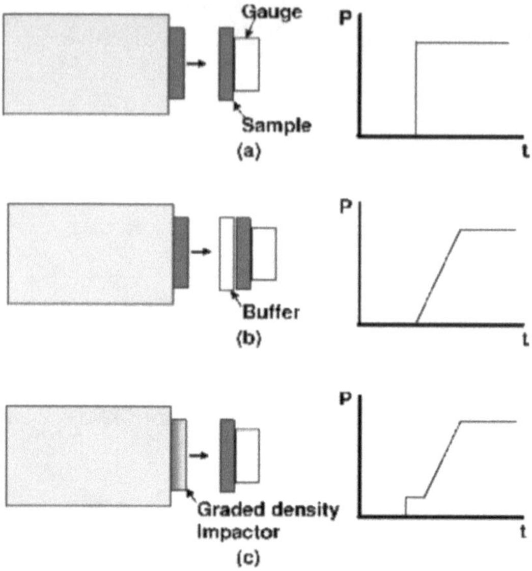

Fig. 10.2. (**a**) Typical impact condition for shock loading of planar samples; (**b**) fused silica buffer for producing shockless loading; (**c**) graded density impactor for shock/shockless loading

non-linear with convex curvature. Pyroceram, another material that exhibits concave curvature in its stress-strain response to ~20 GPa, was used by Benedick and Asay (1976) for ramp studies on aluminum. However, no other material is known to exhibit concave curvature in the stress-strain response to higher stresses. Thus, the technique of using a suitable ramp generator for these studies is limited to stresses of ~20 GPa.

To overcome this stress limitation, Barker developed a projectile impact method using graded density impactors that produce a small shock upon impact with a planar specimen, followed by a gradual increase in loading pressure at the input to the sample [13]. He referred to this as the ICE technique, an acronym for "isentropic compression experiment." The concept for this method is shown in Fig. 10.2c, which depicts a low impedance projectile facing that initially produces a low stress shock wave, followed by a continuous increase in pressure that results in quasi-isentropic loading of a solid specimen. The graded density impactor, prepared with particle sedimentation techniques and referred to as a "pillow" [13], has been used in a variety of experiments to study the high pressure EOS and mechanical properties of materials. However, this technique has offered limited success in providing quasi-isentropic loading curves because of the difficulty in precisely controlling the pillow density gradient and subsequent stress history input to the sample, and the inevitable initial shock wave, which is undesirable in many experiments. Variants of the pillow impactor technique have been developed using plasma spraying techniques to produce impactors that result in quasi-isentropic compression in planar impact experiments [72]. Most recently, an improved method of preparing graded-density impactors was developed [71] that promises to provide more controllability in specifying pressure histories to specimens.

In addition to gun techniques for producing continuous pressure loading, an intense laser energy deposition technique has been developed that produces a stagnating plasma on a planar specimen and shocklessly loads it to high pressures. Presently, ramp wave risetimes available with this method are on the order of a few tens of ns and a few tens of GPa of peak pressure. Work is underway to increase both the loading time and the peak pressure for EOS and material property studies [37, 76, 77].

Isentropic compression experiments have also been conducted under cylindrical loading conditions. Reference [46] used an explosive-driven magnetic field generator to radially compress hydrogen to pressures on the order of 200 GPa. This was an attempt to detect metallization, which has been predicted theoretically to occur at pressures of order a few hundred GPa at low temperatures. In Hawke's experiments, x-radiography was used to determine the density produced by radial compression. The resistivity was measured in the compressed hydrogen and the pressure was calculated from volume measurements. Although these experiments indicated that metallization did occur, subsequent experiments with diamond anvil cells to even higher pressures have not confirmed the metallic transition.

This article will focus on recently developed techniques using pulsed magnetic sources to both significantly improve the smoothness of the ramp load introduced into a planar specimen and also to extend the pressure range accessible with smooth ramp loading. This breakthrough in capability first occurred with techniques developed on the Z Accelerator [67] and has resulted in several recent publications [6, 11, 41, 42, 44, 74].

The magnetic ramp loading concept is illustrated in Fig. 10.3a. In this technique, planar specimens of about 6–10 mm diameter are mounted on a flat anode of either aluminum or copper. A direct short between the anode and cathode produces a planar magnetic field between the conductors and a magnetic pressure proportional to the square of the field [6]. The current is applied in these experiments over a time scale of about 200–300 ns, so that a high-pressure ramp wave, similar to that shown in Fig. 10.3b, is introduced into the anode conductor and then into the planar specimen [42], which is typically 0.5–1.5 mm thick. Using this concept, experimental designs have been developed that allow simultaneous studies of several materials on each experiment. We will show that this is particularly useful for relative comparisons of dynamic response. Since this initial development, a large number of applications and spin-offs of the technique have been developed. These will be discussed in the following sections.

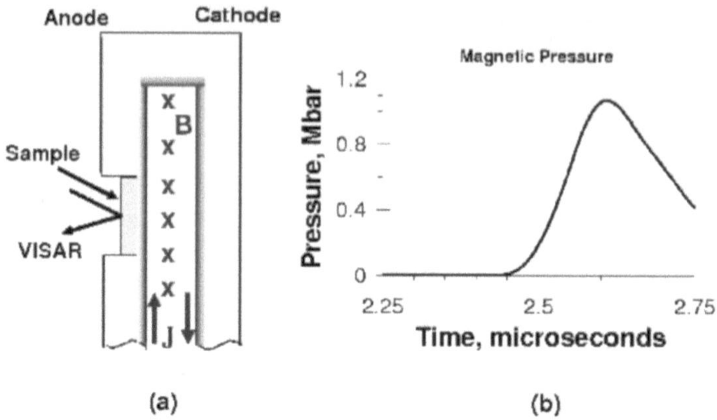

Fig. 10.3. Experimental technique for producing shockless loading with a magnetic drive. (**a**) sample geometry, (**b**) typical input stress-time profile

10.2 Basic Concepts of Ramp Loading

10.2.1 Wave Propagation

Wave propagation in a material is governed by its dynamic properties. In most materials the local sound speed increases with increasing pressure. This is illustrated conceptually in Fig. 10.4a, which shows a convex isentrope and Hugoniot in the P–V plane. For the inviscid fluid response assumed here, continuous compression produces a ramp wave that propagates with the local sound speed at the corresponding pressure in the ramp. Thus this wave tends to steepen as it propagates into the sample, as illustrated in Fig. 10.4. If the propagation distance is sufficiently long, the wave will eventually evolve into a steady, nearly discontinuous jump in pressure, density, energy, temperature, and entropy; this wave is referred to as a steady shock wave, as discussed earlier. Since the wave evolution is governed by the local elastic modulus of the material, measurement of the wave evolution as a function of propagation distance provides information that can be analyzed to determine the continuous isentropic loading response.

In later sections we will discuss the requirements for the accelerator to produce pulses with a specific current history that optimizes propagation distance prior to shock formation. In the idealized optimal case, the ramp will abruptly form a single high-amplitude shock at a specific location in the specimen. At this point, the isentropic pressure will generally drop precipitously from the drive pressure to a state on the Hugoniot for the material. Also at this point, a contact discontinuity in density will occur, which is associated with a forward propagating shock wave into the ambient density material and a backward facing ramp rarefaction wave into the isentropically compressed material. All states to the left of the contact discontinuity are isentropically

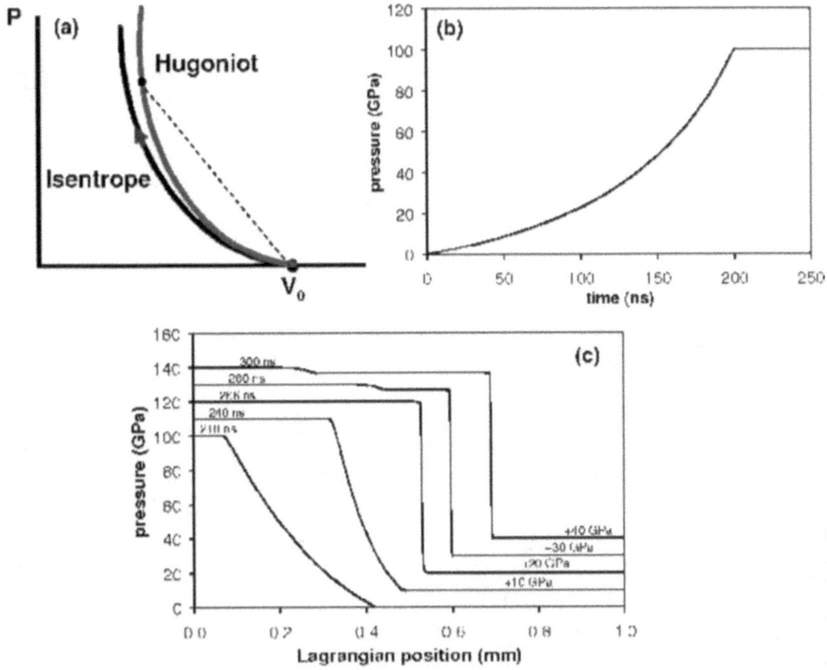

Fig. 10.4. (**a**) Isentrope and Hugoniot for a fluid, (**b**) ideal pressure history applied to the fluid that optimizes the distance for shock formation, (**c**) resulting ramp and shock wave evolution produced in the specimen. The pressure profiles are offset for clarity

compressed and all states to the right are shock compressed. The contact discontinuity is the location where the density changes abruptly between these two states.

Figure 10.4a shows schematically an isentrope and a Hugoniot curve for a fluid that is isentropically compressed to a high pressure state with the idealized input pressure history shown in Fig. 10.4b. Figure 10.4c illustrates snap shots in time of the resulting ramp and shock waves that are formed later in time. For the applied pressure history and specific material represented in this figure, a single shock wave is formed at approximately 0.55 mm into the material. At this point, a rarefaction fan is observed to propagate back into the compressed material and a shock is formed that propagates to the right. The specific current history necessary to produce this optimal pressure history will be discussed in Sect. 10.3.3.

The governing equations for planar shock or ramp loading are summarized as follows for a solid initially at rest [1,8,29].

For a shock wave, we have

$$\sigma - \sigma_0 = \rho_0 U_s, u_p \tag{10.1}$$

$$V/V_0 = 1 - (u_\mathrm{p}/U_\mathrm{s}) \tag{10.2}$$

$$E - E_0 = \frac{1}{2}(\sigma - \sigma_0)(V_0 - V) \tag{10.3}$$

and for a ramp wave we have

$$\frac{\partial \sigma}{\partial h} = -\rho_0 \frac{\partial u_\mathrm{p}}{\partial t} \tag{10.4}$$

$$\frac{\partial V}{\partial t} = V_0 \frac{\partial u_\mathrm{p}}{\partial h} \tag{10.5}$$

$$\frac{\partial V}{\partial t} = V_0 \frac{\partial u_\mathrm{p}}{\partial h} \ . \tag{10.6}$$

Here, σ is longitudinal stress, u_p is particle velocity, V is specific volume, ρ is density $(1/V)$, h is Lagrangian position, and E is specific internal energy. Zero subscripts refer to the initial state. For simple waves where there is no rate dependence and the flow is isentropic ($S = S_0 =$ constant) (10.4)–(10.5) reduce to

$$d\sigma = \rho_0 \, C_\mathrm{L} \, du_\mathrm{p} \tag{10.7}$$

$$d\sigma = \rho_0 \, C_\mathrm{L} \, du_\mathrm{p} \tag{10.8}$$

$$dE = -\sigma(S_0) \, dV \ . \tag{10.9}$$

Measurement of any two of the shock variables in (10.1)–(10.3) allows determination of all five quantities. The locus of end states produced by steady shock compression defines the Hugoniot curve illustrated schematically in Fig. 10.1. Typically, the shock velocity and one other variable are measured in a shock wave experiment. In plate impact experiments where the impacting plate and the sample are identical, measurement of the projectile velocity precisely defines the particle velocity in the shocked state, which is exactly one half of the projectile velocity. In ramp wave experiments, it is also necessary to measure two variables for the complete ramp loading history. This is typically accomplished by measuring the velocity profile of either a free or windowed surface for at least two different specimen thicknesses.

A variety of shock wave diagnostics have been developed for making these measurements [18,23]. Velocity interferometry is one of the most useful techniques for measuring wave profiles at high stresses because of the high precision in particle velocity and the high time resolution that can be achieved. The VISAR (Velocity Interferometer System for Any Reflector) technique developed by Barker and Hollenbach [15] has been particularly useful and has been employed in all of the experiments discussed in this chapter. In this technique, the measured particle velocity, u_p, is related to the wave length of the laser light used, λ, the delay time of the interferometer, τ, and the measured fringe count in the interferometer as follows:

$$u_\mathrm{p}(t) = \frac{\lambda}{2\tau}(1+\delta)^{-1}(1+\Delta\nu/\nu_0)^{-1}F(t) \ . \tag{10.10}$$

Here, δ is an optical correction that accounts for wave length dispersion in the optical etalons used in the interferometer [17], and the correction factor $1 + (\Delta\nu/\nu 0)$ accounts for the change in refractive index of the window due to compression. The VISAR has been particularly useful for analyzing ramp wave profiles because of the small samples required in these experiments and the concomitant high time resolution needed. However, it is necessary to evaluate the correction factor, $1 + (\Delta\nu/\nu_0)$, under ramp loading, as will be discussed later.

Lagrangian and Backward Analysis Techniques

Given the velocity profiles for at least two sample thicknesses subjected to ramp wave loading, the general equations of motion in (10.4)–(10.6) can be applied for non-simple waves [1] or (10.7)–(10.9) can be used for simple waves [1, 42, 74]. In either case, the histories of ramp compression at various sample thicknesses must be precisely measured because of the short transit times obtained from the necessarily small samples required in magnetic loading experiments. Typically, the particle velocity must be measured to ~1% or better and the relative timing of wave profiles for the various sample thicknesses (typically of order 0.5–1 mm) to within ~0.5 ns in order to achieve accuracies on the order of 2–3% in stress and comparable accuracies in density from experimental records in the pressure range below ~100 GPa.

Ideally, in situ wave profile measurements are needed to determine accurate wave velocities to implement the Lagrangian wave analysis technique. This is difficult to achieve while maintaining good time resolution, so the method has not yet been applied to ramp loading experiments with the Z Accelerator. Optical windows can be used in laser interferometer experiments to achieve the required time resolution, but the impedance mismatch between a specimen and window results in wave reflections and interactions that perturb the ramp wave. This causes the measured profile to be considerably different than the profile that a true in situ gauge would record, as illustrated in Fig. 10.5.

For cases where the yield stress is negligible compared to the peak stress and the waves are shown to be self-similar (simple), Hayes has developed a numerical integration technique that is particularly useful [48]. This method uses the measured wave profiles at two or more sample thicknesses, either from a free surface or a sample/window interface measurement, and a "backward" integration technique to determine the stress–time profile at a specified location within the sample that resulted in the measured wave profiles. By iteratively changing the assumed stress–volume material response used in the backward integration such that the stress–time profiles determined from the measurements at each of the sample thicknesses converge, it is possible to obtain a unique stress–volume curve for the material, within the assumptions stated above. Accuracies of a few percent have been quoted for this approach for compressive response over a modest stress range in this approach. A notable advantage of the technique is that it exactly corrects for wave perturbations caused by impedance mismatch at free surfaces or sample/window interfaces for inviscid liquids and is a very good approximation for materials with low yield strength [48]. The backward analysis technique and the Lagrangian wave analysis method are illustrated in Figs. 10.5a and 10.5b.

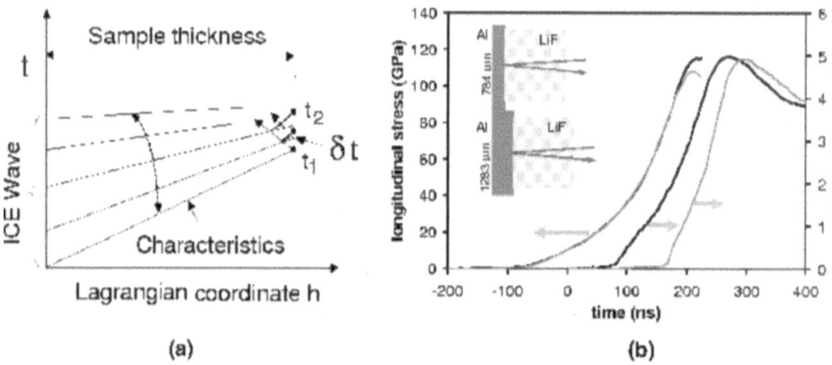

Fig. 10.5. (**a**) Illustration of the affect of wave interactions at an interface. The reflected waves perturb the oncoming compression wave, causing a slight change in the characteristics. This results in a significant difference in the observed wave profile as compared to the in situ response. (**b**) Backward integration method for determining compression states in ramp loading experiments. The EOS for aluminum is iteratively adjusted until the wave profiles measured from two sample thicknesses (784 and 1283 μm) integrate back to a common loading profile at $x = 0$. The inferred pressure profiles for the two sample thicknesses differ in peak pressure as a result of wave interactions, which must be carefully considered in analyzing the results. In particular, data above 100 GPa in this experiment are not usable

Forward Analysis Techniques

The backward integration technique and the Lagrangian analysis method are useful for analyzing continuous material response in several different applications. However, both methods have limitations in certain cases. For example, the integration technique is strictly applicable for rate-independent response of fluids where the material flow should be isentropic. The presence of elastic–plastic response in solids with high compressive yield strengths violates this condition and results in non-unique solutions. The Lagrangian analysis technique is powerful for analyzing generalized wave propagation for either rate-independent or rate-dependent response, but it is generally necessary to measure wave profiles at several sample positions to accurately apply the method. Further, the wave profiles should be measured with both stress and particle velocity gauges if possible.

In many material studies, it is not possible to meet either of these requirements; for example, either only a single continuous wave profile can be measured at a specific sample thickness or the compressive material strength of the sample being investigated is extremely high. In these situations, an alternate method of probing material response is attractive and has proven to be quite useful. The technique makes use of measurements of the particle velocity history directly on the drive plate, which is assumed to have a low yield strength, and simultaneously on the back surface of a sample mounted on the drive plate. Using the measured velocity history from the drive plate, the backward integration technique can be used to estimate the input stress history to the plate, which is also the input stress history applied to the sample mounted on the plate. Once this history is known, a forward calculation can be used to model the measured wave profile at the back surface of the specimen. A major advantage of this approach is that accurate simulations can be obtained for arbitrary material response, including rate-dependent behavior or excessively high yield strengths of the specimen.

The technique is illustrated in Fig. 10.6a, which depicts a sample of molten tin contained on a drive plate. Measurement of the wave profile on the back surface of the drive plate, either with a laser window or on the free surface of the plate, as shown in the figure, determines the input particle velocity or, equivalently, the stress history to the drive plate. As mentioned above, the drive plate should be a conductor with low compressive yield strength. Reference [48] have shown that accurate backward calculations can be performed for either copper or aluminum drive plates at low pressures since they both have low yield strengths. For a specific experiment that will be discussed later, this configuration produces the measured back surface plate velocity shown in Fig. 10.6b and the calculated input particle velocity history to the specimen, which is also shown in Fig. 10.6b. This input history can be used in a forward computer simulation of the drive plate/sample configuration to accurately account for the drive plate/sample impedance mismatch. A calculation and comparison with experimental data obtained on liquid tin is shown in Fig. 10.6c. This result and conclusions regarding kinetic effects of the phase transformation that occurred in the tin are discussed in Sect. 10.4.2.

This technique allows evaluation of material response using the standard methods of comparing numerical simulations and measured wave profiles through a forward calculation in time with a known input history. This is a powerful method that can be used on a variety of time-dependent and arbitrary material response problems. However, it is important to note that the experiment must be properly designed to prevent wave interactions between the ramp wave and the actual input history (in this case, calculated) which can easily occur because of the long risetime of the input ramp wave.

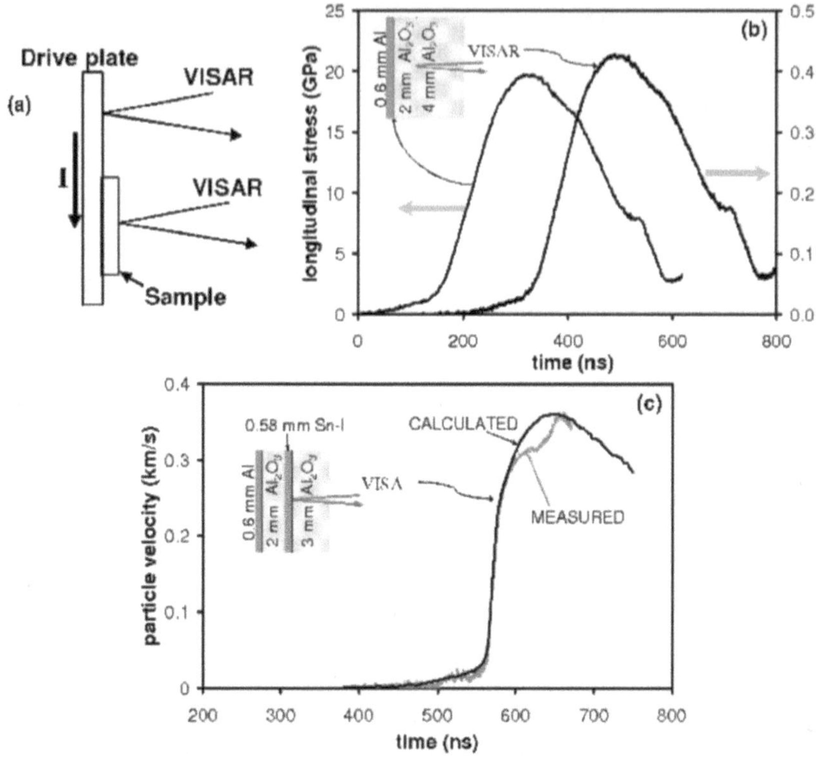

Fig. 10.6. Forward integration method for determining the dynamic response of a specimen. (**a**) experimental configuration showing velocity measurements on both a drive plate and a tin specimen. (**b**) Measured free surface velocity history on the back surface of the drive plate (*right*) and the calculated input velocity history to the drive plate and specimen determined with a backward calculation from the free surface velocity (*left*). (**c**) Forward calculation of the sample response using the calculated input history and an assumed equation of state for the tin sample. Also shown is an experimental profile to compare with the numerical simulation

10.3 Experimental Technique

10.3.1 Experimental Configuration

The examples discussed in later sections were obtained using pulsed magnetic fields achievable with the Z Accelerator. The accelerator is a low inductance pulsed power generator capable of capacitively storing ~11.5 MJ of electrical energy [67]. The machine uses a combination of fast switches and transmission lines to deliver a ~20 MA, ~100–300 ns rise-time current pulse through parallel anode and cathode plates to generate a time-varying magnetic field between the plates. The detailed experimental configuration used for the magnetically driven ICE technique is shown schematically in Fig. 10.7. The

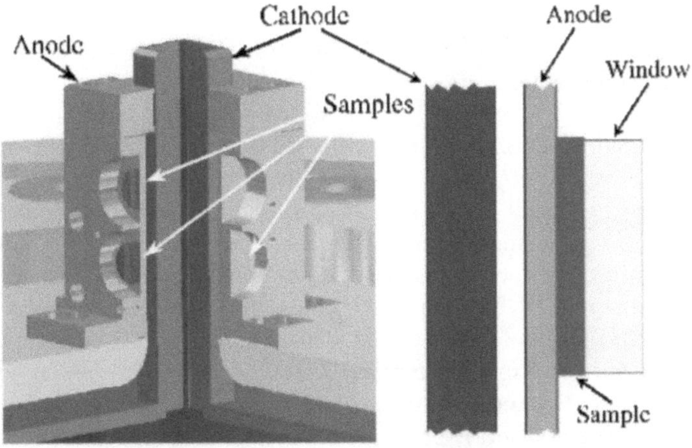

Fig. 10.7. Experimental configuration for Magnetic drive, in this case the square short circuit geometry. The square cathode and flat plate anode produce a planar magnetic field within the A–K gap. The sample of interest (which may be backed by a window) is mounted onto the anode plate. The typical dimension of the short circuit load is ~20–35 mm square (outside dimension of the cathode) by ~30–50 mm in height

sample holder comprises the anode of a square or rectangular coaxial short circuit load at the center of the Z Accelerator. The interaction of the current density and the magnetic field produced in the insulating gap results in a time-dependent magnetic pressure, $P(t)$, through the Lorentz force exerted on the inner surfaces of the anode and cathode plates. The magnitude of this loading is given by

$$P(t) = B^2/2\mu_0 = \mu_0 J^2(t)/2 = \mu_0 I^2(t)/2S^2 . \tag{10.11}$$

Here, $J(t)$ is the time-dependent current density (amps/unit length) at the sample location, B is the corresponding magnetic field strength, μ_0 is the magnetic permeability of free space, and S is a scale factor (generally time dependent) used to convert the current, I, into current density, J. The scale factor S is typically between 2 and 10 cm, and is both load geometry dependent and time dependent. The magnetic pressure applied to the anode surface induces a hydrodynamic stress wave of equal magnitude and rise-time that propagates into the anode material.

Typically, four anode panels are arranged about a central stainless steel cathode post to form a symmetric anode–cathode (A–K) gap. A short circuit is created between the anode panels and cathode post through a shorting cap at the top of the coaxial load. Each anode panel becomes an experimental platform with the capability of holding up to five samples that are all subjected to essentially identical loading. Two dimensional magnetohydrohynamic (MHD) calculations have been performed for this configuration and show that magnetic field uniformity of about 0.5% can be obtained over the central horizontal region of the anode plate [74]. These numerical simulations were confirmed with a line-imaging VISAR technique [86] that measures spatially resolved particle velocity [42]. In addition, similar interferometer measurements obtained along the vertical direction of the magnetic drive indicate considerably better uniformity along the entire length of the coaxial load. As many as twenty samples can be subjected simultaneously to identical, planar loading histories in a single firing of the accelerator.

The magnitude of the peak magnetic field in the A–K gap, thus the peak magnetic pressure, can be adjusted by changing the geometry of the short circuit load and therefore the current density. Two typical short circuit geometries are routinely used in materials testing: a square coaxial geometry, in which four anode panels of equal width surround a square cathode stalk, and a rectangular geometry, in which two sides of the square geometry are reduced to roughly half the width. The square geometry produces magnetic pressures up to ~100 GPa. The rectangular or slab geometry produces up to ~300–400 GPa of magnetic pressure through increased current density, but at the expense of available sample area. In addition to varying the short circuit geometry, the current density for a given geometry can also be lowered by reducing the charge voltage of the accelerator. Since the peak current scales approximately linearly with the charge voltage, the magnetic pressure can be reduced by as much as ~45% using this approach. These two parameters, the geometry and the charge voltage, allow the peak magnetic pressure to be continuously varied from a few tenths of a GPa to several hundreds of GPa.

Due to significant joule heating and resulting ablation of the conducting plate near the current carrying surface as well as magnetic field penetration into the metal associated with this technique, the specific details of ramp loading are extremely complicated [63, 64]. Computer simulations, which include all the experimental parameters and the necessary physical properties of the conducting plate, are of great value in designing the experimental configuration. This is particularly true for high pressure ICE experiments in which there is significant deformation of the anode and cathode plates during the rise-time of the current pulse. This deformation results in significant increases in load region inductance with time and affects the shape of the current rise and the peak current. Furthermore, the load deformation alters the scale factor, S, used in converting current to magnetic field strength and thus to magnetic pressure acting on the anode. This further modifies the shape and peak of the resulting pressure pulse.

10.3.2 MHD Modeling

Both one- and two-dimensional Eulerian simulations have been performed using the finite element, arbitrary Lagrangian–Eulerian, MHD code ALEGRA [85] to solve the MHD equations for a compressible material with material strength. In these simulations, a complete equation of state valid for a wide range of pressures, densities, and temperatures is used for the conductor materials, in addition to physical models for thermal and electrical conductivities [30, 32]. Material density and internal energy are used in the EOS to obtain pressure and temperature at each time step in the calculation. The calculated density and temperature are used in the conductivity model to obtain electrical and thermal conductivities. The form of Ohm's law used in ALEGRA is

$$\mathbf{J} = \sigma (\mathbf{E} + \mathbf{v} \times \mathbf{B}) \qquad (10.12)$$

where $\mathbf{J}$ is the current density, σ is electrical conductivity, $\mathbf{E}$ is the resistive electric field, $\mathbf{v}$ is velocity of a charge element, and $\mathbf{B}$ is the applied magnetic field. More details concerning the MHD modeling can be found elsewhere [63, 64].

Results of one-dimensional MHD simulations using measured current drives as the input are in very good agreement with experiment and have provided significant insight into the details of the magnetic loading [63]. As an example, snapshots at three different times from a one-dimensional MHD simulation are shown in Fig. 10.8. These snapshots nicely illustrate the physics involved in the magnetically driven ICE experiment. In particular, note the time evolution of the hydrodynamic pressure wave, the magnetic diffusion front, and the material density profile. Several points are evident. First, there is significant diffusion of the current and field into the electrode during the timescale of the experiment; the location of the current is bounded on the left by the density profile (the current must reside in the anode material) and on the right by the magnetic pressure profile. However, the predictions of the magnetic diffusion rates, of order 2–3 km/s, depending on the current density, are significantly lower than the wave speeds of the hydrodynamic front. This is significant since the stress waves, which typically propagate at speeds of ~4–7 km/s, outrun the lagging diffusion wave; this ensures that the hydrodynamic wave propagation occurs in pristine material and is not yet affected by magnetic field effects and Joule heating. The predicted diffusion rates are in good agreement with experimental results [63, 83]. Second, note that there is not a well-defined boundary at which the magnetic pressure is applied. Instead, due to the significant diffusion and Joule heating, the magnetic pressure broadens with time and propagates into the electrode. Finally, there is significant ablation of the anode material, evident by the broadening of the density profile at the rear of the anode. This is significant since wave interactions emanating from either a free surface or window interface will interact with this rather ill-defined pressure boundary earlier than would be expected had the pressure boundary been at the initial power-flow surface. Careful attention to this effect is required when designing ICE experiments.

Two-dimensional MHD simulations enable determination of the extent of pressure gradients due to magnetic field gradients within the A–K gap as well as the effect of load deformation on the resulting current pulse. The two-dimensional nature of the square or rectangular short circuit load results

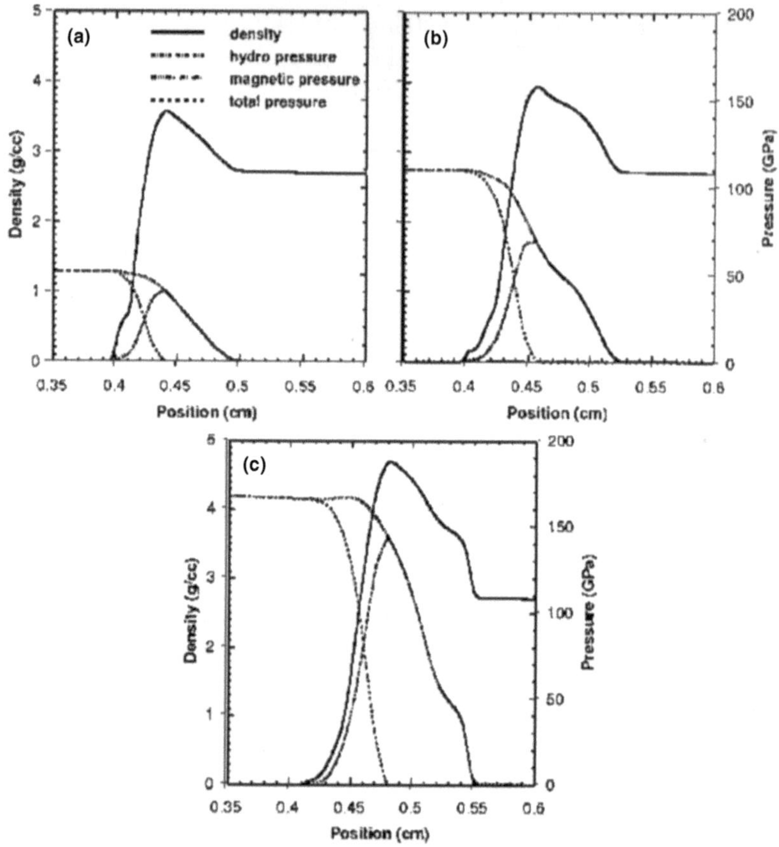

Fig. 10.8. Snapshots of a one-dimensional MHD calculation at three different times (2.46, 2.51, and 2.56 μs) during loading; in this case an ICE experiment on aluminum to ~170 GPa. Indicated in the figure are the density profile, and the hydrodynamic, magnetic, and total pressure

in magnetic pressure gradients across the horizontal direction of the anode plate (perpendicular to the axis of the load). The electromagnetic (EM) code QUICKSILVER [81] was used to determine these gradients for the initial geometry of the anode and cathode. As a specific example, the simulations indicate that magnetic pressure is uniform to better than 1% over the central 6 mm of a 15-mm wide anode plate. However, the applied pressure drops quite rapidly toward the corners of the cathode due to magnetic field gradients that cause slight hydrodynamic pressure variations across the anode surface. The pressure variations result in an unequal expansion of the A–K gap, which, independent of additional edge effects, gives rise to further pressure gradients across the anode. Consequently, the area over which the magnetic pressure is uniform decreases as a function of time. Nevertheless, two-dimensional MHD simulations that account for the time-dependent deformation of the A–K gap suggest that the central ∼4–5 mm width of the original 15-mm wide anode remains at a nearly uniform pressure to late times [64].

In contrast, the magnetic pressure gradient is rather independent of height along the coaxial short circuit load. There are variations as a function of height near the transition of the cathode from a radial feed to a coaxial feed; however, the current density is quickly redistributed and becomes substantially uniform over the coaxial region of interest. This suggests there should be little to no gradient along the vertical direction; i.e., the principal curvature effects are limited to the horizontal direction. This has been verified experimentally using a line imaging VISAR diagnostic [42].

As mentioned above, power flow into the ICE load is affected by the dynamics of the electrodes caused by the magnetic pressure. Deformation of the load conductors, material motion, and Joule heating all result in an increase in the inductance and resistance of the load during the current pulse. Simulations must account for the coupling between the accelerator and the load in a self-consistent manner in order to accurately predict the drive current and the time-dependent loading profile of the stress wave. This is accomplished by coupling an accurate circuit model of the accelerator to the two-dimensional simulation of the load [64].

The specific circuit, shown in Fig. 10.9, includes equivalent resistances and inductances for the accelerator, as well as a time-dependent model of current loss and a time-dependent resistance that emulates a short circuit, both of which occur upstream of the MHD load. Values for the inductances L_{m} , $L_{\mathrm{convolute}}$, and L_{im} are determined from geometrical considerations. The inductance of the MHD load is calculated self-consistently in the simulation using an effective transverse length representative of the actual experimental load. The circuit is driven by a time-dependent accelerator voltage obtained from experimental measurements of accelerator performance in which similar short circuit loads were used.

Two-dimensional simulations using such a circuit model produce currents upstream (load current) and downstream current (current in the magnetically insulated transmission lines, MITL, the feed lines inside the vacuum section of the accelerator) of Z_{flow}, denoted I_{u} and I_{d}, respectively, that accurately match the experimentally measured values shown in Fig. 10.10. It is the load current, I_{d}, along with the time-varying scale factor, S, that effectively determines the shape of the pressure pulse that drives the ICE load. These MHD simulations are invaluable for providing an accurate means to predict peak pressures and magnetic diffusion for arbitrary short circuit configurations and are critical for designing ICE experiments. This recent advance in the ability to predict the performance of arbitrary load geometries from known input conditions and material properties has allowed the identification of optimal anode and cathode geometries, which have resulted in peak stresses of nearly 400 GPa applied to planar samples.

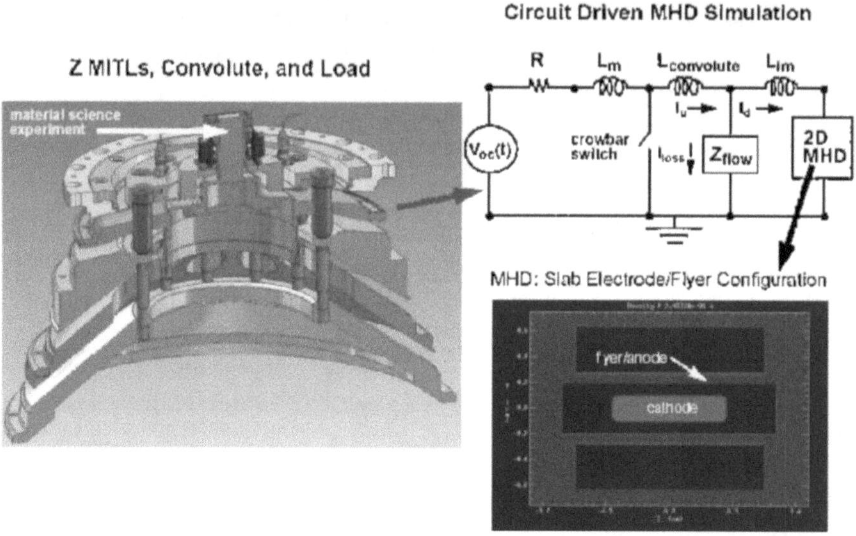

Fig. 10.9. Circuit model of the Z accelerator with a two-dimensional simulation as one of the circuit elements. In this case, the two-dimensional simulation is of a rectangular geometry short circuit load

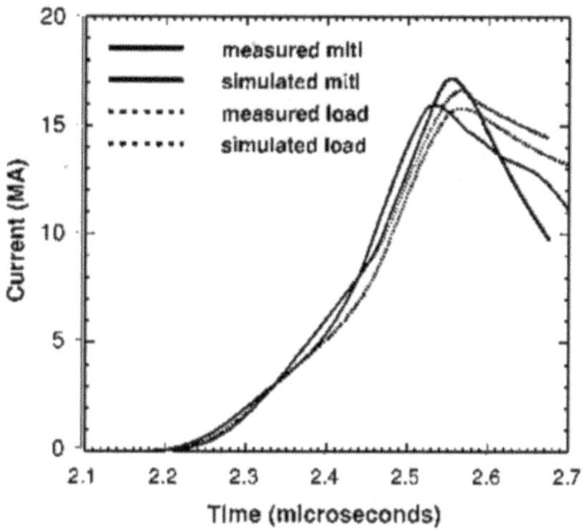

Fig. 10.10. Comparison of calculated (*gray*) and measured (*black*) load (*dashed line*) and magnetically insulated transmission line (*mitl*) (*solid line*) currents

10.3.3 Pulse Shaping

In the normal firing configuration for the accelerator, all of the 36 individual transmission lines are triggered simultaneously. This firing configuration results in essentially a linearly increasing current pulse with a rise-time of approximately 200 ns. In most materials, the ensuing ramp wave will tend to form a shock at the base of the wave after propagation of only a few hundred microns. It is possible to substantially increase the propagation distance without shock formation by tailoring the current pulse.

As was briefly discussed in Sect. 10.2, the ideal wave profile will result in an abrupt shock formation at a single plane in the sample. Such an ideal profile maximizes the propagation distance for a given rise time prior to shock formation. This ideal profile can be determined by considering a position-time diagram, as illustrated in Fig. 10.11a. It is desired that all of the characteristics converge at a single time and position in the sample. Projecting these characteristics, which have a slope given by the inverse of the Lagrangian wave speed at the particular pressure, backwards in both space and time to a given location in the sample results in the desired optimal pressure–time profile. Note that the rise time of the resulting pulse depends upon the desired thickness of the sample; i.e., a longer rise time is required to delay shock formation for propagation through a thicker sample. As an example, the optimal pressure profile determined in this way for a 200 GPa ramp wave in aluminum is shown in Fig. 10.11b. In this case the rise time is 300 ns, and the shockless propagation distance is ~2 mm.

For completeness, we note there are slight modifications necessary to account for MHD effects and the significant deformation of the load. Nevertheless, this profile accurately represents the form of the required current pulse. Note that the shape differs significantly from the essentially linearly increasing current pulse produced by the accelerator in the normal firing configuration. Current tailoring, or pulse shaping, is required to reproduce the desired current shape. This is accomplished through proper staggered triggering of the individual transmission lines. To do this, individual laser triggers have been installed at the accelerator to enable individual triggering of the

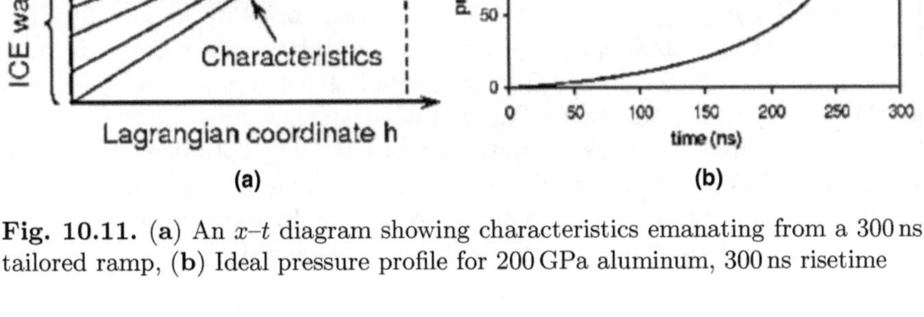

Fig. 10.11. (**a**) An x–t diagram showing characteristics emanating from a 300 ns tailored ramp, (**b**) Ideal pressure profile for 200 GPa aluminum, 300 ns risetime

transmission line gas switches. This modification to the accelerator enables significant flexibility and allows arbitrary current profiles to be produced. In particular, it has enabled a more concave current profile to be produced, which is necessary to delay shock formation to a few millimeters propagation distance for most materials. This type of pulse shaping was critical in performing many of the experiments that are described below.

10.4 Applications

10.4.1 Isentropic EOS Measurements

Although the smooth compression profile shown in Fig. 10.3b is extremely useful for material property studies, several issues must be addressed to obtain high quality data with this technique because small deviations in material response produce noticeable inflections in the transmitted wave profiles. One issue is that the high magnetic field (of order a few MG) produced in the A–K gap between the conductors rapidly diffuses into the anode as discussed in Sect. 10.3.2. This requires an optimal design for both the conductor and sample geometries with MHD simulations as described above in order to prevent magnetic disturbances on the measured velocity profile [63, 64] and resulting errors in interpreting material response. In addition, it is necessary to shape the current history to prevent premature formation of shock waves in the samples [41, 63], as discussed in Sect. 10.3.3. Finally, in order to apply the various analysis methods discussed earlier, it is necessary to estimate the minimum rise time or peak strain rate where dissipative effects become important and to determine the conditions for thermal conduction across the ramp wave that result in non-isentropic loading. Theoretical work to evaluate these limitations is in progress for aluminum quasi-isentropic compression experiments [33] and should lead to general rules that can be applied to experimental configurations and loading conditions sufficient to obtain accurate loading data for any material.

A recent example illustrating the use of the ICE technique to estimate the high pressure compression response of aluminum is shown in Fig. 10.12 [27]. In this experiment, pairs of planar aluminum alloy 6061-T6 samples were mounted on aluminum drive plates as illustrated in Fig. 10.12a and subjected to a shaped current pulse that produced a peak pressure about 170 GPa. Thicknesses of the aluminum samples ranged from about 0.77 mm to 1.56 mm. For thicknesses greater than 1.56 mm a small shock was observed to form at the foot of the ramp wave. This implies that the input profile, though designed to minimize shock formation, was not fully optimized to form a single, 170 GPa shock, as we discussed earlier. Single crystal LiF windows (< 100 > orientation) were used to record the ramp wave profiles at these different thicknesses. The average strain rate produced in this experiment

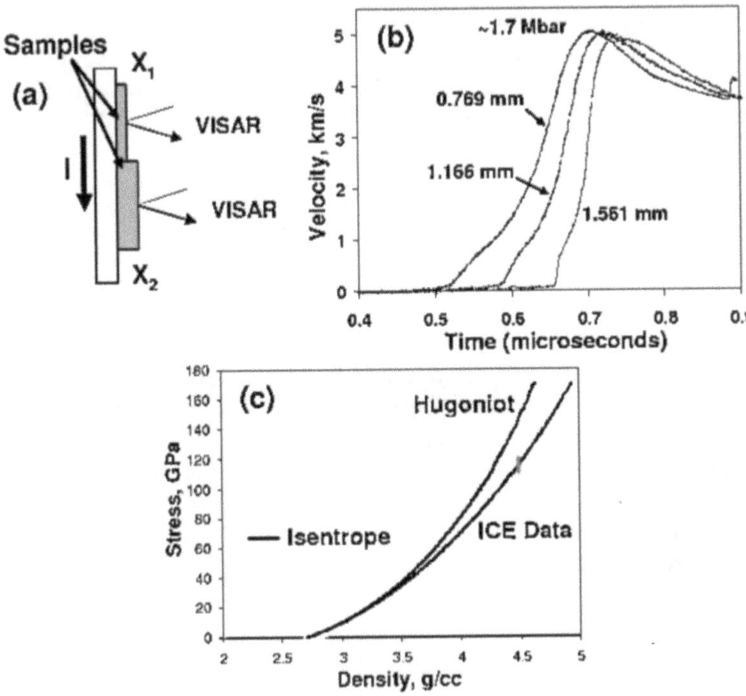

Fig. 10.12. Ramp loading of aluminum alloy 6061-T6 to 170 GPa. (**a**) Experimental geometry, (**b**) measured wave profiles, (**c**) stress-volume response

ranged from about 4–8 × 10^6 s^{-1}. Careful analysis of the wave profiles indicated that self-similar response was achieved, to a good approximation, and therefore that the material response is quasi-isentropic to within experimental error. Using the reported equation of state for LiF and the high-pressure optical properties of LiF determined for shock compression, the backward integration technique was used to estimate the high-pressure loading response of aluminum (the validity of this of this assumption is described below).

The resulting compression curve is shown in Fig. 10.12b, along with a calculated isentrope and the Hugoniot curve using the Sesame 3700 equation of state [52,53]. The experimental compression curve agrees with the calculated isentrope to within the experimental error of about ±7% in stress over the stress range to about 170 GPa and further confirms the assessment of simple flow from the wave profiles. It should also be noted that there is a significant difference between the isentrope and the Hugoniot near the peak stress level, as observed in Fig. 10.12b. The good agreement observed between the calculated and experimental isentropes shown in Fig.10.12 implies that this method should be valuable for off-Hugoniot measurements of EOS properties for a wider variety of materials, once all sources of errors in the assumed isentropic response are evaluated. We note the data are still preliminary, since the high pressure isentropic response of single crystal < 100 > LiF has not been experimentally evaluated at these stress levels and the refractive index must also be established in this stress range. Experiments are underway to obtain these data, which should allow refinement of this analysis and reduction in experimental uncertainty.

With these caveats, it is useful to put the Hugoniot and isentropic curves in the context of an equation-of-state surface. Figure 10.13 depicts the EOS surface for aluminum and illustrates the thermodynamic paths followed by shock loading and for isentropic loading from room temperature. The large departure of these curves at the higher pressures illustrates the EOS range that can be covered with these combined loading techniques and further emphasizes the importance of ICE experiments for developing EOS theories.

The observed correspondence between theoretical and experimental isentropes in Fig. 10.12 also provides preliminary information on the high-stress mechanical properties of aluminum by placing an upper limit on the magnitude of compressive yield strength for the specific strain-rate loading history applied in these experiments. This analysis indicates that the compressive yield strength of aluminum can be no more than 12 GPa at a stress level of

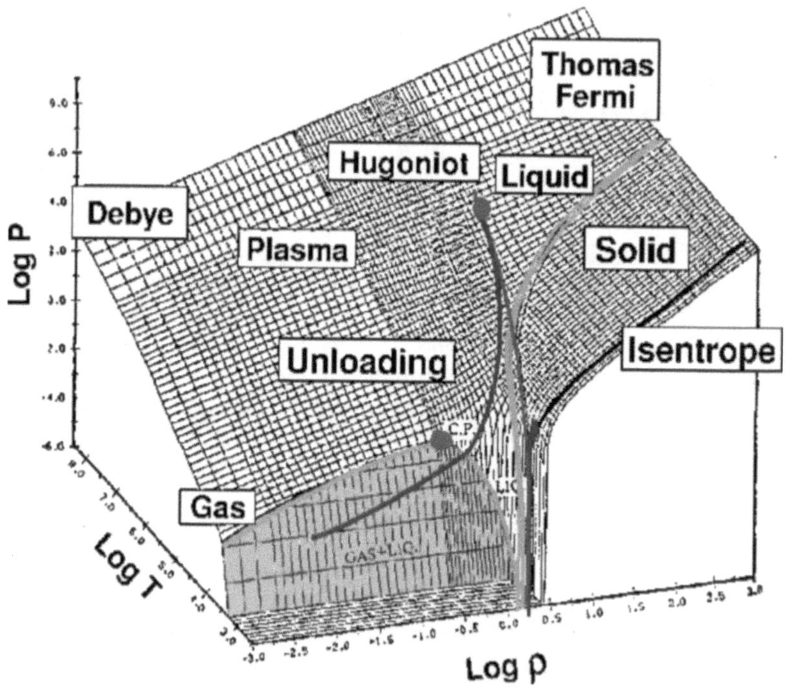

Fig. 10.13. Equation-of-state surface for aluminum showing the approximate location of the isentrope, Hugoniot and the unloading path from a Hugoniot state of about 700 GPa (Courtesy of V.E. Fortov 2004)

170 GPa. Additional loading and unloading experiments from the compressed state are needed to provide a more direct measurement of the compressive strength during ramp loading [22]. Experiments are presently in progress on the Z accelerator to evaluate this method [88].

A second example of using this technique for EOS measurements is illustrated by recent ramp loading experiments on energetic materials. The ICE technique has particular significance in this case since the considerably lower thermal energy produced by quasi-isentropic compression in comparison with shock compression allows investigation of the EOS of energetic materials to higher stresses. In contrast, shock loading at significantly lower stresses will induce energetic reaction. ICE experiments were recently conducted on LX-04, which is an energetic material (an explosive) consisting of 85% HMX (octahydro 1,3,5,7-tetranitro-1,3,5,7-tetrazocihne) and 15% Viton-A (vinylidene fluoride/hexafluoropropylene copolymer) [45]. The unreacted response of this material under compression is necessary for developing models of energetic initiation, but previous experiments using shock wave techniques were unable to obtain this information in the required stress range. Ramp compression experiments were conducted on LX-04 to about 20 GPa. These experiments used an aluminum drive plate to conduct the current from the accelerator. Several samples of LX-04 were bonded to the back side of the drive plates and NaCl windows were used to monitor the velocity history at the rear of the LX-04 samples, as shown in Fig. 10.14a. A shaped current pulse was used to minimize the formation of shocks over propagation distances on the order of 0.4 to 0.7 mm and the Lagrangian wave analysis method described in Sect. 10.2.1 was used to determine the stress–volume response. Figure 10.14b shows the resulting compression curve for LX-04, which is assumed to be quasi-isentropic, along with a calculated Hugoniot and other data. These measurements represent the first dynamic data on the unreacted response of this material over this pressure range and illustrate the potential for mining new information – in this case an unreacted compression curve that is not accessible with other methods.

10.4.2 Phase Transitions

Since ramp loading produces continuous loading curves, the measured wave profiles and resulting stress–strain response are sensitive to very small changes in material response; in fact, the wave profiles represent, essentially, the derivative of the loading curve, as is apparent from inspection of (10.7)–(10.9). First order polymorphic phase changes produce a discontinuity in the loading curve with a typical volume discontinuity of a few percent [36]. The effect of a first order phase change is illustrated schematically in Fig. 10.15a. For this example, the stress–volume loading response is assumed to be continuous to point A, where a transition to a higher density phase occurs. If the material is shocked to stress B, the loading curve, referred to as the Rayleigh

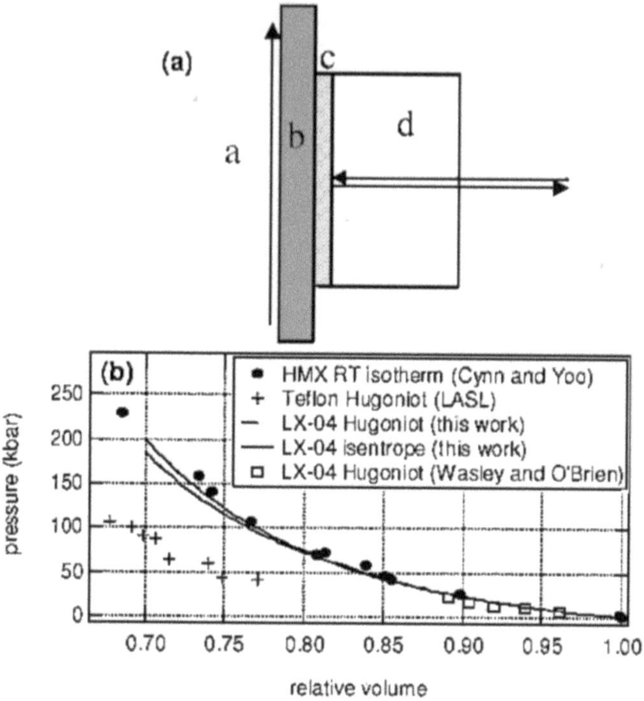

Fig. 10.14. Determination of the unreacted isentrope in LX-04. (**a**) Experimental technique showing the current drive a, aluminum drive plate b, the LX-04 sample c, and a NaCl window d, (**b**) Calculated unreacted isentrope (*lower solid curve*) and Hugoniot (*upper solid curve*). From [45]

line, is the path followed for steady shock compression. For the case illustrated, a single shock wave would be observed so that the phase change at point A would not be detected. However, if the material is loaded with a ramp wave to point B, it will be continuously compressed along this path and the discontinuity at point A is easily observed as a disturbance in the propagating wave. Several recent ramp wave experiments performed on phase transforming materials have illustrated the potential of the method. Three examples are provided in the following discussion.

ICE experiments to demonstrate the ability of the technique to detect phase transitions were initially performed on Armco iron, since iron has well-established phase transition and kinetic properties [16]. The high-pressure Hugoniot response of iron to about 30 GPa is shown in Fig. 10.15b and illustrates a transition from the α (bcc) phase to the ε (hcp) phase beginning at approximately 13 GPa. At this pressure, a volume discontinuity of about 3% occurs.

Ramp wave profiles obtained on several samples simultaneously loaded to 30 GPa with an input ramp compression are shown in Fig. 10.16a [26]. Sample

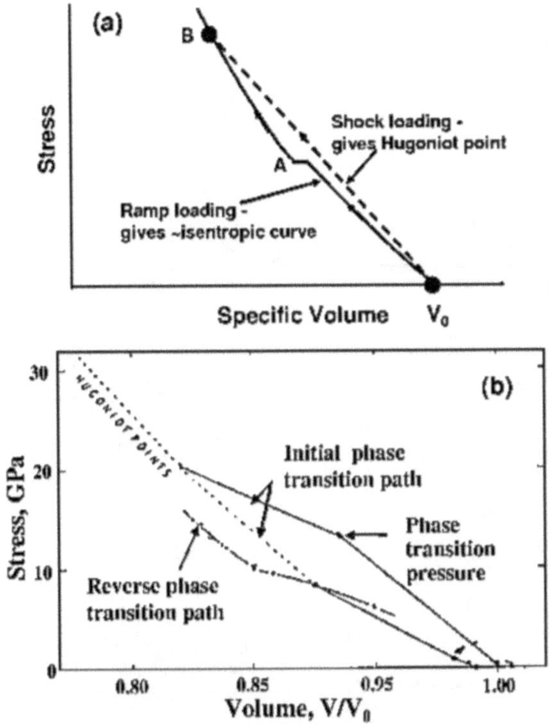

Fig. 10.15. Phase transition studies with ramp waves. (**a**) Stress-volume paths followed by isentropic and shock loading, (**b**) Stress-volume response for iron in the vicinity of the α–ε phase transition

thicknesses ranged from 0.3–0.9 mm. In this experiment Z-cut sapphire interferometer windows were placed on the back side of the iron specimens to approximate in situ wave profile measurements. For the set of experiments shown, the base of the ramp waves exhibit evidence of elastic–plastic compression, as expected. Following initial elastic compression, the iron is compressed plastically to the final drive stress. The resulting wave profiles show a clear break at an interface particle velocity of about 0.35 km/s, corresponding to a stress of about 13 GPa and onset of the α–ε transition. Further compression occurs in the high-pressure phase. These data are providing important information on the kinetic properties of the α–ε phase transition in iron [26].

Another example of using the ICE technique to detect very small volume change phase transitions, and also to evaluate the effects of impurities on phase transition kinetics, is illustrated by experiments on highly characterized zirconium samples. In these experiments, two different samples of zirconium with differing impurity levels (principally oxygen) were subjected to identical loading to a peak stress of ~25 GPa. Figure 10.16b shows the resulting wave profiles obtained with single crystal LiF windows [78]. As noted in the figure,

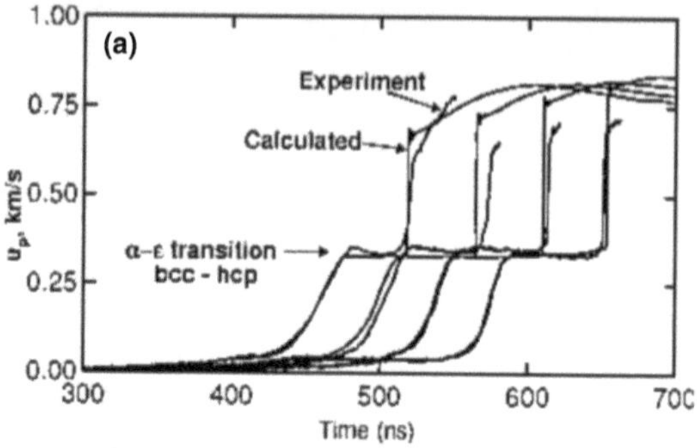

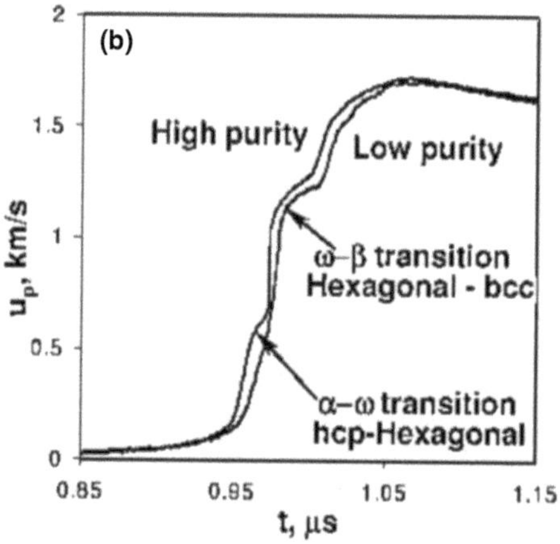

Fig. 10.16. Examples of using ramp waves for detecting polymorphic phase transitions. (**a**) iron [26], (**b**) zirconium [78]

two phase transitions, evident as small disturbances in the wave profiles at interface velocities of about 0.6 and 1.2 km/s, are observed during compression. The lower stress perturbation occurring at the α–ω transition is of particular interest. As observed in the figure, the signature of this transition on the wave profiles is significantly different for the two samples; the signature is much more pronounced and appears to occur at a higher interface velocity for the high-impurity sample. In contrast, the signature at the higher stress ω–β transition appears to be similar for both samples. Presently, it is unclear whether these differences are due to different kinetic effects in the α–ω transition or to an increased transition stress induced by the impurities. It is also not clear why only the α–ω transition appears to be affected by the differing impurity levels and not the ω–β transition. In both cases, the volume change and therefore the effect on the ramp wave profile, is very small. This would make the transitions difficult to observe in shock wave experiments unless the transition pressure is well known and the shock amplitude is carefully chosen to not produce an over-drive effect, as illustrated in Fig. 10.15a.

Quasi-isentropic loading can also be used to study other types of phase transitions occurring under dynamic compression. In particular, the possibility of freezing on dynamic time scales has been of interest from both a theoretical and an experimental perspective for several years. Whether dynamic freezing can occur in shock wave experiments has been a long-standing controversy, but there has been no definitive evidence for the dynamic resolidification under shock compression [34]. Recently, however, [35] demonstrated that freezing of water does occur on the 100 ns time scale for step-wise loading, which results in quasi-isentropic loading. The ICE wave technique developed on the Z Accelerator can be used to study such effects with more precision, since the technique produces a smooth compression loading history to extremely high pressures, in contrast to the step-wise loading produced in reverberating wave experiments. Therefore, this is a potentially new approach for probing these transitions. The graded density impactor techniques being developed by [71] and the laser drive technique by [76,77] are also potentially useful for these studies.

Ramp loading techniques have been used to study dynamic resolidification of both liquid tin [27] and liquid bismuth [71]. Results for liquid tin are discussed here since they illustrate several important requirements for performing these investigations. Reference [27] have described experimental techniques for performing freezing experiments on the Z Accelerator. In these experiments, two to four aluminum or copper drive plates, similar to the configuration shown in Fig. 10.7, are used to ramp load preheated cells containing liquid tin. Each cell contains a rear window of sapphire or LiF for measuring the wave evolution properties with a VISAR [16] after a specified propagation distance in the liquid. In addition, the rear surface velocity of the aluminum drive plate is also directly measured with a VISAR, so that the input pressure history to the sample can be precisely determined. The known loading history applied to the tin sample, as determined in this way, is used in a forward numerical simulation method that includes a kinetic transformation capability. Since each liquid cell is subjected to the same nominal input pressure history, this technique allows for the determination of kinetic effects for solidification from the liquid phase by comparing the observed wave evolution with rate-dependent models of the phase transition at different propagation distances. It is particularly important to properly design these experiments to minimize wave interactions that can give rise to misinterpretation of the results. In particular, the necessary conditions for obtaining good data related to liquid–solid transition are: (i) the input pressure history must be precisely known and shaped in time to minimize premature shock formation, (ii) each liquid sample must be subjected to identical pressure histories, and (iii) ramp wave histories should be measured at several propagation distances in order to assess the importance of kinetic effects.

Figure 10.17a illustrates the path followed in the temperature–pressure plane for isentropic and shock loading of tin. Under the assumption that

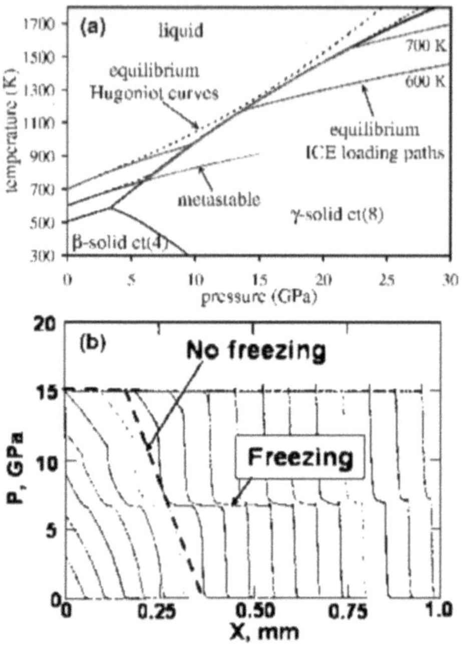

Fig. 10.17. (**a**) Phase diagram for tin showing the paths followed by shock and isentropic loading [27]. The *lower dashed line* in the solid phase represents a metastable response (no phase transition) of liquid tin. (**b**) The calculated wave profiles versus propagation distance for a ramp input at constant pressure of 15 GPa

thermodynamic equilibrium is achieved at all times during the loading process, isentropic compression produces a final state in the solid (bct) phase for initial temperatures of 600 and 700 K. Note that shock loading produces end states either on the melt boundary or in the liquid state at high pressure for both cases, and is thus not useful for probing dynamic resolidification of tin. If freezing is suppressed for the isentropic loading case, the metastable isentrope extends into the stability field of the solid phase as an extension of the liquid isentrope. Figure 10.17b illustrates the wave profiles expected for both the cases of metastable and equilibrium loading of liquid tin to 15 GPa from an initial temperature of 700 K. For a given ramp input condition, a single ramp wave output will be observed for metastable response of the liquid, a structured two-wave ramp for equilibrium response, and a complex evolving ramp wave for time-dependent freezing. For propagation distances near 1 mm, the structured ramp representing equilibrium freezing shown in Fig. 10.17b is observed to evolve into a shock, illustrating the need to measure profiles at different propagation distances, especially for thin samples.

Experimental ramp profiles obtained from pure tin samples preheated to 530 K and loaded into the stability field of the solid phase for propagation

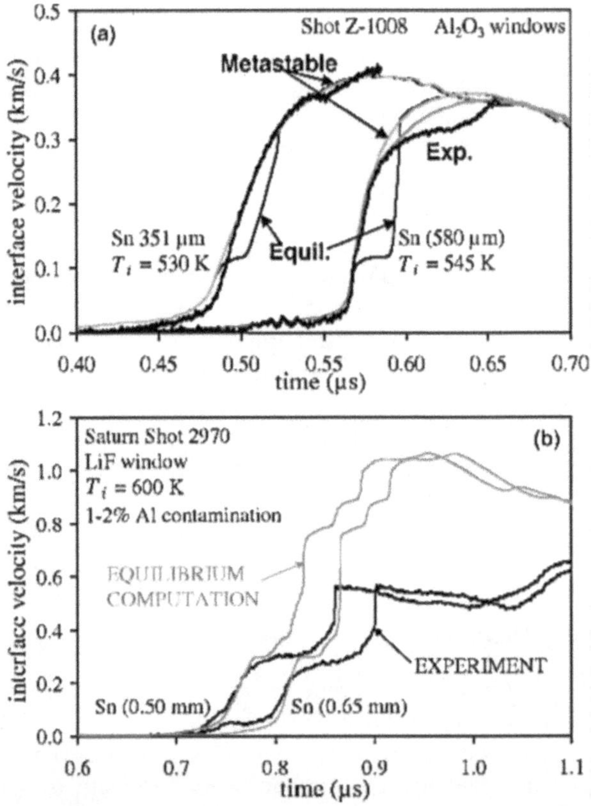

Fig. 10.18. (**a**) Ramp profiles observed in pure tin continuously loaded to 13 GPa at an initial temperature of 530 K. (**b**) Ramp profiles observed in liquid tin containing impurities for ramp loading at an initial temperature of 600 K [27]

distances of 531 and 580 μm are shown in Fig. 10.18 [27]. Comparison of the experimental and calculated profiles in the figure suggests that equilibrium freezing does not occur on the timescale of the experiment for this driving stress and propagation distance. Furthermore, the experimental profiles do not agree with simulations assuming metastable response, i.e., liquid response on this time scale, as also illustrated in the figure. The experimental data strongly suggest that kinetic effects are important for this particular loading condition, although further work is necessary to substantiate this hypothesis and to systematically evaluate the kinetic effects [27].

Figure 10.18b shows a similar ICE experiment on liquid tin that was known to contain large amounts of aluminum contaminants, some of which were likely present as suspended solid particulates in the liquid tin, presumably coming from the sample holder of aluminum used to contain the liquid tin. A significant observation from this experiment is that the measured wave structures are much more compatible with those expected for equilibrium response, suggesting transformation to the solid phase is occurring over this time scale. These tentative results indicate the possibility that contaminants in the liquid phase may play a significant role in the freezing process, possibly by providing nucleation sites for solidification. Further work is necessary to investigate this hypothesis [27].

Similar discontinuities shown for the impure tin samples in Fig. 10.8b have been observed in water and preheated bismuth subjected to quasi-isentropic compression [71], although the purity of these materials was not reported. In contrast, recent experiments have shown that freezing in ultra-pure water involves an incubation time and is in general of a heterogeneous nature [34,35]. These effects will have definite consequences on continuum wave profile measurements, as discussed in this section. In particular, wave profile predictions based on one-dimensional, homogeneous and instantaneous freezing will likely not provide an accurate representation of the freezing process. Systematic studies using ramp loading to explore dynamic freezing at different propagation distances and for different peak pressures should be useful in resolving these issues.

10.4.3 Optical Properties

The optical properties of transparent materials under dynamic compression are of interest in a variety of shock physics applications. For velocity interferometer measurements of arbitrary wave profiles, it is necessary to know how the refractive index depends on the compressed density or uniaxial stress state produced in the window material. Optimal materials that have been qualified for shock or ramp measurements are presently extremely limited. Most interferometer wave profile measurements have been performed with either sapphire, quartz or LiF windows. Specifically, previous refractive index measurements under shock loading have been made on sapphire, polymethylmethacrylate, fused quartz, single crystal quartz, and LiF to determine the correction factor, $1 + (\Delta\nu/\nu_0)$ appearing in (10.10). This factor accounts for the change in refractive index with density occurring as a shock or arbitrary wave propagates through the window, and is required for accurately inferring particle velocity from VISAR measurements [16,51,82,90]. The refractive index change induces an additional Doppler shift in the VISAR measurement of particle velocity. The correction factor is usually assumed to be independent of stress or particle velocity. Reference [47] has shown that this approximation is valid if the refractive index is linear in material density and a function only of density (i.e., temperature independent). For this special case, the index determined from shock experiments can be applied to unloading wave profile measurements or for any other arbitrary wave profile.

Only a limited number of experiments have been performed to test this hypothesis, primarily because of the complexity and time involved in making these measurements. Techniques have been developed using ramp loading that result in a more efficient way to evaluate the refractive index under compressive loading. Hayes and colleagues have developed an experimental approach for evaluating this property continuously during ramp loading [47]. The configuration for these experiments is illustrated in Fig. 10.19a, in which the refractive index of Z-cut sapphire was determined continuously to about 20 GPa. In these experiments, two or more thicknesses of sapphire were placed on a drive plate and subjected to ramp loading. Velocity interferometers (VISARs) were used to measure the free surface velocity simultaneously with the interface velocity between the window and the drive plate. As was discussed earlier, the drive plate measurement defines the initial pressure input to the sapphire window using the backward integration technique, and thus determines the actual interface velocity, u_A, at the window interface. The measured interface velocity produces an apparent particle velocity, u_a, which differs from the actual particle velocity because of changes in the index under compression. The apparent and free surface velocity measurements for

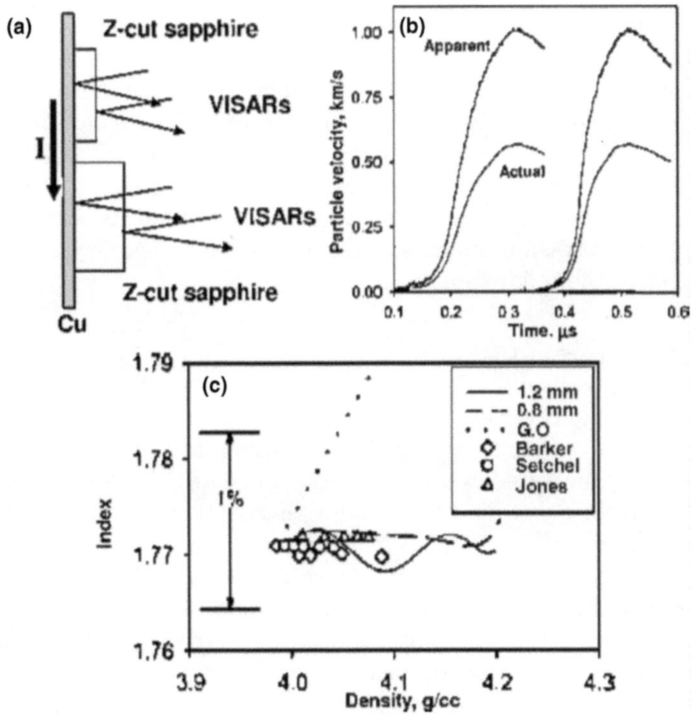

Fig. 10.19. (**a**) Experimental configuration used to study optical effects under ramp loading in Z-cut sapphire, (**b**) VISAR measurements of the interface and free surface velocities, (**c**) index properties of sapphire versus density [49]. The dotted line designated GD corresponds to the Gladstone–Dale dependence of index on density

an experiment on Z-cut sapphire are shown in Fig. 10.19b. These combined measurements determine the flow variables during ramp loading in a rate-independent material – specifically, the stress and density during compressive loading.

Reference [47, 49] has shown that the apparent and actual velocities can be determined continuously during ramp loading from the relationship

$$du_{\mathrm{a}}/du_{\mathrm{A}} = n - \rho\,(dn/d\rho)\,, \tag{10.13}$$

where n is the refractive index and ρ is corresponding density. By using the free surface velocity history to determine the actual velocity history and comparing that with the measured apparent velocity at each point in the profile, (10.13) can be applied incrementally to determine the refractive index as a function of density in a single experiment. The results from two experiments on Z-cut sapphire are shown in Fig. 10.19c, along with shock wave measurements of the refractive index [14, 51, 82]. The excellent agreement between measurements obtained with shock and ramp loading indicates that the refractive index of Z-cut sapphire is essentially independendt of density over the stress range of 0–20 GPa and further illustrates that the correction factor for Z-cut sapphire previously determined from shock measurements [14, 51, 82] can be used to determine the correction factor for arbitrary wave profile measurements. It should be noted that this correlation may not hold at higher stresses where the Hugoniot differs substantially from the isentrope and temperature effects may become more important. Experiments are in progress to evaluate the optical properties of LiF at elevated stress levels [28], which should be useful for addressing this issue.

10.4.4 Constitutive Properties

In the ICE technique, the loading time can be precisely controlled, resulting in applied strain rates from about 10^5/s to 10^7/s, so that material properties can be continuously measured over a large stress range to a few hundred GPa and at variable strain rates. Since the loading history is known to substantially influence thermomechanical and physical states produced in high-rate loading [9, 70], and therefore the constitutive properties, ramp loading provides a way to change these states for a given stress level. In particular, the dynamic yield strength is thought to depend on the loading history [9]. The use of ramp waves for exploring the range of constitutive properties that can be studied is only beginning. However, a few examples will be presented here to illustrate the potential of the method.

It is known that grain size affects the quasi-static yield strength of materials, but the effects under dynamic loading are less well understood. Reference [73] has demonstrated the possibility of using ramp loading to determine small changes in mechanical properties. Again, the ability to simultaneously load multiple samples in one experiment makes the technique attractive for this type of high-fidelity A–B material comparison. The ramp waves produced in different materials, all for the identical input history, follow the actual loading curve for the corresponding strain rate. Thus, independent of whether the material is rate dependent or not, a comparison of the ramp profiles and resulting stress–strain paths enables a critical evaluation of differences in mechanical properties.

Figure 10.20 illustrates the results of ramp loading on two polycrystalline uranium alloys with average grain size that varied by about a factor of two. Microscopic photographs of the materials studied in a single experiment are shown in Fig. 10.20a. The stress–density curves obtained for each of the materials to stresses of about 20 GPa are shown in Fig. 10.20b. As illustrated, there is no observable difference in the constitutive response, at least to within experimental uncertainty. Also shown in the figure is a calculated hydrostat for both materials, which provides an estimate of the compressive yield strength from measurement of the difference between the longitudinal stress and the hydrostatic pressure [39]. For this case, there is no difference in mechanical response, including yield strength, to within experimental error. These

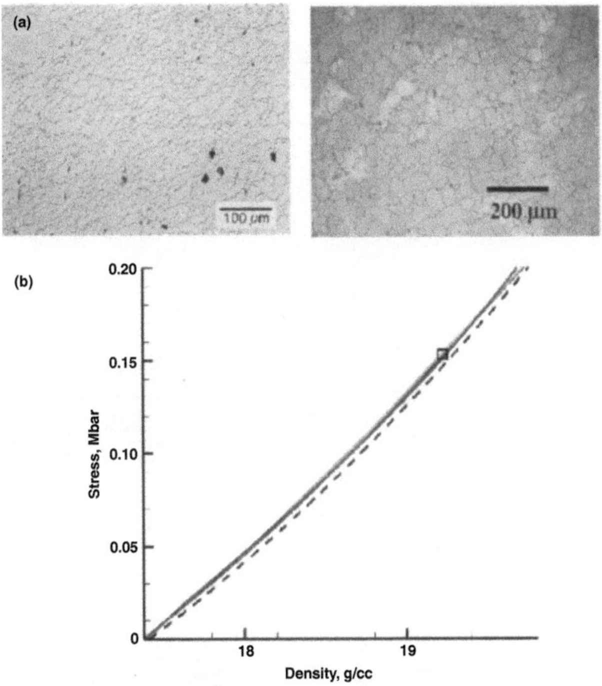

Fig. 10.20. Effects of grain size for ramp loading of polycrystalline uranium alloy. (**a**) Grain structures studied; the bar in the graph on the left is 100 μm and 200 μm in the graph on the right. (**b**) Stress-strain response (Courtesy of Reisman 2003)

experiments illustrate the power of the ICE technique for quickly making relative comparisons of dynamic material response.

A second example concerns the effects of atomic-scale porosity that can be induced in metals through radiation-produced defects. Reisman and colleagues [75] performed a single experiment on various samples of stainless steel subjected to different levels of irradiation, resulting in different levels of atomic porosity. Four sets of steel samples were prepared, ranging from the as-received, un-irradiated material to samples that contained 0.4%, 1.1% and 2.7% levels of radiation-induced porosity. Two sample thicknesses of each concentration were mounted on four separate panels and subjected to magnetic pressure that continuously increased to 16 GPa in 200 ns, as shown in Fig. 10.21a. The resulting wave profiles for the four samples are shown in Fig. 10.21b. For the three samples containing porosity, the initial part of the wave profile up to a particle velocity of about 0.15 (~4 GPa) overlaps with that of the virgin material. At higher stress levels, the wave profiles diverge for higher particle velocities. In particular, there is a plateau region in the wave profiles similar to effects previously observed in the compaction

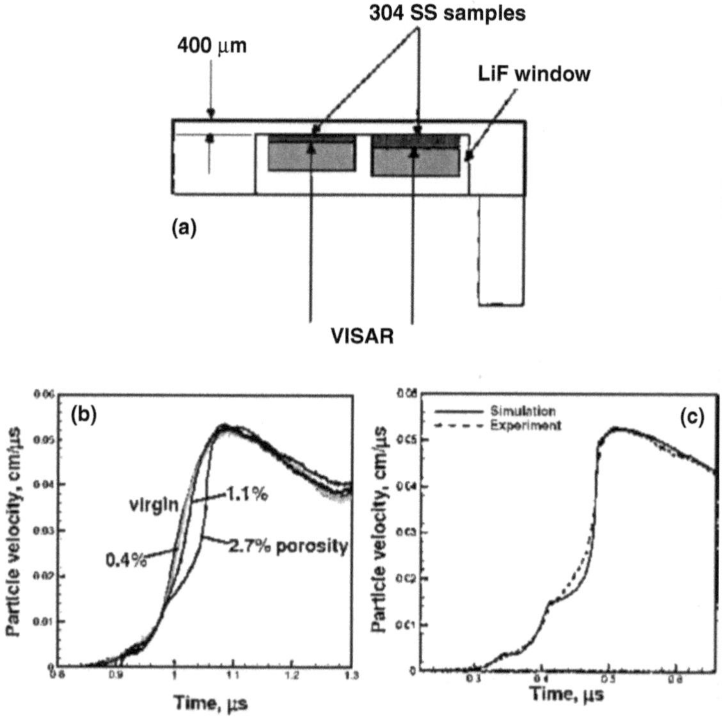

Fig. 10.21. Effects of radiation-induced defects on ramp wave propagation in stainless steel. (**a**) Experimental configuration. (**b**) Wave profiles for different defect volumes. (**c**) Calculated and experimental wave profiles [75]

of porous materials. As shown, these data make it possible to easily discriminate the dynamic response for the different defect concentrations, and to detect differences in mechanical response from the virgin material for defect concentrations as low as 0.4%.

Reference [75] have developed a model of the atomic void collapse process based on discrete atomistic effects. These include void shrinkage due to the emission of vacancies and vacancy clusters, and the absorption of self-interstitials or interstitial clusters. In this model, the emission of vacancy-type dislocation loops results in the shrinkage of voids during compression, which produce an overall volume reduction of material containing voids. When combined with the EOS and constitutive behavior for the material, a compaction model is obtained that can be used to simulate the ramp wave experiments in a one-dimensional wave code. Figure 10.21c shows the resulting wave profile predicted with this model for the 2.7% porosity sample [75]. Good agreement is observed in comparison with experimental results – in particular, predictions of the crush up response between 4 and 8 GPa. The slight discrepancy noted between simulations and experiments could result from several factors, including a good description of porous material release and the crush strength [75]. Evaluation of these small variations would be very difficult using conventional methods that would literally require tens of experiments.

The ability to perform this evaluation in a single experiment is another demonstration of the powerful capabilities provided by ramp waves for quickly discriminating changes in the dynamic response of materials. It is expected that these methods will also be useful for evaluating the effects of several other applications, including the aging of materials resulting from changes in chemical or mechanical effects or the effects of impurities on mechanical, chemical or optical properties. For example, it might be possible to determine the correlation between defect concentration and dynamic elastic yielding that heretofore has required a large number of shock experiments [10].

Likewise, this technique should be able to identify subtle changes in mechanical response due to small changes in dynamic strength. In using this approach, it would be desirable to determine the absolute magnitude of dynamic compressive strength of a standard material under ramp loading that could then be used for making relative comparisons of strength as discussed for the uranium alloy. Self-consistent test methods have been developed for determining dynamic compressive strengths for shock loading followed by either unloading or reloading from the shocked state [5, 7]. In principle, this approach could also be used with ramp waves and would allow the determination of material flow strength under isentropic or much cooler conditions than possible with shock experiments. The input pressure profiles produced by magnetic loading, as shown in Fig. 9.3b, allow evaluation of the continuous unloading response from the peak stress state. Implementing the approach of Asay and Chhabildas would require determination of the loading and unloading response in the vicinity of the peak stress, which would provide a measure of dynamic material strength. Reference [88] is presently exploring analytic techniques for making these estimates for ramp loading of aluminum. The results of his work may lead to new capabilities for estimating material strength under quasi-isentropic loading.

10.4.5 Magnetically Driven Flyer Plates

The ramp loading technique can also be used to launch flyer plates to high velocities [43,58,59]. In this application, the impulsive pressure produced by the loading provides sufficient momentum over a specified time duration to launch a flyer plate to high velocity. The configuration for launching flyer plates in this way is illustrated in Fig. 10.22. In most cases, the current-carrying aluminum anode plate also becomes the flyer plate. However, it is also possible to mount a different material, such as a titanium plate, onto the aluminum drive plate to create a composite flyer. It is also extremely desirable to design the flyer plate configuration so that shock formation in the flyer plate, resulting from evolution of the driving ramp wave, is minimized and so that magnetic diffusion through the plate is also minimized. These are conflicting requirements since minimization of shock formation requires a minimum plate thickness for a given current profile, while minimization of magnetic

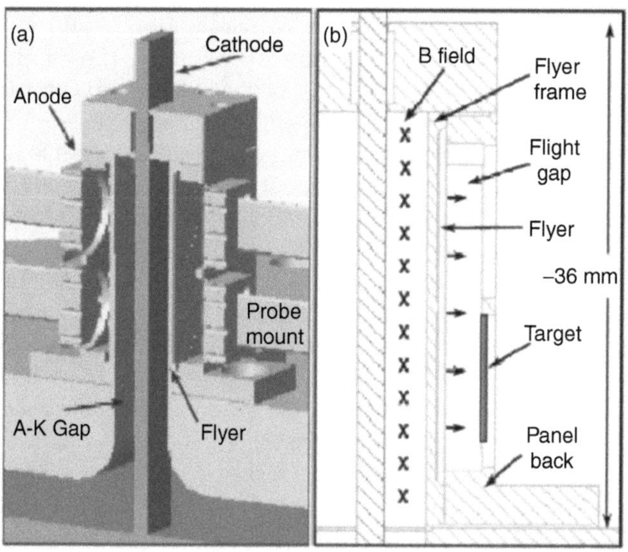

Fig. 10.22. Schematic of the load used to launch flyer plates on the Z accelerator, in this case the rectangular short circuit geometry. (**a**) Section view of the coaxial short circuit load. Also shown are the probe mounts used to hold the fiber optics used to diagnose the sample. (**b**) Section view showing more detail. The height of the anode panel is ~36 mm

diffusion is accomplished with a thicker plate. This is a highly coupled magnetohydrodynamic problem, since successful plate launch requires properly shaping the current pulse in order to minimize shock formation, while also limiting magnetic diffusion. Nevertheless, these obstacles have been overcome through combined experimental and computational studies and magnetically launched flyer plates are now routinely used on the Z Accelerator. Design of the flyer plate configuration is accomplished using the fully self-consistent two dimensional MHD simulations described in a previous section that include accurate equations of state and conductivity for each component in the assembly and accurate representation of the accelerator performance, including prediction of the significant deformation occurring in the anode load during the pulse rise time [63, 64].

One-dimensional solutions can be used to address shock formation and magnetic diffusion effects, while two-dimensional representations are necessary to model the lateral geometries of the flyer plate in order to optimize flatness of the flyer plate during launch. Figure 10.23 illustrates use of a line imaging VISAR [86] to measure shock breakout along the vertical and horizontal directions of stepped aluminum target plates that were impacted with the flyer plate. Shock breakout is indicated by loss of light when the shock wave reaches the back surface of the stepped aluminum plate. The near simultaneous breakout of the shock at each step in Fig. 10.23a indicates both the effect of low impact tilt of the flyer plate on the target and the extremely good uniformity of magnetic loading over the vertical dimension of the flyer plate [59]. Similar Line VISAR measurements made across the lateral dimensions of the flyer plate, as in Fig. 10.23b, indicate curvature near the edges of the plate with a near planar portion in the center 3–4 mm of the plate. Both experimental observations agree with two dimensional MHD simulations of the experiment [63, 64], as shown in the figure.

Figure 10.24 shows an experimental measurement of flyer plate velocity versus time for an aluminum flyer plate accelerated to over 25 km/s. In this case, design of the launch geometry resulted in a flyer plate that was continuously accelerated without shock formation, so the flyer plate impacting surface remained essentially in the ambient thermodynamic state. For symmetric impact experiments, i.e., the flyer plate and target are the same material, measurement of the final flyer velocity precisely determines the particle velocity needed in the solution of (10.1)–(10.3). Measurement of shock velocity then allows solution of the three equations relating five independent variables of shock compression. It is important to note that these experiments must be carefully designed with MHD simulations, as with the ICE experiments previously discussed, in order to specify the initial impacting state of the flyer plate and thus the effective shock impedance of the flyer plate before impact. MHD calculations are also necessary to determine the optimal flight distance to minimize the flyer plate acceleration (and therefore residual stress) at impact. If shock formation and magnetic diffusion are eliminated,

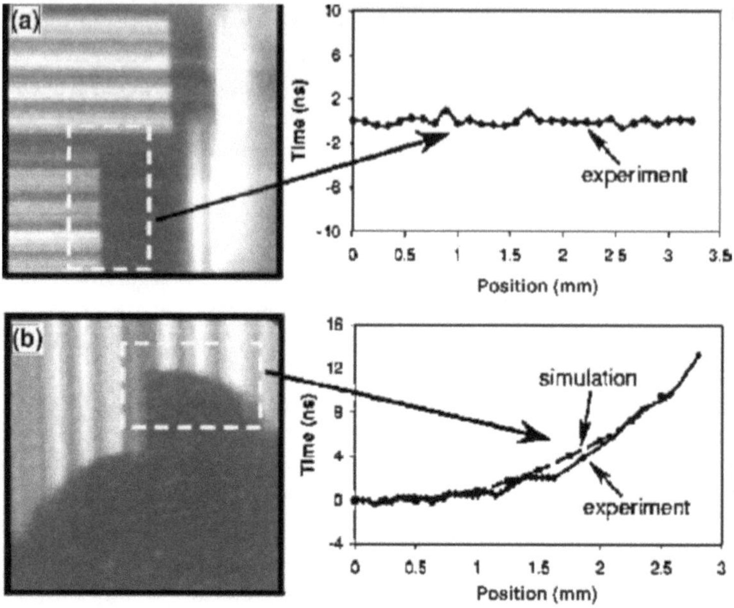

Fig. 10.23. Streaked line-imaging VISAR records of shock break out in stepped aluminum targets impacted by an aluminum flyer plate at ~20 km/s. Orientations of the streak camera slit relative to the target image was (**a**) parallel to the coaxial load and (**b**) perpendicular to the coaxial load. In each case the spatial extent of the image was ~7 mm

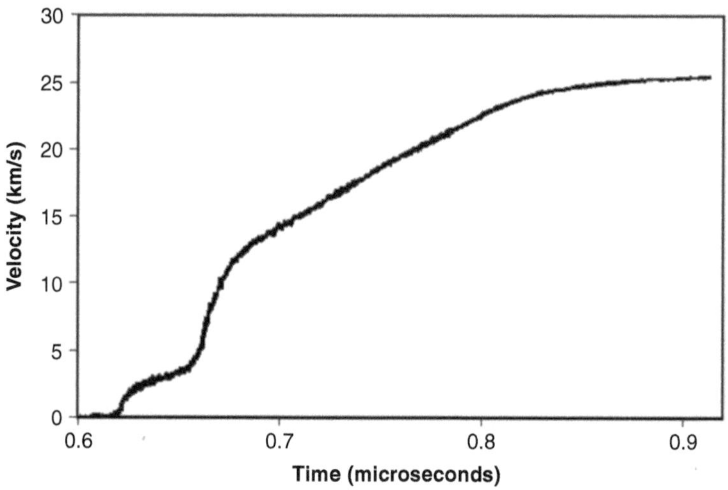

Fig. 10.24. VISAR measurement of an aluminum flyer plate accelerated to over 25 km/s

the front surface of the flyer plate remains at the ambient thermodynamic state so that the particle velocity produced in the target for a symmetric impact experiment is precisely one half of the measured impact velocity.

The flyer plate technique has been used to determine Hugoniot curves in several materials. Two examples will be shown to illustrate the potential of the magnetically-driven flyer plate method. The first example involves symmetric impact experiments of aluminum alloy 6061-T6 samples [58]. The experimental configuration is shown in Fig. 10.25a. Symmetric impact of an aluminum flyer plate onto an aluminum target results in near-absolute Hugoniot data for aluminum if the velocity of the flyer plate is measured, along with the shock velocity in the aluminum sample and the physical state can be inferred as discussed above. In the early experiments with this technique, current shaping was not available so shock formation occurred in some of the impact experiments. Because of this, it was necessary to make a small correction for the change in density and shock impedance of the flyer plate before impact [58]. When these corrections are made, the state of the impactor can be determined and Hugoniot data to nearly 500 GPa are obtained, as shown in Fig. 10.25b. Since only small corrections to the measured states are necessary to obtain these data from the measured flyer plate velocity and shock velocity, the Hugoniot data shown in the figure represent nearly absolute measurements.

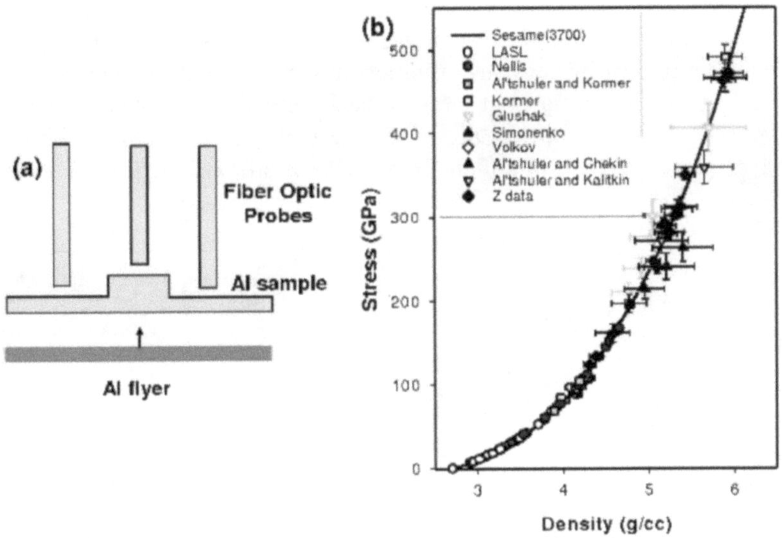

Fig. 10.25. Magnetically driven plate impact technique for near-absolute Hugoniot experiments on aluminum. (**a**) Experimental configuration. (**b**) Hugoniot results for aluminum

Several points are apparent from the figure. First, Hugoniot data obtained over the stress range of about 120–500 GPa with the magnetic flyer technique overlap with gas gun data at the lower stresses where previous data have been obtained. Gas gun data are generally considered to have the highest accuracy. Agreement of the magnetic-flyer-generated data with these data provides a partial validation of the technique. At higher stresses, agreement is obtained with Hugoniot results from several other investigators, including an absolute Hugoniot measurement obtained by Russian investigators in a nuclear-driven shock wave experiment. Good agreement is also obtained with a well-established theoretical EOS for aluminum over the full range of data taken with magnetic drive technique [59], as shown in the figure.

The magnetically driven flyer plate technique was also used to investigate the high pressure EOS of liquid deuterium. There is considerable interest in the equation of state of hydrogen and its isotopes for scientific applications. Prior gas gun data on this material were limited to about 20 GPa [69]. These data agreed with existing theoretical models, which extrapolated to density compression ratios of about four at higher pressures [56]. However, Hugoniot data obtained with a laser-shock technique disagreed with these theoretical estimates and indicated substantially higher compressibility at pressures of 200–300 GPa [24, 25].

The experimental configuration shown in Fig. 10.26a was used to study the Hugoniot response of liquid deuterium over the pressure range of about 30–100 GPa with magnetically driven flyer plates. In these experiments, either aluminum or titanium flyer plates were used to impact a cryogenic cell of liquid deuterium initially at a temperature of about 21 K. The liquid was confined by an aluminum front plate and a Z-cut sapphire window back plate that was also used for optical access, as illustrated in Fig. 10.26a. Impact of the flyer plate on the aluminum drive plate produced a constant pressure

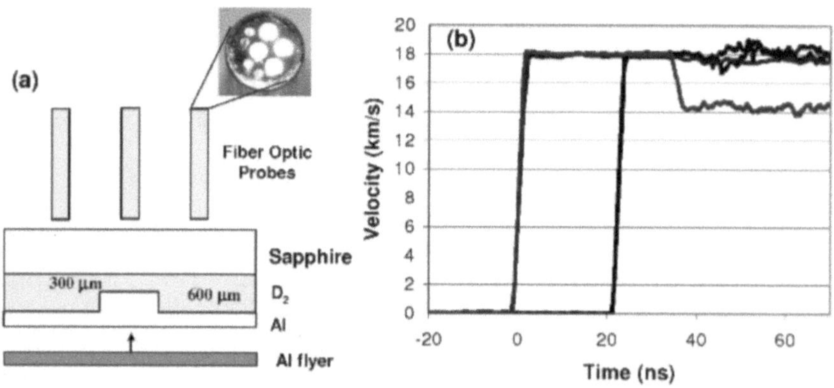

Fig. 10.26. Impact technique for Hugoniot experiments on liquid deuterium. (**a**) Experimental configuration. (**b**) Direct shock velocity measurements

drive for 30–40 ns, which, in turn, propagated a steady shock first into the aluminum drive plate and then into the deuterium cell. Shock steadiness was verified by directly measuring the shock velocity in the deuterium cell at two different cell thicknesses using the VISAR diagnostic. This was possible because the shock front in liquid deuterium becomes reflective for shock stresses exceeding about 30 GPa allowing direct measurement of shock velocity with a VISAR [56]. The measured constant velocity of the shock wave in deuterium shown in Fig. 10.26b indicates that the drive pressure is constant for at least 30 ns (at which point the shock reaches the sapphire window), with the result that the shock pressure is steady for the timescale of the experiment.

The Hugoniot of liquid deuterium was determined from an impedance matching technique in which the deuterium sample was placed behind an aluminum drive plate and the shock velocities measured in both the driver plate and the deuterium sample. The experimental configuration for using an impedance matching technique to obtain Hugoniot data in liquid deuterium is shown in Fig. 10.27a. Impact of an aluminum or titanium flyer plate on the aluminum drive plate produces a shock state in aluminum that is determined from measurement of the flyer velocity and the shock velocity in the drive plate. Since aluminum has a well-established Hugoniot curve for shock loading over the range studied [58], measurement of the flyer velocity accurately determines the impact stress state. Determination of the shocked state in deuterium is thus established by measuring the shock velocity in liquid deuterium and using the unloading response of aluminum to impedance match to the impact state in deuterium, as illustrated in the figure. This approach requires an accurate knowledge of the unloading response of aluminum from high impact stresses. Knudson et al. [55] have experimentally determined the unloading response of aluminum from various impact stresses to 500 GPa in

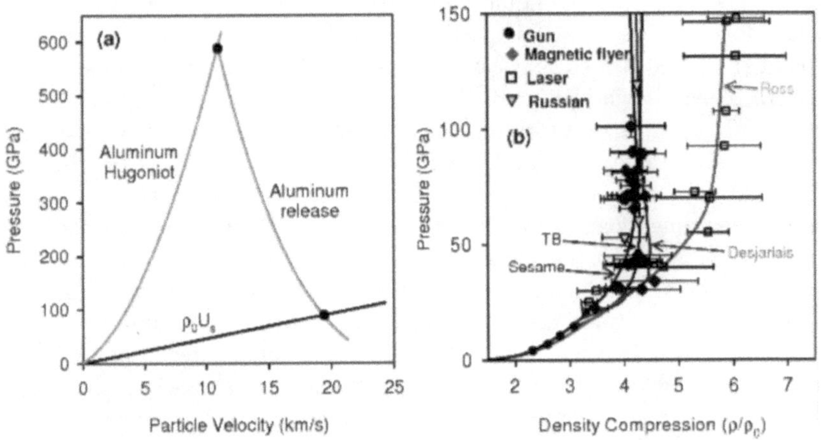

Fig. 10.27. (**a**) Impedance matching technique used to determine the Hugoniot of deuterium, and (**b**) experimental Hugoniot for liquid deuterium

order to establish the required accuracy of aluminum unloading that allows accurate measurements of low impedance materials such as liquid deuterium by impedance matching.

Figure 10.27b presents the Hugoniot response of liquid deuterium to about 100 GPa, along with experimental results using light-gas guns [69], a laser drive [24,25] and a multi-stage explosive device [19,20,87]. The magnetically driven flyer plate results agree well with results from the gas gun and explosive drive techniques in the region of data overlap. These combined data do disagree with results obtained using a laser shock drive technique and a laser backlighter to determine absolute Hugoniot states [24]. Reference [55–57] discussed potential reasons for this discrepancy and provide further data to support the stiffer response of deuterium determined by the magnetic flyer results.

Also shown in Fig. 10.27b are theoretical curves that include a tight binding model for deuterium [65], a chemical mixing model by Kerley [54], a chemical mixing model by Ross [79], and an ab-initio quantum mechanical model based on density functional theory developed by Desjarlais [31]. It is noted that except for the laser Hugoniot results, good agreement is achieved with data from the other experimental approaches and with the theories for deuterium mentioned above. This good agreement gives some confidence that the shock compression response of deuterium over the pressure range to 100 GPa is well established.

The relatively large size of the flyer plates launched using this technique also allows high-pressure shock and release experiments to be performed. Due to the significant temperature increase associated with shock compression, the release states achieved are relatively hot and expanded. In particular, release states from Hugoniot pressures of ~1000 GPa in aluminum (corresponding to ~30 km/s flyer plate impact) are predicted to intersect the vapor dome very near the liquid-vapor critical point. This region of the EOS surface for most materials is not well understood. Such experiments further expand the EOS surface that can be accessed through experiments on the Z accelerator, as shown in Fig. 10.13.

As a first step to performing these types of experiments, release experiments in aluminum were performed for initial shock stresses ranging from ~250–500 GPa, using a low density (200 mg/cm^3) silica aerogel. We note that these measurements are also relevant to the deuterium data described above since the release into the aerogel simulates the unloading of aluminum in the deuterium impedance match experiments. We also note that this technique is similar to that used by Holmes to measure the aluminum release from ~80 GPa with an accuracy of ~1% in u_p [50]. Direct impact experiments were performed to generate Hugoniot data for aerogel in the range of ~30–75 GPa. Experiments were then performed in which a shock was transmitted from an aluminum drive plate into the silica aerogel, as illustrated in Fig. 10.28a. The measured shock velocity in the aerogel in the release

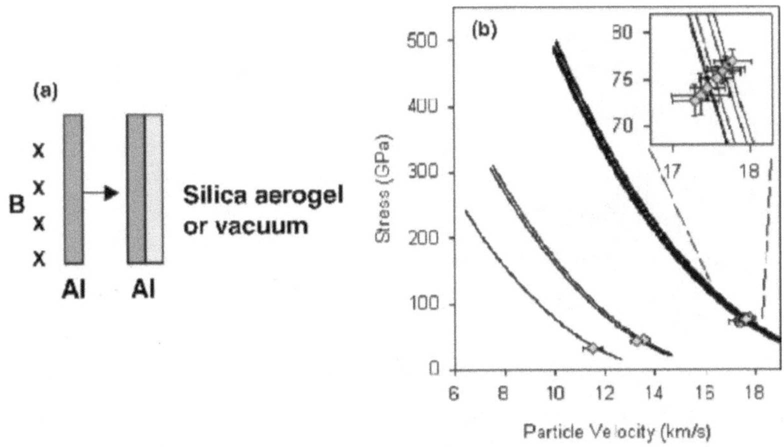

Fig. 10.28. Aluminum unloading experiments. (**a**) Experimental configuration. (**b**) Release data obtained using low density silica aerogel

experiment, along with the measured aerogel Hugoniot, determines a point in $P - u_p$ space through which the aluminum release isentrope must pass. A series of experiments was performed, in which release points in aluminum were measured from initial shock states in the range of ~250–500 GPa. The results of these experiments are plotted in Fig. 10.28b, along with aluminum release calculations from the Sesame 3700 model for aluminum [52, 53]. The agreement between experiment and calculation is within experimental uncertainty over this entire range. Thus, these measurements validate the Sesame 3700 release behavior over the pressure range of interest in this study and indicate that no significant errors in the inferred deuterium density are a result of the aluminum EOS. In particular, these release experiments make a strong case for the use of the Sesame 3700 EOS model for aluminum in performing impedance matching experiments.

We have also performed preliminary full release experiments for aluminum. For these experiments a LiF window was placed a known distance, typically of order 1 mm, from the rear of the aluminum target and the response of the aluminum was diagnosed by looking through the LiF window. When the shock reached the rear surface of the target, the shocked aluminum fully released to zero pressure and accelerated toward the LiF window. Two distinct impacts at the LiF window were observed for initial shock states in the aluminum of over ~250 GPa (particle velocity of >7 km/s). The results of three experiments at ~250, ~300, and ~500 GPa are shown in Fig. 10.29. The data are plotted as the measured release velocity less twice the particle velocity of the shocked state (this velocity is referred to as the expansion velocity since it represents the increment of velocity above that obtained by approximating the release path as the reflected Hugoniot) as a function of the particle velocity of the shocked state. Notice that the slower feature observed

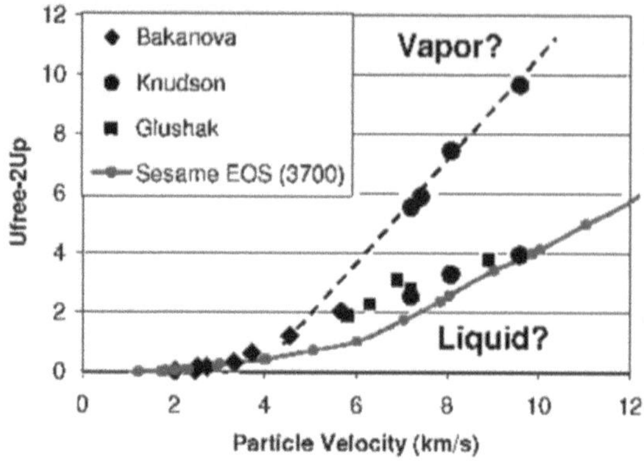

Fig. 10.29. Full release experiments. Plotted are the measured expansion velocities ($u_{\mathrm{free}} - 2u_{\mathrm{p}}$) as a function of the particle velocity of the shocked state

in our experiments is in good agreement with data obtained by Russian investigators in experiments where the shocked aluminum was released into an atmosphere of air. It appears that the faster feature observed in our experiments was suppressed in their experiments. This is likely due to the fact that the air tamps the release somewhat. This apparent two-phase flow is being interpreted as a fast vapor component followed by a slower, possibly liquid component and is preliminary evidence that these release adiabats are intersecting the vapor dome (as indicated in Fig. 10.13). However, further experiments are required to investigate this hypothesis.

10.5 Conclusion and Future Applications

The magnetic loading technique discussed here provides an experimental capability for accessing states of matter previously inaccessible in the laboratory. In particular, the ability to continuously load a planar specimen to peak pressures of several hundred GPa is unique and allows access to thermodynamic states of matter that approach planar isentropic compression. In this regard, the technique provides an important bridge between techniques such as the diamond anvil cell that produce high-pressure data under isothermal conditions and shock loading data that produce high temperature states associated with the high compression. The ability to obtain EOS data intermediate between these two extremes enables a critical evaluation of EOS theories. When combined with the ability to explore highly expanded regions of the thermodynamic surface using combined magnetically-driven flyer impact and unloading, as illustrated in Fig. 10.13, the method provides a powerful capability for EOS studies. The continuous loading technique is also useful for identifying various phase transformations that can occur under dynamic loading. Examples presented in the text on iron, zirconium and tin illustrate the ability to effectively determine transition stresses associated with a transformation and bound the associated kinetic effects.

In addition to thermodynamic information, the ramp loading technique is also useful for evaluating the dynamic mechanical response of materials at loading rates intermediate to those obtained in shock experiments and with techniques such as Hopkinson bars. The ICE technique is especially useful for detecting relatively small differences in mechanical behavior that could result from differences in grain sizes or by defects introduced into materials. Examples presented in the text include effects of grain size on the dynamic mechanical response of two different uranium alloys and the effects of radiation-induced defects on the high-stress mechanical response of stainless steel. Differences of 1% or so in stress can easily be detected. More importantly, the tests on several different grain sizes or levels of radiation damage can be conducted in a single experiment. With conventional methods this could literally require tens of experiments.

The magnetic compression technique is also potentially useful in a variety of other material property measurements. We have illustrated use of the technique to simultaneously obtain the mechanical and optical properties of laser windows used for dynamic wave profile measurements. The ability to obtain this information in a single experiment can considerably reduce the time and expense of acquiring this information compared to conventional methods.

An important spin-off of the technique is the continuous acceleration of flyer plates to high velocities for conventional plate impact shock wave experiments. In this approach, the continuously varying magnetic field is used to accelerate flyer plates on the order of 1 mm thick and 10 or more mm in width to velocities exceeding 20 km/s. With the magnetic drives presently available, flyer velocities of 25 km/s have been achieved. This capability has several important applications, one of which is to produce essentially absolute Hugoniot data through symmetric impact of a flyer plate onto a sample of the same material to access stresses considerably higher than possible with conventional techniques. An example is presented for aluminum shock-compressed to stresses of about 500 GPa, which is almost three times higher than that possible with light gas guns. The comparable accuracy to gas guns and other well established techniques possible with the magnetic drive technique helps to establish its validity as a useful tool for acquiring high-pressure Hugoniot data. Another example presented in the text concerns use of the flyer technique to obtain relative Hugoniot data on liquid deuterium to pressures about five times higher than conventional techniques, but with accuracies comparable to those techniques. These results are helping to validate first principles calculations of the high pressure response of hydrogen and its isotopes and are providing important ancillary data to other techniques such as diamond anvil cells in order to establish the EOS of hydrogen over extended EOS ranges.

Several additional applications have also been identified that illustrate further potential for the magnetic loading technique. It has been recently shown that the Z Accelerator can be used in a cylindrical loading mode to characterize porous materials, such as ceramic powders [62]. In this configuration, the square load shown in Fig. 10.3 is replaced by a cylindrical cathode and anode. The anode is used for the cylindrical drive for the cylindrical sample which is placed in contact with the anode. This produces a one dimensional cylindrically diverging wave in the sample that can be diagnosed at different propagation distances to evaluate constitutive response for cylindrical loading. Initial experiments performed to demonstrate the technique in porous powders have been very promising [62]. An important factor in using the approach for highly compressible materials, such as porous materials, is that edge perturbations, which limit sample thicknesses in planar impact experiments, are essentially eliminated, resulting in longer times to perform experimental measurements.

Another possible application of the magnetic isentropic compression technique concerns soft recovery and material synthesis applications. In typical plate impact experiments, soft recovery of the target is exacerbated by the requirement that the projectile used to launch the impacting plate must be separated from the recovered target in order to prevent late-time damage to the target. This requirement is essentially eliminated in the magnetic drive technique. It is only necessary to configure the sample so the momentum produced by the magnetic field is dissipated in the sample assembly. Similar principles can be used in the synthesis of novel materials through magnetic compression. In addition, it should be possible to compact several samples simultaneously using isentropic compression techniques. An advantage of this approach over shock compaction using conventional plate impact techniques is that the additional heating associated with shock loading is not present so the sample should be considerably cooler.

Acknowledgments

We would like to acknowledge the large number of people who have been involved in the development and application of the Isentropic Compression Technique and associated magnetically-driven flyer plates. Because of the involvement by a large team of people many of whom are identified in the references, we cannot list everyone by name, but we would like to acknowledge several individuals who played a crucial role in the development of this technology. They include Rick Spielman, who provided the opportunity and the engineering and moral support for initiating isentropic compression techniques on the Z accelerator; Clint Hall, who engineered the original configuration used for isentropic loading of multiple samples simultaneously; David Reisman, who determined the magnetic conditions necessary to achieve accurate planar continuous loading of specimens and was instrumental in using this technique for several innovative applications; Ray Lemke and Mike Desjarlais, who were instrumental in developing and defining magnetic loading conditions necessary to shocklessly accelerate flyer plates to high velocities; David Hanson, who developed the cryogenic system that allowed accurate studies of liquid deuterium; Dennis Hayes, who has been instrumental in developing computer simulation techniques for analyzing ICE wave profiles and for investigating the optical properties of laser windows under isentropic loading; Randy Hickman, who was instrumental in developing the large array of diagnostic techniques at the Z accelerator that are necessary to measure wave profiles under ramp loading; the superb operations and diagnostic team at the Z facility whose dedication and constant support through several difficult iterations of the technique were critical to successful development of the method; Don Cook, who was the director of the Pulsed Power Sciences Center that had the foresight to initiate the shock physics program at the Z accelerator; and finally, the current management of the Pulsed Power Sciences Center, in particular Jeff Quintenz, Keith Matzen and Chris Deeney, who have been fully supportive of these developments over the past several years and who provided the considerable resources necessary to develop the technique.

Sandia is a multiprogram laboratory operated by Sandia Corporation, a Lockheed Martin Company, for the United States Department of Energy's National Nuclear Security Administration under Contract DE-AC04-94AL85000.

References

1. Aidun, J.B., and Y.M. Gupta (1991), "Analysis of Lagrangian gauge measurements of simple and non-simple plane waves," *J. Appl. Phys.* **69**, p. 6998.
2. Al'tshuler, L.V., and B.S. Chekin (1974), "Metrology of high pulsed pressures," in: Proc. of 1 All-Union Pulsed Pressures Simposium, VNIIFTRI, Moscow vol. 1, pp. 5–22. (in Russian).
3. Al'tshuler, L.V., S.B. Kormer, A.A. Bakanova, and R.F. Trunin (1960), "The isentropic compressibility of aluminum, copper, lead, and iron at high pressures," *Zh. Eksp. Teor. Fiz.* **38**, p. 790. (in Russian) *Sov. Phys. JETP* **11**, p. 573.
4. Al'tshuler, L.V., N.N. Kalitkin, L.V. Kuz'mina, and B.S. Chekin (1977), "Shock adiabats for ultrahigh pressures," *Zh. Eksp. Teor. Fiz.* **72**, p. 317 (in Russian) *Sov. Phys. JETP* **45**, p. 167.
5. Asay, J.R. (1997), "The use of shock-structure methods for evaluating high-pressure material properties," *Int. J. Impact Engng.* **20**, p. 27.
6. Asay, J.R. (2000), "Isentropic compression experiments on the Z accelerator," in: *Shock Compression of Condensed Matter – 1999*, (eds. M.D. Furnish, L.C. Chhabildas, and R.S. Hixson) American Institute of Physics, New York, p. 261.
7. Asay, J.R., and L.C. Chhabildas (1981), "Determination of the shear strength of shock compressed 6061-T6 aluminum," in: *Shock Waves and High-Strain-Rate*

Phenomena in Metals (eds. M.A. Meyers and L.E. Murr) Plenum Press, New York, p. 417.

8. Asay, J.R., and G.I. Kerley (1987), "The response of materials to dynamic loading," *Int. J. Impact Engng.* **5**, p. 69.
9. Asay J.R., and L.C. Chhabildas (2003), "Paradigms and Challenges in Shock Wave Research," in: *High-Pressure Shock Compression of Solids VI* (eds. Y. Horie, L.Davison, and N.N. Thadhani) Springer-Verlag, New York, p. 57.
10. Asay, J.R., G.R. Fowles, G.E. Duvall, M.H. Miles, and R.F. Tinder (1975), "Effects of point defects on elastic precursor decay in LiF," *J. Appl. Phys.* **45**, p. 2132.
11. Asay, J.R., C.A. Hall, K.G. Holland, M.A. Bernard, W.A. Stygar, R.B. Spielman, S.E.Rosenthal, D.H. McDaniel, and D.B. Hayes (2000), "Isentropic compression of iron with the Z accelerator," in: *Shock Compression of Condensed Matter – 1999*, (eds. M.D. Furnish, L.C. Chhabildas, and R.S. Hixson) American Institute of Physics, New York, p. 1151.
12. Benedick, W.B., and J.R. Asay (1976), *Bull. Amer. Phys. Soc.* **21**, p. 1298.
13. Barker, L.M. (1984), "High-pressure quasi-isentropic impact experiments," in: *Shock Compression of Condensed Matter – 1983*, (eds. J.R. Asay, R.A. Graham, and G.K.Straub) North–Holland, Amsterdam, p. 111.
14. Barker, L.M., and R.E. Hollenbach (1970), "Shock-wave studies of PMMA, fused silica and sapphire." *J. Appl. Phys.* **41**, p. 4208.
15. Barker, L.M., and R.E. Hollenbach (1972), "Laser interferometer for measuring the high velocities of any reflecting surface," *J. Appl. Phys.* **43**, p. 4669.
16. Barker, L.M., and R.E. Hollenbach (1974), "Shock wave studies of the α–ε phase transition in iron," *J. Appl. Phys.* **45**, p. 4872.
17. Barker, L.M., and K. Schuler (1974), "Correction to the velocity-per-fringe relationship for the VISAR interferometer," *J. Appl. Phys.* **45**, p. 3692.
18. Barker, L.M., M. Shahinpoor, and L.C. Chhabildas (1983), "Experimental Diagnostic Techniques," in: *High-Pressure Shock Compression of Solids* (eds. J.R. Asay and M. Shahinpoor) Springer-Verlag, New York, p. 43.
19. Belov, S.I., G.V. Boriskov, A.I. Bykov, R.I. Il'kaev, N.B. Luk'yanov, A.Ya. Matveev, O.L. Mikhailova, V.D. Selemir, G.V. Simakov, R.F. Trunin, I.P. Trusov, V.D. Urlin, V.E. Fortov, and A.N. Shuikin (2002), "Shock compression of solid deuterium," *Sov. Phys. JETP Lett.* **76**, p. 433.
20. Boriskov, G.V., A.I. Bykov, R.I. Il'kaev, V.D. Selemir, G.V. Simakov, R.F. Trunin, V.D. Urlin, V.E. Fortov, and A.N. Shuikin (2003), "Shock-wave compression of solid deuterium at a pressure of 120 GPa," *Dokl. Phys.* **48**, p. 553.
21. Boslough, M.B., and J.R. Asay (1993), "Basic principles of shock compression," in: *High-Pressure Shock Compression of Solids* (eds. J.R. Asay and M. Shahinpoor) Springer-Verlag, New York, p. 7.
22. Chhabildas, L.C., and J.R. Asay (1982), "Time-resolved wave profile measurements in copper to megabar pressures," in: High Pressure in Research and Industry – 8th AIRAPT Conf. VI (eds. C.M. Backman, T. Johanisson, and L. Tegner) Arkitektkopia, Sweden, p. 183.
23. Chhabildas, L.C., and R.A. Graham (1987), "Techniques and theory of stress measurements for shock wave applications," in: (eds. R.R. Stout, F.R. Norwood, and M.E. Fourney) AMD 83. American Society of Mechanical Engineers, New York, p. 1.

24. Collins, G.W., L.B. DaSilva, P. Celliers, D.M. Gold, M.E. Foord, R.J. Wallace, A. Ng, S.V. Weber, K.S. Budil, and R. Cauble (1998), "Measurements of the equation of state of deuterium at the fluid insulator-metal transition," *Science* **281**, p. 1178.
25. Da Silva, L.B., P. Celliers, G.W. Collins, K.S. Budil, N.C. Holmes, T.W. Barbee Jr., B.A. Hammel, J.D. Kilkenny, R.J. Wallace, M. Ross, R. Cauble, A. Ng, and G. Chiu (1997), "Absolute equation of state measurements on shocked liquid deuterium up to 200 GPa (2 Mbar)," *Phys. Rev. Lett.* **78**, p. 483.
26. Davis, J.-P. (2004), "Quasi-isentropic compression of iron," Private communication.
27. Davis, J.-P., and D.B. Hayes (2003), "Isentropic compression experiments on dynamic solidification in tin," in: *Proceedings of the 13th APS topical group conference on shock compression science of condensed matter – 2003*, July 20–25, 2003, Portland Oregon.
28. Davis, J.-P. et al. (2004), "Quasi-isentropic response of aluminum," Private communication.
29. Davison, L., and R.A. Graham (1979), "Shock compression of solids," *Physics Reports* **55**, p. 257.
30. Desjarlais, M.P. (2001), "Practical improvements to the Lee–More conductivity near the metal-insulator transition," *Contrib. Plasma Phys.* **41**, p. 267.
31. Desjarlais, M.P. (2003), "Density functional calculations of the liquid deuterium Hugoniot, reshock and reverberation timing," *Phys. Rev. B* **68**, p. 064204.
32. Desjarlais, M.P., J.D. Kress, and L.A. Collins (2002), "Electrical conductivity for warm, dense aluminum plasmas and liquids," *Phys. Rev. E* **66**, p. 025401.
33. Ding, J.L., J.R. Asay, and M.D. Knudson (2003), "Thermal and mechanical analysis of material response under ramp and shock loading," *Bull. Amer. Phys. Soc.* **48**, p. 83.
34. Dolan, D.H., III (2003), "Time-dependent freezing of water under multiple shock wave compression," Ph.D. thesis, Washington State University.
35. Dolan, D.H., III, and Y.M. Gupta (2003), "Time dependent freezing of water under dynamic compression," *Chem. Phys. Lett.* **374**, p. 608.
36. Duvall, G.E., and R.A. Graham (1977), "Phase transitions under shock wave loading," *Rev. Mod. Phys.* **49**, p. 523.
37. Edwards, J., K.T. Lorenz, B.A. Remington, S. Pollaine, J. Colvin, D. Braun, B.F. Lasinski, D. Reisman, J.M. McNaney, J.A. Greenbough, R. Wallace, H. Louis, and D. Kalantar (2004), "Laser-driven plasma loader for shockless compression and acceleration of samples in the solid state," *Phys. Rev. Lett.* **92**, p. 075002-1.
38. Fortov, V.E. (2004), "Aluminum phase diagram," Private communication.
39. Fowles, G.R. (1961), "Shock wave compression of hardened and annealed 2024 aluminum," *J. Appl. Phys.* **32**, p. 1475.
40. Glushak, B.L., A.P. Zharkov, M.V. Zhernokletov, V.Ya. Ternovoi, A.S. Filimonov, and V.E. Fortov (1989), "Experimental investigation of the thermodynamics of dense plasmas formed from metals at high energy concentrations," *Zh. Eksp. Teor. Fiz.* **96**, p. 1301. (in Russian) [trans. in *Sov. Phys. JETP* **69**, p. 739].
41. Hall, C.A. (2003), "Isentropic compression experiments on the Sandia Z accelerator," *Phys. Plasmas* **7**, p. 2069.

42. Hall, C.A., J.R. Asay, M.D. Knudson, W.A. Stygar, R. Spielman, T.D. Pointon, D.B. Reisman, A. Toor, and R.C. Cauble (2001), "Experimental configuration for isentropic compression of solids using pulsed magnetic loading," *Rev. Sci. Instrum.* **72**, p. 3587.
43. Hall, C.A., M.D. Knudson, J.R. Asay, R. Lemke, B. Oliver (2002a), "High velocity flyer plate launch capability on the Sandia Z accelerator," *Int. J. Impact Engng.* **26**, p. 275.
44. Hall, C.A., J.R. Asay, M.D. Knudson, D.B. Hayes, R.W. Lemke, J.-P. Davis, and C. Deeney (2002b), "Recent advances in quasi-isentropic compression experiments on the Sandia Z accelerator," in: *Shock Compression of Condensed Matter – 2001* (eds. M.D. Furnish, N. Thadhani, and Y. Horie) American Institute of Physics, New York, p. 1163.
45. Hare, D.E., D.B. Reisman, F. Garcia, L.G. Green, J.W. Forbes, M.D. Furnish, C. Hall, and R.J. Hickman (2003), "The isentrope of unreacted LX-04 to 170 kbar," in: *Proceedings of the 13th APS topical group conference on shock compression science of condensed matter–2003*, July 20–25, 2003, Portland, Oregon.
46. Hawke, R., D.E. Duerre, J.G. Huebel, J.G. Klapper, D.J. Steinberg, R.N. Keeler (1972), "Method of isentropically compressing materials to several megabars," *J. Appl. Phys.* **43**, p. 2734.
47. Hayes, D.B. (2001), "Unsteady compression waves in interferometer windows," *J. Appl. Phys.* **89**, p. 6484.
48. Hayes, D.B., and C.A. Hall (2002), "Correcting free surface effects by integrating the equations of motion backward in space," in: *Shock Compression of Condensed Matter – 2001* (eds. M.D. Furnish, N. Thadhani, and Y. Horie) p. 1177.
49. Hayes, D.B., C.A. Hall, J.R. Asay, M.D. Knudson (2003), "Continuous index of refraction measurements to 20 GPa in *Z*-cut sapphire," *J. Appl. Phys.* **94**, p. 2331.
50. Holmes, N.C. (1994). "Shock compression of low-density foams," in: *High-Pressure Science and Technology – 1993* (eds. S.C. Schmidt, J.W. Shaner, G.A. Samara, and M. Ross). American Institute of Physics, New York, p. 153.
51. Jones, S.C., M.C. Robinson, and Y.M. Gupta (2003), "Ordinary refractive index of sapphire in uniaxial tension and compression along the c axis," *J. Appl. Phys.* **93**, p. 1023.
52. Kerley, G.I. (1987), "Theoretical equation of state for aluminum," *Int. J. Impact Engng.* **5**, p. 441.
53. Kerley, G.I. (1998), *Equations of State for Aluminum and Beryllium*, Kerley Publishing Services Report No. KPS98-1.
54. Kerley, G.I. (2003), *Equations of state for hydrogen and deuterium.* Sandia National Laboratories Report SAND2003-3613.
55. Knudson, M.D. (2004), "Unloading response of aluminum to impact stresses of 500 GPa," manuscript in preparation.
56. Knudson, M.D., Hanson DL, Bailey JE, Hall CA, Asay JR, Anderson WW (2001) Equation of state measurements in liquid deuterium to 70 GPa. Phys Rev Lett 87:225501-1.
57. Knudson, M.D., D.L. Hanson, J.E. Bailey, C.A. Hall, and J.R. Asay (2003a), "Use of a wave reverberation technique to infer the density compression of shocked liquid deuterium to 75 GPa," *Phys. Rev. Lett.* **90**, p. 035505-1.

58. Knudson, M.D., R.W. Lemke, D.B. Hayes, C.A. Hall, C. Deeney, and J.R. Asay (2003b), "Near-absolute Hugoniot measurements in aluminum to 500 GPa using a magnetically accelerated flyer plate technique," *J. Appl. Phys.* **94**, p. 4420.
59. Knudson, M.D., C.A. Hall, R. Lemke, C. Deeney, and J.R. Asay (2003c), "High velocity flyer plate launch capability on the Sandia Z accelerator," *Int. J. Impact Engng.* **29**, p. 377.
60. Knudson, M.D., D.L. Hanson, J.E. Bailey, C.A. Hall, J.R. Asay, and C. Deeney (2004), "Principal Hugoniot, reverberating wave, and mechanical reshock measurements of liquid deuterium to 400 GPa using plate impact techniques," accepted for publication in *Phys. Rev. B.*
61. Kormer, S.B., A.I.Funtikov, V.D. Urlin, and A.N. Kolesnikova (1962), "Dynamical compression of porous metals and the equation of state with variable specific heat at high temperatures," *Zh. Eksp. Teor. Fiz.* **42**, p. 626 (in Russian) *Sov. Phys. JETP* **15**, p. 477].
62. Lawrence, R.J., D.E. Grady, and C.A. Hall (2003), "The response of ceramic powders to high-level quasi-isentropic dynamic loads," in: *Proceedings of the 13th APS topical group conference on shock compression science of condensed matter – 2003*, July 20–25, 2003, Portland, Oregon.
63. Lemke, R.W., M.D. Knudson, C.A. Hall, T.A. Haill, M.P. Desjarlais, J.R. Asay, and T.A. Mehlhorn (2003a), "Characterization of magnetically accelerated flyer plates," *Phys. Plasmas* **10**, p. 1092.
64. Lemke, R.W., M.D. Knudson, A.C. Robinson, T.A. Haill, K.W. Struve, J.R. Asay, and T.A. Mehlhorn (2003b), "Self-consistent, two-dimensional, magnetohydrodynamic simulations of magnetically driven flyer plates," *Phys. Plasmas* **10**, p. 1867.
65. Lenosky, T.J., J.D. Kress, and L.A. Collins (1997), "Molecular-dynamics modeling of the Hugoniot of shocked liquid deuterium," *Phys. Rev. B* **56**, p. 5164.
66. Lindl, J. (1995), "Development of the indirect-drive approach to inertial confinement fusion and the target physics basis for ignition and gain," *Phys. Plasmas* **2**, p. 3933.
67. Matzen, M.K. (1997), "Z pinches as intense x-ray sources for high-energy density physics applications," *Phys. Plasmas* **4**, p. 1519.
68. Mitchell A.C., and W.J. Nellis (1981), "Shock compression of aluminum, copper, and tantalum," *J. Appl. Phys.* **52**, p. 3363.
69. Nellis, W.J., A.C. Mitchell, M. van Thiel, G.J. Devine, and R.J. Trainor (1983), "Equation-of-state data for molecular hydrogen and deuterium at shock pressures in the range 2–76 GPa (20–760 kbar)," *J. Chem. Phys.* **79**, p. 1480.
70. Nesterenko, V. (2001), *Dynamics of Heterogeneous Materials*, Springer-Verlag, New York.
71. Nguyen, J. (2003), "Dynamic compression by design at LLNL gas gun," *Bull. Amer. Phys. Soc.* **48**, p. 23.
72. Perez, M. (1990), "The MIVAR: a ramp-wave generator fabricated by the plasma spray technique," in: *Shock Compression of Condensed Matter – 1989* (eds. S.C. Schmidt, J.N. Johnson, and L.W. Davison) North–Holland, Amsterdam, p. 751.
73. Reisman, D.B. (2003), "Ramp loading of uranium alloys," Private communication.
74. Reisman, D.B., A. Toor, R.C. Cauble, C.A. Hall, J.R. Asay, M.D. Knudson, and M.D.Furnish (2001), "Magnetically driven isentropic compression experiments on the Z accelerator," *J. Appl. Phys.* **89**, p. 1625.

75. Reisman, D.B., W.G. Wolfer, A. Elsholz, and M.D. Furnish (2003), "Isentropic compression of irradiated stainless steel on the Z accelerator," *J. Appl. Phys.* **93**, p. 8952.
76. Remington, B.A., G. Bazan, J. Belak, E. Bringa, M. Caturia, J.D. Colvin, M.J. Edwards, S.G. Glendinning, D. Ivanov, B. Kad, D. Kalantar, M. Kumar, B.F. Lasinski, K.T. Lorenz, J.M. McNaney, D.D. Meyerhofer, M.A. Meyers, S.M. Pollaine, D. Rowley, M. Schneider, J.S. Stolken, J.S. Wark, S.V. Weber, W.G. Wolfer, and B. Yaakobi (2003a), "Materials science under extreme conditions of pressure and strain rate," submitted to *Metall. And Materials Transactions A.*
77. Remington, B.A., R.M. Cavallo, M.J. Edwards, B.F. Lasinski, K.T. Lorenz, H.E. Lorenzana, J. McNaney, S.M. Pollaine, D.P. Rowley (2003b), *Materials science at the extremes of pressure and strain rate*, Lawrence Livermore National Laboratory Report UCRL-JC-152288.
78. Rigg, P.A., W. Greeff, M.D. Knudson, D.B. Hayes, R.S. Hixson, and G.T. Gray III (2003), "Investigation shock-induced phase transitions in zirconium," in: *Proceedings of the 13th APS topical group conference on shock compression science of condensed matter – 2003*, July 20–25, 2003, Portland, Oregon.
79. Ross, M. (1998), "Linear-mixing model for shock-compressed liquid deuterium," *Phys. Rev. B* **58**, p. 669.
80. Saumon, D., and T. Guillot (2004), "Shock compression of deuterium and the interiors of Jupiter and Saturn," Accepted for publication *Astrophysical Journal.*
81. Seidel, D.B., M.L. Kiefer, R.S. Coats, T.D. Pointon, J.P. Quintenz, and W.A. Johnson (1991), "The 3D, electromagnetic, particle in cell code, QUICKSILVER," in: *CP90 Europhysics conference on computational physics* (ed. A. Tenner) World Scientific, Amsterdam, pp. 475–482.
82. Setchell, R.E. (2002), "Refractive index of sapphire at 532 nm under shock compression," *J. Appl. Phys.* **91**, p. 2833.
83. Sharp, G. (2002), *Magnetic diffusion in conductors at ultra-high current density*, Ph.D. thesis, University of New Mexico.
84. Simonenko, V.A., N.P. Voloshin, A.S. Vladimirov, A.P. Nagibin, V.N. Nogin, V.A. Popov, V.A. Sal'nikov, and Yu.A. Shoidin (1985), "Absolute measurements of shock compressibility of aluminum at pressures $\geq$ 1 TPa," *Zh. Eksp. Teor. Fiz.* **88**, p. 1452. (in Russian) *Sov. Phys. JETP* **61**, p. 869.
85. Summers, R.M., J.S. Peery, M.W. Wong, E.S. Hertel Jr, T.G. Trucano, and L.C. Chhabildas (1997), "Recent progress in Alegra development and applications to ballistic impacts," *Int. J. Impact Engng.* **20**, p. 779.
86. Trott, W.M., J.N. Castaneda, J.J. O'Hare, M.R. Baer, L.C. Chhabildas, M.D. Knudson, J.-P. Davis, and J.R. Asay (2001), "Dispersive velocity measurements in heterogeneous materials," *Bull. Amer. Phys. Soc.* **46**, p. 31.
87. Trunin, R.F., W.J. Nellis (2003), *Shock compression of liquid deuterium at 54 GPa*, Lawrence Livermore National Laboratory Report UCRL-JC-152886.
88. Vogler, T.J (2004), "Determining compressive strength for ramp loading," Private communication.
89. Volkov, L.P., N.P. Voloshin, A.S. Vladimirov, V.N. Nogin, and V.A. Simonenko (1981), "Shock compressibility of aluminum at pressure 10 Mbar," *Pis'ma Zh. Eksp. Teor. Fiz.* **31**, p. 623. (in Russian) *Sov. Phys. JETP Lett.* **31**, p. 588].
90. Wise, J.L., and L.C. Chhabildas (1986), "Laser interferometer measurements of refractive index in shock-compressed materials," in: *Shock Waves in Condensed Matter* (ed. Y.M. Gupta) Plenum Publishing Corp., New York, p. 441.

FSC
www.fsc.org
MIX
Papier | Fördert
gute Waldnutzung
FSC® C083411

Kolibri 360 GmbH
Ferdinand-Jühlke-Straße 7
99095 Erfurt, Deutschland
produktsicherheit@kolibri360.de